ORGANIC CHEMISTRY

Structure and Mechanisms

Research Progress in Chemistry

ORGANIC CHEMISTRY

Structure and Mechanisms

Harold H. Trimm, PhD, RSO

Chairman, Chemistry Department, Broome Community College;
Adjunct Analytical Professor, Binghamton University,
Binghamton, New York, U.S.A.

Apple Academic Press

Research Progress in Chemistry Series

Organic Chemistry: Structure and Mechanisms

First Published in the Canada, 2011
Apple Academic Press Inc.
3333 Mistwell Crescent
Oakville, ON L6L 0A2
Tel. : (888) 241-2035
Fax: (866) 222-9549
E-mail: info@appleacademicpress.com
www.appleacademicpress.com

The full-color tables, figures, diagrams, and images in this book may be viewed at www.appleacademicpress.com

ISBN 978-1-926692-60-9

Harold H. Trimm, PhD, RSO

Printed at Thomson Press (I) Ltd.
9 8 7 6 5 4 3 2 1

Cover Design: Psqua

Library and Archives Canada Cataloguing in Publication Data
CIP Data on file with the Library and Archives Canada

CONTENTS

INTRODUCTION

Chemistry is the science that studies atoms and molecules along with their properties. All matter is composed of atoms and molecules, so chemistry is all encompassing and is referred to as the central science because all other scientific fields use its discoveries. Since the science of chemistry is so broad, it is normally broken into fields or branches of specialization. The five main branches of chemistry are analytical, inorganic, organic, physical, and biochemistry. Chemistry is an experimental science that is constantly being advanced by new discoveries. It is the intent of this collection to present the reader with a broad spectrum of articles in the various branches of chemistry that demonstrates key developments in these rapidly changing fields.

Organic chemistry is the study of compounds that contain the element carbon. Carbon is unique among the elements for its ability to bond to itself to form long chains and a myriad of structures. Of the approximately 11 million known chemical compounds, about 10 million are organic. New advances in organic chemistry have allowed the production of better polymers with specific properties, such as biodegradable plastics. The elucidation of new drug structures from plants and the synthesis of improved pharmaceuticals is another area of great interest. Organic chemists are also researching the reactions that occur in living systems and understanding the molecular causes of disease.

These chapters will allow the reader to keep up with the latest methods and applications being used in the twenty-first century, as well as other developments in the field of organic chemistry.

— Harold H. Trimm, PhD, RSO

Mechanistic Aspects of the Isomerization of Z-Vinylic Tellurides Double Bonds in the Synthesis of Potassium Z-Vinyltrifluoroborate Salts

Hélio A. Stefani, Rafael C. Guadagnin, Artur F. Keppler, Giancarlo V. Botteselle, João V. Comasseto and Carlos A. Suganuma

ABSTRACT

Through direct transmetalation reaction of Z-vinylic tellurides with nBuLi was observed the unexpected isomerization of double bonds leading to potassium E-vinyltrifluoroborates salts in low to moderate yields. Using EPR spin trapping experiments the radical species that promoted the stereoinversion of Z-vinylic organometallic species during the preparation of potassium vinyltrifluoroborate salts was identified. The experiments support the proposed mechanism, which is based on the homolytic cleavage of the TenBu bond.

Background

Boronic acids and boronate esters are the most commonly used derivatives in Suzuki-Miyaura cross-coupling reactions. Recently, Molander et al.[1] and our group [2] have explored the use of potassium organotrifluoroborate salts as an alternative to the usual organoboron reagents in alkenyl-alkenyl,[3] aryl-aryl,[4] alkenyl-alkynyl,[5] and alkenyl-aryl [6] cross-coupling reactions.

Distinct from the most commonly explored hydrometallation reactions, the hydrotelluration of alkynes exclusively forms Z-vinylic tellurides. [7] Vinylic tellurides have the ability to undergo tellurium-metal exchange reactions with several different commonly used, commercially available, or easily prepared organometallic reagents, leading to Z-vinyllithiums and Z-vinylcyanocuprates. In reactions promoted by Pd or Ni, these compounds undergo stereospecific coupling with a wide range of organic species. [8] The vinylic organometallic species obtained in this way can also react with carbonyl compounds, α, β-unsaturated systems, or epoxides [9-11] with complete retention of the double-bond stereochemistry

Taking advantage of the regio- and stereocontrol of the preparation of Z-vinylic tellurides,[12] and of the unique features of the transmetallation with complete retention of the original double bond geometry, we report herein the synthesis of potassium vinyltrifluoroborate salts by means of the Te-Li exchange reaction. To the best of our knowledge, this is the first reported preparation of potassium E-vinyltrifluoroborate salts from Z-vinylic tellurides.

Results and Discussion

Functionalized Z-vinylic tellurides were prepared by hydrotelluration of alkynes. [13] Using phenyl vinyl telluride, we performed a series of test reactions to establish the best reaction conditions for the lithium-boron exchange step (Table 1; ii, Scheme 1). Optimum yield was obtained with $B(OiPr)_3$ as the electrophile and ether as the solvent (entry 6).

Scheme 1. Synthetic route used to prepare vinyl BF3K salts.

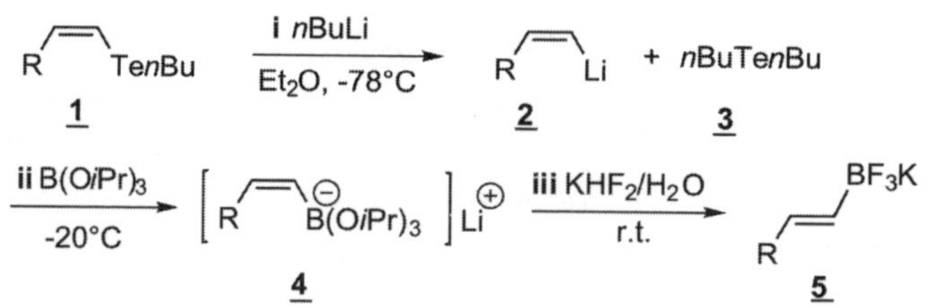

Table 1. Lithium-Boron Test Reaction Conditions.

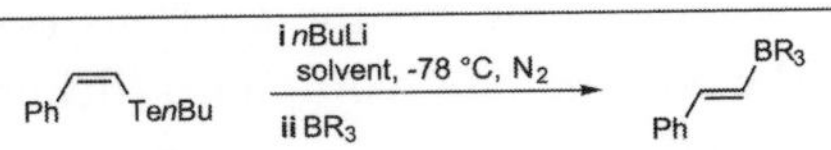

Entry	Electrophile (eq)[a]	Solvent	Yield (%)
1	$B(OMe)_3$ (1.5)	THF	18
2	$B(OiPr)_3$ (1.5)	THF	47
3	$BF_3.OEt_2$ (1.5)	THF	-
4	$B(OiPr)_3$ (1.5)	THF/HMPA	25
5	$B(OiPr)_3$ (1.5)	THF/TMEDA	-
6	$B(OiPr)_3$ (1.5)	Et_2O	51
7	$B(OiPr)_3$ (1.5)	Et_2O	15

Using the optimized conditions (Table 1, entry 6), all the Z-vinylic tellurides were, to our surprise, transformed into potassium E-vinyltrifluoroborate salts exclusively (Figure 1).

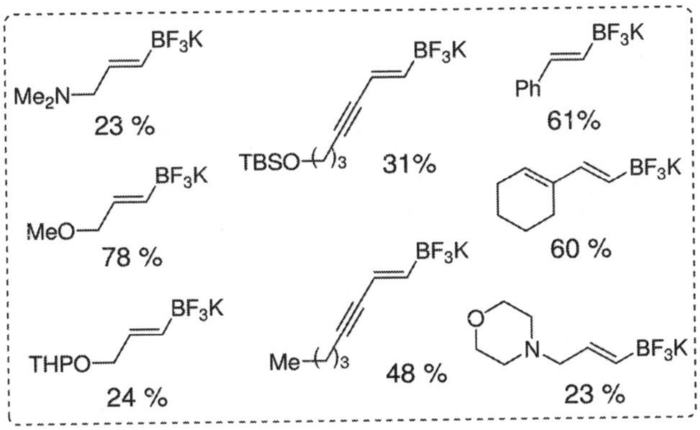

Figure 1. Isolated vinyl BF3K salts.

The 1H NMR spectra of the products showed the presence of the salt nBuBF3K as a by-product (30–50% of the total yield). Use of 1.0 equiv. of nBuLi instead of 1.5 equiv. as in the optimized protocol gave the same proportion of nBuBF3K.

With 1H NMR, we tried to observe the coupling constants of the vinylic hydrogens for each intermediate of the reaction route. Using this approach, we could prove the cis geometry of the vinylic hydrogens of the intermediate 2(Scheme 1), which presented a coupling constant of 18.09 Hz. [14,15] Unfortunately, the boronic "ate" complex 4(Scheme 1) is an insoluble species and no 1H NMR spectra were obtained. However, these results indicated that the double bond geometry isomerization occurred only after the formation of the intermediate 4 (Scheme 1).

We suggest that homolytic cleavage of the Te-Bu bond, from 3 (i, Scheme 1), generates nBu•, which is responsible for the cis-trans isomerization. The butyl radical attack occurs at the boronic "ate" complex 4 (Scheme 1),[16] yielding the nBuBF3K salt as a final product.

In order to verify the presence of radical species in the reaction mixture, we performed EPR spin trapping experiments using 3,5-dibromo-4-nitrosobenzene-sulfonate (DBNBS), which is an appropriate spin trap for tellurium centered radicals. [17] Radical species were detected at the i and ii steps of the proposed route. In the first step (i, Scheme 1), the detected spectra contained a mixture of DBNBS radical adducts (Figure 2A). The triplet of triplets (aN = 21.6 G, aH = 0.7 G,) is the DBNBS/•TenBu radical adduct [17] and the broadened triplet (aN = 9.1 G, aH = 1.0 G) can be attributed to another DBNBS radical adduct. The intensity of the broadened triplet started to decay after 5 min incubation, and was barely detected in the 15 min incubation spectrum (Fig. 2B). The DBNBS/•TenBu signal maintained its intensity during the course of the EPR analysis.

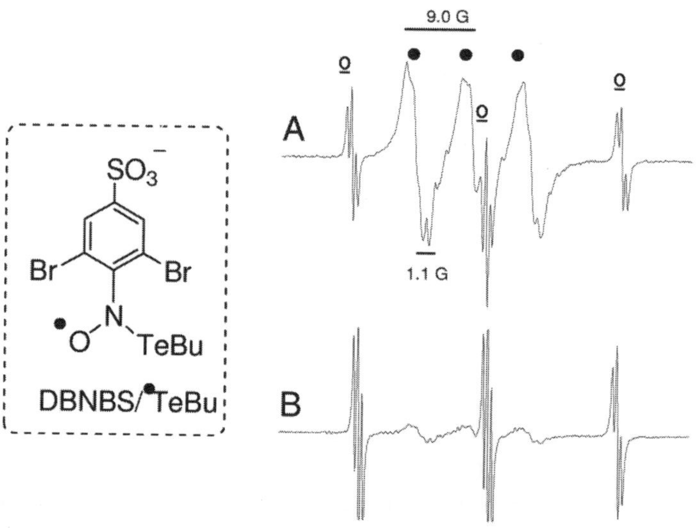

Figure 2. Representative EPR spectra of DBNBS radical adducts obtained during the Te-Li exchange reaction. (A) EPR spectrum obtained after 1 min incubation of the reaction mixture with the DBNBS aqueous solution, (B) EPR spectrum obtained after 15 min incubation of the reaction mixture with the DBNBS aqueous solution; (o) DBNBS/•TenBu radical adduct and (•) transient DBNBS radical adduct.

After the addition of the B(OiPr)3 (ii, Scheme 1), the reaction mixture produced a complex EPR spectra that can be attributed to a mixture of radical species (Figure 3). The addition of the boron reagent generated different radical species from those observed in the previous reaction step (Figure 2).

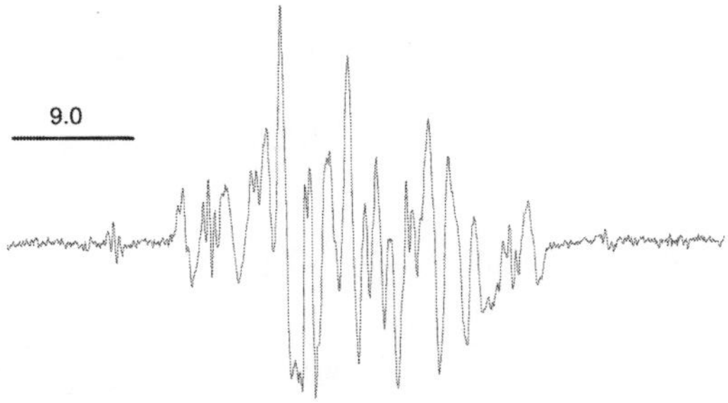

Figure 3. Representative EPR spectrum of DBNBS radical adducts obtained during the Li-Boron exchange reaction. EPR spectrum obtained after 15 min incubation of the reaction mixture with the DBNBS aqueous solution.

We performed control experiments to exclude the possibility of radical generation by the combination of the boron reagent with O2 [18] or by the self-radical generation of the nBuTenBu reagent. Incubation of nBuTenBu, nBuLi and B(OiPr)3 with DBNBS produced no EPR signals (Table 2, entries 3–5). Equimolar solutions of nBuTenBu, nBuLi and DBNBS (Table 2, entry 6) produced a radical signal with similar parameters as those detected during the Te-Li exchange (i, Scheme 1). In the absence of the reducing reagent (nBuLi), an equimolar solution of nBuTenBu, B(OiPr)3 and DBNBS also did not produce EPR signals (Table 2, entry 8).

Table 2. Reactions and Control Experiments Performed

Entry	Reactions	DBNBS radical adducts EPR Hyperfines (G)			
		a_N	a_H	a'_N	a'_H
1	BuTeCH = CHPh+^{n}BuLi+DBNBS	21.6	0.7	9.1	1.0
2	LiCH = CHPh+B(O^{i}Pr)$_3$+DBNBS		complex signal		
3	BuTeBu+DBNBS		no signal		
4	^{n}BuLi+DBNBS		no signal		
5	B(O^{i}Pr)$_3$+DBNBS		no signal		
6	BuTeBu+^{n}BuLi+DBNBS	-	-	9.1	1.0
7	BuTeBu+^{n}BuLi+B(O^{i}Pr)$_3$+DBNBS		complex signal		
8	BuTeBu + B(OiPr)$_3$+ DBNBS		no signal		

To test our proposed mechanism, we repeated the reaction using (Z)-β-bromostyrene, to achieve the desired Z-vinyllithium, the experiments were performed using tBuLi in a solution composed of THF/Et2O/petrol ether, at -120°C, with and without nBuTenBu, instead of Z-vinylic tellurides to examine the effect of the nBuTenBu as the source of the butyl radical. From this reaction, the expected potassium vinyltrifluoroborate salt was not isolated, probably because it is necessary to use experimental conditions [19] that differ from those that were selected to perform the synthesis of the BF3K salts. To maintain the same reaction conditions, other control experiments were performed (Scheme 2).

Scheme 2. Experimental conditions. i: 1 equiv nBuLi, Et2O, -78°C, 30 minutes.ii: 0.8 equiv B(OiPr)$_3$, -20°C, 60 minutes.iii. 3 equiv KHF2 in aqueous solution, -20°C to r.t., 30 minutes.

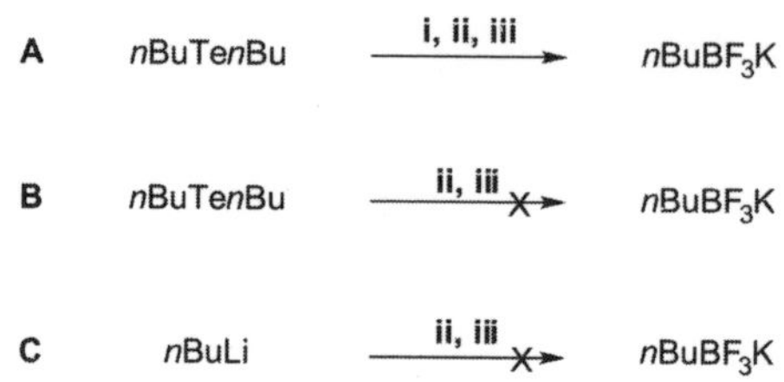

Instead of having the double bond isomerization as a radical pathway model, evidence of nBuTenBu radical behavior came from the detection of nBuBF3K as a product only from experiment A (Scheme 2). With the control experiments (Scheme 2), it was proven that the generation of nBuBF3K salt is dependent on the presence of nBuTenBu, as well as that that occurs during the reaction to prepare the alkenyltrifluoroborate salts.

The results presented above support a free radical pathway for the trans-cis double bond isomerization. Scheme 3 was proposed to account for the E-vinyl and nBuBF3K salts. In the first step, the butyl radical 5 is formed by homolytic cleavage of the nBu-Te bond of the compound 3, caused by the lithium species present in the reaction medium. The second step consists of an attack of 5at the boronic "ate" complex 4, leading to the vinylic radical, which undergoes self-isomerization to the most stable isormer 8. In the third step, the vinylic radical 8 attacks a B(OiPr)3 species, yielding an anionic vinyl boronic "ate" radical. The boron-centered radical is then reduced by a •TenBu radical 6, leading to the E-vinyltrifluoroborate salt 9 after the reaction work up with aqueous KHF2.

Scheme 3. Proposed mechanism of the reaction.

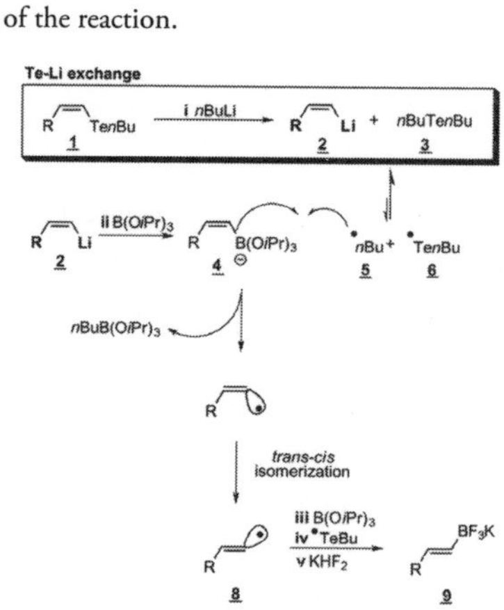

Conclusion

In conclusion, we have identified the radical species that promoted the stereoinversion of vinylic compounds during the preparation of potassium vinyltrifluoroborate salts. The above experiments support the proposed mechanism, which is based on the homolytic cleavage of the TenBu bond.

Acknowledgements

The authors wish to thank FAPESP (Grants 05/59141-6 and 03/01751-8 and the scholarship 04/13978-1-AFK, 03-13897-7-RC), and CNPq agencies for financial support.

References

1. Molander GA, Ellis N: Acc Chem Res. 2007, 40:275.
2. Stefani HA, Cella R, Vieira AS: Tetrahedron. 2007, 63:3623.
3. Cella R, Orfão ATG, Stefani HA: Tetrahedron Lett. 2006, 47:5075.
4. Cella R, Cunha RLOR, Reis AES, Pimenta DC, Klitzke CF, Stefani HA: J Org Chem. 2006, 71:244.
5. Stefani HA, Cella R, Dörr FA, Pereira CMP, Zeni G, Gomes M Jr: Tetrahedron Lett. 2005, 46:563.
6. Stefani HA, Cella R: Tetrahedron. 2006, 62:5656.
7. Zeni G, Lüdtke DS, Panatieri RB, Braga AL: Chem Rev. 2006, 106:1032.
8. Zeni G, Braga AL, Stefani HA: Acc Chem Res. 2003, 36:731.
9. Barros SM, Comasseto JV, Barriel JN: Tetrahedron Lett. 1989, 30:7353.
10. Hirro T, Kambe N, Ogawa A, Miyoshi N, Murai S, Sonoda N: Angew Chem, Int Ed Engl. 1987, 11:1187.
11. Marino JP, Tucci FC, Comasseto JV: Synlett. 1993, 761.
12. Barros SM, Dabdoub MJ, Dabdoub VB, Comasseto JV: Organometallics. 1989, 8:1661.
13. Zeni G, Formiga HB, Comasseto JV: Tetrahedron Lett. 2000, 41:1311.
14. Nesmeyanov AN, Borisov AE: Tetrahedron. 1957, 1:158.
15. Seyferth D, Vaughan LG: J Am Chem Soc. 1964, 86:883.
16. Pozzi D, Scanlan EM, Renaud P: J Am Chem Soc. 2005, 127:14204.

17. Keppler AF, Cerchiaro G, Augusto O, Miyamoto S, Prado F, Di Mascio P, Comasseto JV: Organometallics. 2006, 25:5059.
18. Cadot C, Cossy J, Dalko PI: Chem Commun. 2000, 1017.
19. Neumann H, Seebach D: Tetrahedron Lett. 1976., 52: 4839.

The Role of Heterogeneous Chemistry of Volatile Organic Compounds: A Modeling and Laboratory Study

Gregory R. Carmichael and Vicki H. Grassian

Overview

The outputs of this research have been reported annually via the RIMS system. This report serves as an update and final report. The focus of our DOE BES funded project is on the importance of heterogeneous reactions in the troposphere. The primary objectives of our study were to: (i) Evaluate the extent to which heterogeneous chemistry affects the photochemical oxidant cycle, particularly, sources and sinks of tropospheric ozone; and (ii) Conduct laboratory studies on heterogeneous reactions involving NO_y, O_3 and VOCs on aerosol surfaces. These objectives were pursued through a multidisciplinary approach that combines modeling and laboratory components as discussed in more detail below. In addition, in response to the reconfiguring of the Atmospheric Science Program to focus on aerosol radiative forcing of climate, we also began to investigate the radiative properties of atmospheric aerosol.

Laboratory Studies

We investigated heterogeneous reactions involving ozone, nitrogen oxides and volatile organics on mineral dust and carbonaceous aerosol using a variety of techniques that we have developed over the course of the grant period. In addition to investigating heterogeneous reactions, we begun to measure the optical properties of these tropospheric aerosol so as to further our understanding of the scattering and absorption of aerosol in the context of climate change. Of particular interest here is the relationship between heterogeneous chemistry and changes in the optical properties of mineral dust and carbonaceous aerosol. This helps us understand how aerosol chemistry and climate are interconnected.

Heterogeneous Reactions of O_3 on Mineral Dust

We studied the initial reactive uptake of ozone on several mineral oxides, clays and authentic dust samples. In those studies, a Knudsen cell was used and it was determined that the initial uptake coefficient was 2.0 +/- 0.3x10^{-4} for alpha-Fe_2O_3, 1.2 +/- 0.4x 10^{-4} for α-Al_2O_3, 6.3 +/- 0.9 x 10^{-5} for SiO_2 and 3 +/- 1 x 10^{-5} for kaolinite. The γ_0,BET for authentic dusts, Saharan sand and China loess was determined to be 6 +/- 2 x 10^{-5} and 2.7 +/- 0.8 x 10^{-5}, respectively. It was also shown that after long ozone exposure experiments demonstrated that the mineral oxide surfaces did not passivate, rather they approached non-zero steady-state uptake values, which were lower by approximately 80% from the initial values. However, the mineral oxide powders and dusts exhibited catalytic behavior with the destruction of many more ozone molecules than the total number of surface sites present on the particles. Furthermore, we discussed that the process of ozone adsorption onto mineral oxide surfaces showed a weak temperature dependence, consistent with a low energy of activation.

More recently, we began to consider the effect of coatings on the particle surface. For example, as particles age in the atmosphere they become coated with organic and inorganic coatings. Therefore, laboratory measurements of the reactive uptake on these "clean" powders may be somewhat different from the particles encountered in the atmosphere. In our most recent published study, laboratory experiments were done to simulate changes in the reactivity of mineral dust particles after being processed or aged in the atmosphere. Initial uptake coefficients of ozone on processed and unprocessed dust was measured with a Knudsen cell reactor and the relative reactivities are compared. In particular, the reactive uptake of ozone with mineral oxide particles that had been pretreated by exposure to nitric acid, sulfur dioxide, and organics are compared to particles that had not been pretreated. In some cases, it was found that the reactivity of ozone with pretreated particles was significantly reduced whereas in other cases the reactivity was enhanced. For example, the reactive uptake of ozone decreased by approximately 70% for α-Al2O3 particles coated with a layer of nitrate from reaction of nitric acid compared to particles that did not have a nitrate coating, whereas pretreatment of α-Al2O3with sulfur dioxide showed a 33% enhancement toward ozone reactivity. For organic coatings, it was determined that SiO_2 particles functionalized with a C_8-alkene displayed enhanced reactivity toward ozone by 40% relative to untreated SiO_2, while SiO_2 particles functionalized with a C_8-alkane exhibited decreased reactivity by approximately 40% relative to untreated SiO_2particles. The reaction mechanism of ozone uptake with these particles is attributed to the presence and blocking of particular sites on the particles surface causing both a decrease and increase in the reactivity of the particles. Clearly the interaction of these particles with gases prior to exposure to ozone produces changes in the reactivity of the particles toward ozone. Atmospheric models describing the chemistry of the troposphere should not only take into account heterogeneous reactions of ozone, but also adjust the values used according to the history of the particles. Mineral dusts that have accumulated a coating of nitrate or aliphatic material will affect the ozone less than mineral dusts that have accumulated coatings of sulfite or olefins, and the differences can potentially influence the partitioning of ozone in significant ways.

Another important coating is that of adsorbed water. The Knudsen cell apparatus used in the experiments discussed above were limited to much drier conditions than typically found in the atmosphere. We have begun a series of experiments using an environmental aerosol chamber to investigate the destructive uptake of ozone as a function of relative humidity. Thus far, we have determined that for ozone destruction on iron oxide, the most reactive oxide studied, adsorbed water can influence the rate of ozone destruction. Our studies as well as literature data indicate that water adsorbs onto the most reactive sites on the surface, namely Lewis acid sites. These studies are ongoing and will be completed in the near future.

Heterogeneous Reactions of Organic and Inorganic Acids with Calcium Carbonate as a Function of Relative Humidity: Studies Using Single Particle Analysis and an Environmental Aerosol Chamber

Calcium carbonate is a common component of mineral dust and may in fact be a very reactive component of the aerosol present in the Earth's atmosphere. The heterogeneous chemistry of individual calcium carbonate particles with nitric acid at

293 K has been followed using Scanning Electron Microscopy (SEM) and Energy Dispersive X-Ray (EDX) analysis as a function of time and relative humidity (RH). The rate of calcium carbonate to calcium nitrate conversion is significantly enhanced in the presence of water vapor. As discussed in the last report, the SEM images clearly show that solid $CaCO_3$ particles are converted to spherical droplets as the reaction proceeds. The change in phase of the particles and the significant reactivity of nitric acid and $CaCO_3$ at low RH is a direct result of the deliquescence of the product, calcium nitrate, below 20% RH. The reaction efficiency is enhanced at higher relative humidity. We have done a series of experiments in the last year at the EMSL at Pacific Northwest National Laboratory on reactions of calcium carbonate with organic acids as well as reaction of authentic dust samples with nitric acid. Reaction of calcium carbonate with acetic acid does not show similar morphology changes as that with nitric acid. This is thought to be due to the fact that calcium acetate undergoes deliquescence at much higher relative humidity. Reaction of four different authentic dust samples from different arid regions around the world shows that the reaction with nitric acid with the carbonate component of the dust reacts in the same way as commercial samples.

Complementary experiments using our environmental aerosol chamber which is able to measure the absorption and scattering properties of tropospheric aerosol over a wide range of wavelengths show that the nitric acid reacted carbonate particles scatter infrared radiation differently from the unreacted particles. The observed changes in the optical properties of the aerosol are consistent with the changes in morphology and chemical composition of the aerosol as it reacts. We have been able to model the absorption and scattering properties of unreacted calcium carbonate as well as the reacted particles using Mie theory. Mie theory does a fairly good job in fitting the experimental data, however we are currently working on using other scattering theories to determine if they provide a better fit the experimental data.

Modeling Analysis

We used these field experiments as a way to evaluate the importance of the new chemistry arising from the laboratory. The high dust environments of Asia provide the best real environment for looking at the role of heterogeneous chemistry. Our recent results are summarized below.

Dust influences on regional gas-phase chemistry can be classified into heterogeneous influences and radiative influences, and both were studied. In our analysis we introduced four heterogeneous reactions involving O_3, NO_2, SO_2 and HNO_3 reactions on dusts, with reaction rates determined based on Prof. Grassian's laboratory studies. The C-130 flight 6 was significantly affected by heterogeneous and radiative influences. The O_3 heterogeneous uptake on dust had a significant impact on flight 6, accounting for a 20 ppbv decrease in ozone levels. Only when this reaction was included in the model were we able to represent the observed values. This reaction was shown to cause a broad decrease in background O_3. In polluted areas, this low O_3 background reduced NO_2 production, but caused NO enhancement. The impact of the O3 heterogeneous loss on NO_2 was usually stronger than that due to the direct NO2 reaction on dust. As a result of these reactions HONO levels increased by up to 30% in some polluted areas. The radiative influence of dust on photochemistry was largest for HOx. For ozone, radiative influence of aerosols was large, but the contribution due to dust was not as strong as the influences of the heterogeneous reactions.

The presence of dust was also shown to enhance sulfate production by 10 to 40% in dust rich regions, and to result in an increase in sulfate amount in the coarse fraction. Reactions involving NO_2 and nitric acid were shown to result in the accumulation of nitrate into the aerosol, and this occurs mainly in the coarse mode (however, appreciable amounts may also appear in the fine mode).

We have extended the analysis using a version of our model that incorporated an on-line, size-resolved, aerosol thermodynamics model (SCAPE-II). This model was used to further study the aerosol ion distributions, and factors that influence the composition-size relationships, in the East Asia outflow during the TRACE-P and ACE-Asia periods. Results from the model were compared with various observations, and used to study how the inorganic aerosol composition changed as airmasses travel off the continent and out over the western Pacific.

Dust was shown to strongly affect the aerosol ions and their size distributions. Results from this study indicated that dust alters the partitioning of the semi-volatile components between the gas and aerosol phases as well as the size distributions of the secondary aerosol constituents. A main role of dust in the equilibrium process is through the enhancement of the aerosol calcium concentration, which shifts the equilibrium balance to an anion-limited status. This status benefits the uptake of sulfate and nitrate, but repels ammonium. Surface reactions on dust provide an additional mechanism to produce aerosol nitrate and sulfate. The size distribution of dust was shown to be a critical factor. As much of the dust mass resides in the coarse mode (70-90%), appreciable amounts of sulfate and nitrate are in the super-micron particles. For sulfate the observations and the analysis indicate that 10-30% of sulfate was in the coarse fraction. In the case of nitrate more than 80% was found in the coarse fraction. The strength of dust influence was shown to be determined by its fresh ratio and concentration.

The results of this analysis also point out remaining issues and challenges for the model and measurements. From the modeling perspective the results were found to be very sensitive to the dust mass, its size distribution, assumptions about its extent of equilibrium involvement, and the fraction of the aerosol mass available for reaction. Estimating dust mass and size distributions using emission models is fraught with uncertainty. Results obtained in ACE-Asia and TRACE-P have allowed for more rigorous evaluation of model prediction of dust, and provided more details into the dust size distribution. As a result we have improved our ability to estimate dust emissions in East Asia; but quantitative, episodic, and predictive capabilities remain a challenge.

The issue of how to characterize the chemically active portion of the aerosol presents a challenge. Our modeling studies suggest that an appreciable fraction of the calcium in the aerosol in the outflow while internally mixed may be chemically in-active. Quantifying the chemical – activity state from the observations is difficult, and can not be determined by composition measurements by themselves. Microscopy and single particle techniques provide important information. At present the information provided from the single particle statistics derived from particle numbers and the model mass-based predicts do not correspond. Our results show how these can be used together to provide useful information and aid in the interpretation, but ways to facilitate more quantitative comparisons need to be found.

These results suggest that present day atmospheric models have substantial interpretive and diagnostic capabilities. However the results presented here also point out that improvement in our predictive capability will require substantial reductions in uncertainties. Continued integration of models and measurements are clearly needed. More detailed considerations of the possible effects of surface saturation, as well as competition for reactions on other surfaces such as BC need to be considered. These effects, as well as a comprehensive comparison of across the ACE-ASIA measurements are the subjects that we will focus on in 2004.

Products Delivered

Publications

1. A. D. Clarke, Y. Shinozuka, V. N. Kapustin, S. Howell, B. Huebert, S. Doherty, T. Anderson, D. Covert, J. Anderson, X. Hua, K. G. Moore II, C. McNaughton,G. Carmichael and R. Weber, Size distributions and mixtures of dust and black carbon aerosol in Asian outflow: Physiochemistry and optical properties, J. Geophys. Res., 109, D15S09.
2. Cameron S. McNaughton, Antony D. Clarke, Steven G. Howell, Kenneth G. Moore II, Vera Brekhovskikh, Rodney J. Weber, Douglas A. Orsini, David S. Covert, Gintautas Buzorius, Fred J. Brechtel, Gregory R. Carmichael, Youhua

Tang, Fred L. Eisele, R. Lee Mauldin, Alan R. Bandy, Donald C. Thornton, and Byron Blomquist, Spatial distribution and size evolution of particles in Asian outflow: Significance of primary and secondary aerosols during ACE-Asia and TRACE-P, J. Geophys. Res., 109, D19S06.

3. Carlos-Cuellar, S, Christensen, A.P., Burrichter, C., Li, P. and Grassian, V. H. "Heterogeneous Uptake Kinetics of VOCs on Oxide Surfaces Using a Knudsen Cell Reactor: Adsorption of Acetic Acid, Formaldehyde and Methanol on α-Al2O3, SiO2 and α-Fe2O3" J. Phys. Chem. A. 2003, 107, 4250-4261.

4. Carmichael, G. R., Y. Tang, G. Kurata, I. Uno, D.G. Streets, J.-H. Woo, H. Huang, J. Yienger, B. Lefer, R.E. Shetter, D.R. Blake, A. Fried, E. Apel, F. Eisele, C. Cantrell, M.A. Avery, J.D. Barrick, G.W. Sachse, W.L. Brune, S.T. Sandholm, Y. Kondo, H.B. Singh, R.W. Talbot, A. Bandy, A.D. Clarke, and B.G. Heikes, Regional-Scale chemical transport modeling in support of intensive field experiments: overview and analysis of the TRACE-P observations, Journal of Geophysical Research, 108(D21), 8823, doi:10.1029/2002JD003117, 2003.

5. Conant, W.C., J.H. Seinfeld, J. Wang, G.R. Carmichael, Y. Tang, I. Uno, P.J. Flatau, and P.K. Quinn, A model for the radiative forcing during ACE-Asia derived from CIRPAS Twin Otter and R/V Ronald H. Brown data and comparison with observations, Journal of Geophysical Research, 108(D23), 8661, doi: 10.1029/2002JD003260, 2003.

6. Johnson, E. R. and Grassian, V. H. "Environmental Catalysis of the Earth's Atmosphere: Heterogeneous Reactions on Mineral Dust Aerosol" Environmental Catalysis, Ed. Vicki H. Grassian, CRC Publishing, Boca Raton, FL, 2005.

7. Krueger, B. J., Grassian, V. H., Iedema, M. J., Cowin, J. P. and Laskin, A. "Probing Heterogeneous Chemistry of Individual Atmospheric Particles Using Scanning Electron Microscopy" Analytical Chemistry 2003, 75, 5170-5179.

8. Krueger, B. J., Grassian, V. H. and Laskin, A. "Heterogeneous Chemistry of Individual Mineral Dust Particles from Different Dust Source Regions: The Importance of Particle Mineralogy" Atmospheric Environment 2004, 38, 6253-6261.

9. Krueger, B. J., Grassian, V. H., Laskin, A. and Cowin, J. P., "The Transformation of Solid Atmospheric Particles into Liquid Droplets through Heterogeneous Chemistry: Laboratory Insights into the Processing of Calcium Containing Mineral Dust Aerosol in the Troposphere" Geophys. Res. Letts. 2003 30, 48-1 to 48-4.

10. Krueger, B. J., Grassian, V. H., Wietsma, T. W. and Laskin, A. "Heterogeneous Chemistry of Individual Mineral Dust Particles with Nitric Acid. A Combined CCSEM/EDX, ESEM and ICP-MS Study." (submitted to J. Geophys. Res.).

11. Michel, A. E., Usher, C. R. and Grassian, V. H. "Reactive Uptake of Ozone on Mineral Oxides and Mineral Dusts" Atmos. Env. 2003, 37, 3201-3211.

12. Seinfeld, J., G.R. Carmichael, R. Arimoto, W.C. Conant, F.J. Brechtel, T.S. Bates, T.A. Cahill, A.D. Clarke, P. Flatau, B.J. Huebert, J. Kim, S.J. Masonis, P.K. Quinn, L.M. Russell, P.B. Russell, A. Shimizu, Y. Shinozuka, C. Song, Y. Tang, I. Uno, R.J. Weber, J.H. Woo & X.Y. Zhang, Regional Climatic and Atmospheric Chemical Effects of Asia Dust and Pollution, Bulletin of the American Meteorological Society, doi: 10.1175/BAMS-85-3-367, 2004.

13. Tang, Y., G. R. Carmichael, G. Kurata, I. Uno, R. J. Weber, C.-H. Song, S. K. Guttikunda, J.-H. Woo, D. G. Streets, C. Wei, A. D. Clarke, B. Huebert, and T. L. Anderson, The impacts of dust on regional tropospheric chemistry during the ACE-Asia experiment: a model study with observations, Journal of Geophysical Research, doi:10.1029/2003JD003806, 2004.

14. Tang, Y., G. R. Carmichael, I. Uno, J.-H. Woo, G. Kurata, B. Lefer, R. E. Shetter Tang, Y., G.R. Carmichae1, J.H. Seinfeld, D. Dabdub, R.J. Weber, B. Huebert, A.D. Clarke, S.A. Guazzotti, D.A. Sodeman, K.A. Prather, I. Uno, J.-H. Woo, J.J. Yienger, D.G. Streets, P.K. Quinn, J.E. Johnson, C.-H. Song, V.H. Grassian, A. Sandu, R.W. Talbot and J.E. Dibb, Three-dimensional Simulations of Inorganic Aerosol Distributions in East Asia During Spring 2001. Journal of Geophysical Research, 109, D19S23, doi:10.1029/2003JD004201, 2004.

15. Usher, C. R. , Michel, A. and Grassian, V. H. "Reactions on Mineral Dust," Chemical Reviews 2003, 103, 4883 – 4940 (Invited Review Article).

16. Usher, C. R., Michel, A. E. Stec, D. and Grassian, V. H., "Laboratory Studies of Ozone Uptake on Processed Mineral Dust" Atmos. Env. 2003, 37, 5337.

Total Synthesis of the Indolizidine Alkaloid Tashiromine

Stephen P. Marsden and Alison D. McElhinney

ABSTRACT

Background

Tashiromine 1 is a naturally occurring indolizidine alkaloid. It has been the subject of thirteen successful total syntheses to date. Our own approach centres on the stereoselective construction of the indolizidine core by capture of an electrophilic acyliminium species by a pendant allylsilane. The key cyclisation precursor is constructed using olefin cross-metathesis chemistry, which has the potential to facilitate both racemic and asymmetric approaches, depending upon the choice of the allylsilane metathesis partner.

Results

The use of the allyltrimethylsilane cross-metathesis approach enables the rapid construction of the key cyclisation precursor 3 (3 steps from commercial materials), which undergoes acid-induced cyclisation to give the desired bicyclic indolizidine skeleton as a 96:4 mixture of diastereomers. Simple functional group interconversions allowed the completion of the total synthesis of racemic tashiromine in six steps (19% overall yield). Three chiral α-alkoxyallylsilanes (12,14 and 15) were prepared in enantioenriched form and their cross-metathesis reactions studied as part of a putative asymmetric approach to tashiromine. In the event, α-hydroxysilane 12 underwent isomerisation under the reaction conditions to acylsilane 17, while silanes 14 and 15 were unreactive towards metathesis.

Conclusion

A concise, stereoselective total synthesis of racemic tashiromine has been developed. Attempts to translate this into an asymmetric synthesis have thus far been unsuccessful.

Background

Tashiromine (1) is a naturally occurring indolizidine, isolated from an Asian deciduous shrub Maackia tashiroi. [1] As one of the structurally simpler indolizidine alkaloids, [2] tashiromine has been a popular target for synthetic chemists, and to date has succumbed to total synthesis on thirteen occasions. [3-15] A wide variety of reactions have been employed to assemble the core indolizidine structure, including radical cyclisations;[3] nucleophilic addition to imines;[5,14,15] electrophilic alkylation of pyrroles;[7,13] alkylation of enamines,[6] β-amino esters [8] and pyrrolidinyllithiums;[12] stereoselective reduction of enamines [4,9] and pyridinium salts; [11] and titanium-mediated reductive imide-olefin cyclisation. [10] Our own approach [14] utilises an intramolecular addition of an allylsilane to an N-acyliminium ion to deliver the [4.3.0]-azabicyclic (indolizidine) skeleton 2 (Scheme 1), wherein the pendant vinyl group acts as a handle to install the hydroxymethyl sidechain found in tashiromine. The synthesis of azabicyclic assemblies by intramolecular allylsilane/N-acyliminium cyclisations was first studied by Hiemstra and Speckamp,[16] who prepared their functionalised allylsilane cyclisation precursors (such as 3) by alkylation of cyclic imides with reagent 4 (X = OMs). This, in turn, was prepared in four steps by alkylation of an acetylide anion with commercially available iodomethyltrimethylsilane, followed by partial reduction of the alkyne. Alternative synthetic approaches to 4 (X = OMs, I) involve olefination of aldehydes using the Seyferth-Fleming phosphorane [17] or nickel-catalysed 1,2-metallate rearrangement of lithiated dihydropyran. [18] Our approach was informed by the prior work by our own group [19-24] and others [25-38] on the use of olefin metathesis to generate functionalised allylsilanes.

Specifically, cross-metathesis of N-pentenylsuccinimide 5 with allyltrimethylsilane 6 [39] followed by chemoselective partial reduction of the imide would give the cyclisation precursor 3 in short order. Further, the use of chiral allylsilanes as cross-metathesis partners would potentially facilitate an asymmetric approach to the total synthesis of 1. We report herein full details of the successful synthesis of racemic tashiromine 1 by this strategy,[14] as well as our initial attempts towards an asymmetric variant.

Scheme 1. Retrosynthesis for tashiromine

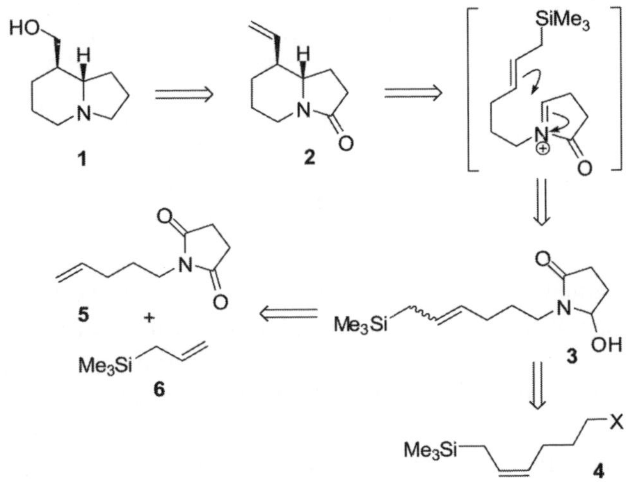

Results and Discussion

Metathesis precursor 5 was prepared by alkylation of the sodium salt of succinimide with 5-bromo-1-pentene in near quantitative yield (Scheme 2). The key cross-metathesis reaction of 5 was carried out using a fourfold excess of allyltrimethylsilane 6 and 5 mol% of Grubbs' second generation catalyst in refluxing dichloromethane. The desired product 7 was formed in 73% yield as an inseparable 3:1 mixture of E- and Z-isomers. Partial reduction with sodium borohydride generated the cyclisation precursor 3 in 86% yield, again as a 3:1 mixture of olefin isomers. Exposure of this mixture to trifluoroacetic acid in dichloromethane at room temperature gave the bicyclic amide 2 in 85% yield as a 96:4 mixture of diastereomers. The identity of the major diastereomer was confirmed by comparison of the spectral data with those of Hiemstra:[16] specifically, the signal for the (ring-fusion) proton at C6 for the major diastereomer appeared as a doublet of triplets with δ = 3.19 ppm, whereas the corresponding signal for the minor diastereomer appeared at δ = 3.67 ppm. The stereochemical outcome of this reaction was rationalised on the basis of the model shown in Scheme 2, whereby nucleophilic addition of the allylsilane to the N-acyliminium ion occurs through a chair-like transition state with the nascent alkene equatorially disposed.

Scheme 2. Stereoselective construction of the indolizidine core 2

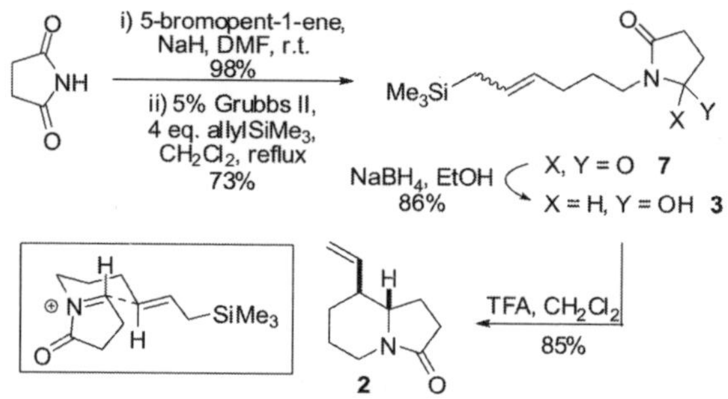

All that remained to complete the synthesis of tashiromine 1 was to effect the oxidative cleavage of the C5 vinyl substituent, then carry out a global reduction of the resulting carbonyl function and the amide. In the event, attempts to form a C5 aldehyde using either ozonolytic or dihydroxylation/periodate alkene cleavage protocols were unsuccessful, with complex mixtures being obtained in both cases. We suspected that the problem lay in the potential for the desired aldehyde to undergo retro-Mannich fragmentation, and so elected to carry out a reductive work-up to the ozonolysis procedure (Scheme 3). The desired alcohol 8 was obtained in a crude form and immediately subjected to reduction with lithium aluminium hydride to give our target tashiromine 1 in 36% yield over two steps. Our stereochemical assignment for the cyclisation of 3 was further corroborated by the agreement of the spectral data for 1 with those previously published in the literature. [3-5,9-12] Additionally, the spectral data for the diastereomeric epi-tashiromine have been reported and differ significantly from those recorded for 1. [10]

Scheme 3. Completion of the total synthesis of tashiromine 1

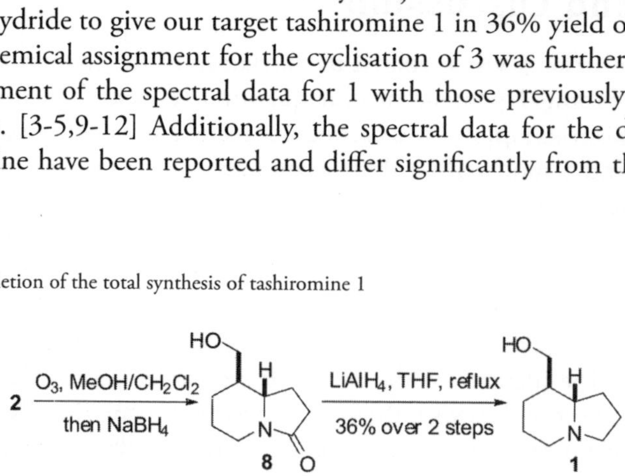

Having completed our target synthesis, our next goal was to investigate an asymmetric approach to tashiromine. Specifically, we envisaged that cyclisation precursors of type 9 ought to be readily available by cross-metathesis of 5 with an appropriate chiral allylsilane followed by chemoselective partial reduction by borohydride. Thereafter, exposure to acid would generate an N-acyliminium ion, which would cyclise through a chair-like transition state with the nascent alkenyl side-chain equatorially disposed, as in the racemic series (Figure 1). The absolute stereochemistry of the newly established asymmetric centres would be controlled by allylic strain arguments, assuming that the well-established precedent for anti-SE2' attack of the iminium on the allylsilane was upheld here. [40] Thus, the predicted major stereoisomer 10 would have (5S, 6S) stereochemistry and an E-configured side-chain, while cyclisation to the predicted minor (5R, 6R) isomer 11 would be disfavoured by A1,3-interactions between the R1 group and vinylic proton (leading to the Z-configured side-chain). This would represent an immolative transfer of chirality approach to tashiromine, since the olefinic side-chains would be cleaved to install the hydroxymethyl side-chain required by the natural product.

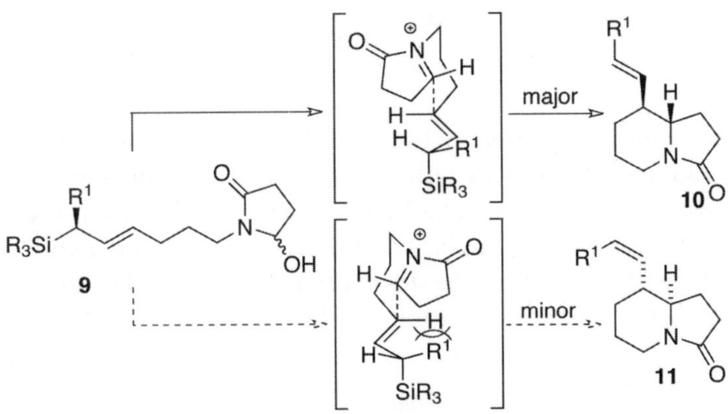

Figure 1. Rationale for stereoselective assembly of the indolizidine core using chiral allylsilanes.

Our approach centred on the readily availability of chiral α-hydroxysilane 12 in enantioenriched format. [41] Protection of the hydroxyl group, either before or after cross-metathesis, would allow access to chiral allylsilanes 9 with R1 being an alkoxy or acyloxy group. Furthermore, this would generate products 10 and/or 11 with a readily oxidised enol-ether/ester side chain for progression to tashiromine. We were, of course, mindful that these functions could potentially act as nucleophiles themselves in the acidic medium of the electrophilic cyclisation, and the investigation of such chemoselectivity issues provided a further impetus for this study. Acylsilane 13 was therefore prepared from propargyl alcohol in four steps then subjected to asymmetric reduction with (-)-DIPCl according to Buynak et al. (Scheme 4). [41] The desired hydroxysilane 12 was obtained in 53% yield and with 91% ee as determined by chiral HPLC analysis. Compound 12 was converted by standard methods to the acetate 14 and the tetrahydropyranyl ether 15. The latter compound was formed as a 1.3:1 mixture of diastereomers which were partially separated by column chromatography – all subsequent reactions were carried out on diastereomerically pure material for ease of analysis.

Scheme 4. Asymmetric synthesis of chiral (alkoxy) allylsilanes

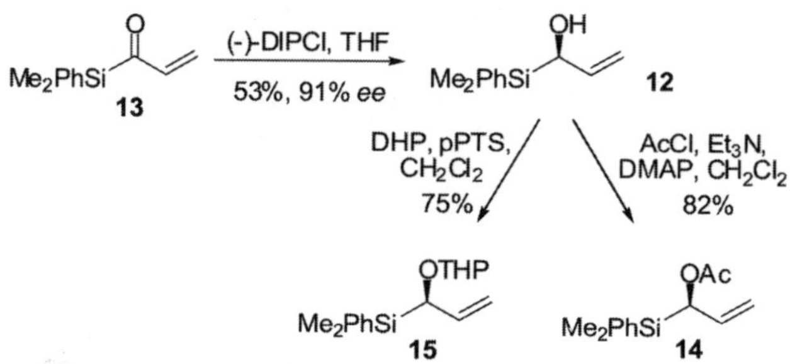

With the requisite enantioenriched allylsilanes in hand, we next investigated their behaviour in olefin cross-metathesis reactions. Unfortunately, neither 14 nor 15 reacted with 5 under the standard cross-metathesis conditions used for trimethylsilane 6; the use of more forcing conditions (elevated temperature and higher catalyst loadings) did not effect the desired transformation, the only product observed being that of homodimerisation of 5 (Scheme 5).

Scheme 5. Attempted cross-metathesis of (alkoxy)allylsilanes

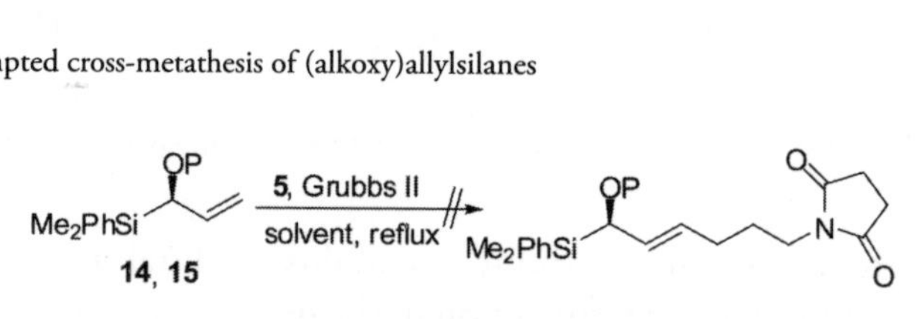

	cat. loading	solvent	outcome
14	5%	CH_2Cl_2	no reaction
14	10%	CH_2Cl_2	no reaction
14	20%	CH_2Cl_2	no reaction
14	10%	toluene	no reaction
15	10%	CH_2Cl_2	**5** homodimerises
15	10%	toluene	**5** homodimerises

Finally, we examined the behaviour of alcohol 12 under cross-metathesis conditions. In the event, two isomerised products were isolated from this reaction (Scheme 6): the internal alkene 16 (formed in 99% yield as a ca. 3:1 mixture of E:Z isomers) and the acylsilane 17. The formation of isomerised alkenes accompanying (or instead of) metathesis processes using ruthenium-based catalysts is well documented, [42-63] as is the formation of carbonyl compounds by isomerisation of the corresponding allylic alcohols. [64-68] At this stage we therefore reluctantly abandoned our investigations into the asymmetric synthesis of tashiromine.

Scheme 6. Competing isomerisation processes in attempted cross-metathesis of (hydroxy) allylsilane 12

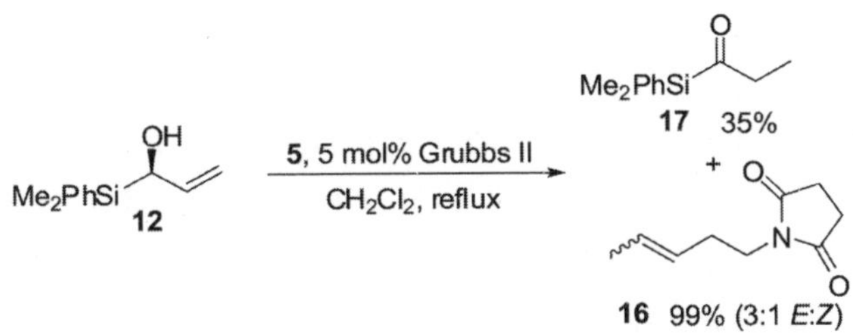

Conclusion

A concise, stereoselective total synthesis of racemic tashiromine has been developed (six steps from succinimide, 19% overall yield) in which the key steps are the preparation of a functionalised allylsilane by olefin cross-metathesis and the construction of the indolizidine core by intramolecular addition of the allylsilane to an N-acyliminium ion. Attempts to translate this into an asymmetric synthesis utilising cross-metathesis reactions of chiral α-alkoxysilanes have thus far been unsuccessful.

Acknowledgements

We thank the EPSRC for a studentship (ADM) and Pfizer and Merck for generous unrestricted research funding.

References

1. Ohmiya S, Kubo H, Otomasu H, Saito K, Murakoshi I: Heterocycles. 1990, 30:537–542.

2. Michael JP: Natural Product Reports. 2007, 24:191–222 and references therein.
3. Beckwith ALJ, Westwood SW: Tetrahedron. 1989, 45:5269–5282.
4. Haddad M, Celerier JP, Haviari G, Lhommet G, Dhimane H, Pommelet JC, Chuche J: Heterocycles. 1990, 31:1251–1260.
5. Nagao Y, Dai W-M, Ochiai M, Tsukagoshi S, Fujita E: J Org Chem. 1990, 55:1148–1156.
6. Paulvannan K, Stille JR: J Org Chem. 1994, 59:1613–1620.
7. Gage JL, Branchaud BP: Tetrahedron Lett. 1997, 38:7007–7010.
8. Ha D-C, Park S-H, Choi K-S, Yun C-S: Bull Korean Chem Soc. 1998, 19:728–730.
9. David O, Blot J, Bellec C, Fargeau-Bellassoued M-C, Haviari G, Celerier J-P, Lhommet G, Gramain J-C, Gardette D: J Org Chem. 1999, 64:3122–3131.
10. Kim S-H, Kim S-I, Lai S, Cha JK: J Org Chem. 1999, 64:6771–6775.
11. Bates RW, Boonsombat J: J Chem Soc, Perkin Trans 1. 2001, 654–656.
12. Dieter RK, Watson R: Tetrahedron Lett. 2002, 43:7725–7728.
13. Banwell MG, Beck DAS, Smith JA: Org Biomol Chem. 2004, 2:157–159.
14. McElhinney AD, Marsden SP: Synlett. 2005, 2528–2530.
15. Belanger G, Larouche-Gauthier R, Menard F, Nantel M, Barabe F: J Org Chem. 2006, 71:704–712.
16. Hiemstra H, Sno MHAM, Vijn RJ, Speckamp WN: J Org Chem. 1985, 50:4014–4020.
17. Paquette LA, Mendez-Andino JL: J Org Chem. 1998, 63:9061–9068.
18. Bergmeier SC, Seth PP: J Org Chem. 1997, 62:2671–2674.
19. Cassidy JH, Marsden SP: Synlett. 1997, 1411–1413.
20. Miles SM, Marsden SP, Leatherbarrow RJ, Coates WJ: J Org Chem. 2004, 69:6874–6882.
21. Miles SM, Marsden SP, Leatherbarrow RJ, Coates WJ: Chem Commun. 2004, 2292–2293.
22. Akindele T, Marsden SP, Cumming JG: Org Lett. 2005, 7:3685–3688.
23. Akindele T, Marsden SP, Cumming JG: Tetrahedron Lett. 2005, 46:7235–7238.
24. Cassidy JH, Farthing CN, Marsden SP, Pedersen A, Slater M, Stemp G: Org Biomol Chem. 2006, 4:4118–4126.

25. Teare H, Huguet F, Tredwell M, Thibaudeau S, Luthra S, Gouverneur V: [http:// content.arkat-usa.org/ ARKIVOC/ JOURNAL_CONTENT/ manuscripts/ 2007/ AK-2285GP as published mainmanuscri pt.pdf]ARKIVOC. 2007, 232–244.
26. Thibaudeau S, Gouverneur V: Org Lett. 2003, 5:4891–4893.
27. Alvarez Corral M, Lopez Sanchez C, Jimenez Gonzalez L, Rosales A, Munoz Dorado M, Rodriguez Garcia I: Beilstein J Org Chem. 2007, 3.
28. Jimenez Gonzalez L, Garcia Munoz S, Alvarez Corral M, Munoz Dorado M, Rodriguez Garcia I: Chem Eur J. 2007, 13:557–568.
29. Jimenez Gonzalez L, Garcia Munoz S, Alvarez Corral M, Munoz Dorado M, Rodriguez Garcia I: Chem Eur J. 2006, 12:8762–8769.
30. Garcia Munoz S, Jimenez Gonzalez L, Alvarez Corral M, Munoz Dorado M, Rodriguez Garcia I: Synlett. 2005, 3011–3013.
31. Jimenez Gonzalez L, Alvarez Corral M, Munoz Dorado M, Rodriguez Garcia I: Chem Commun. 2005, 2689–2691.
32. Vedrenne E, Dupont H, Oualef S, Elkaim L, Grimaud L: Synlett. 2005, 670–672.
33. He A, Yan B, Thanavaro A, Spilling CD, Rath NP: J Org Chem. 2004, 69:8643–8651.
34. BouzBouz S, De Lemos E, Cossy J: Adv Synth Cat. 2002, 344:627–630.
35. Meyer C, Cossy J: Tetrahedron Lett. 1997, 38:7861–7864.
36. Taylor RE, Engelhardt FC, Schmitt MJ, Yuan H: J Am Chem Soc. 2001, 123:2964–2969.
37. Taylor RE, Engelhardt FC, Yuan H: Org Lett. 1999, 1:1257–1260.
38. Chang SB, Grubbs RH: Tetrahedron Lett. 1997, 38:4757–4760.
39. Crowe WE, Goldberg DR, Zhang ZJ: Tetrahedron Lett. 1996, 37:2117–2120.
40. Fleming I, Barbero A, Walter D: Chem Rev. 1997, 97:2063–2192.
41. Buynak JD, Strickland JB, Lamb GW, Khasnis D, Modi S, Williams D, Zhang H: J Org Chem. 1991, 56:7076–7083.
42. Miller SJ, Blackwell HE, Grubbs RH: J Am Chem Soc. 1996, 115:9606–9614.
43. Hoye TR, Promo MA: Tetrahedron Lett. 1999, 40:1429–1432.
44. Maynnard HD, Grubbs RH: Tetrahedron Lett. 1999, 40:4137–4140.
45. Edwards SD, Lewis T, Taylor RJK: Tetrahedron Lett. 1999, 40:4267–4270.

46. Fürstner A, Thiel OR, Ackermann L, Schanz H-J, Nolan SP: J Org Chem. 2000, 65:2204–2207.
47. Cadot C, Dalko PI, Cossy J: Tetrahedron Lett. 2002, 43:1839–1841.
48. Arisawa M, Terada Y, Nakagawa M, Nishida A: Angew Chem, Int Ed. 2002, 41:4732–4734.
49. Sutton AE, Seigal BA, Finnegan DF, Snapper ML: J Am Chem Soc. 2002, 124:13390–13391.
50. Wipf P, Rector SR, Takahashi H: J Am Chem Soc. 2002, 124:14848–14849.
51. Schmidt B: Eur J Org Chem. 2003, 68:816–819.
52. Sworen JC, Pawlow JH, Case W, Lever J, Wagener KB: J Mol Cat A. 2003, 194:69–78.
53. Lehman SE, Schwendeman JE, O'Donnell PM, Wagener KB: Inorg Chim Acta. 2003, 345:190–198.
54. Peczuh MW, Snyder NL: Tetrahedron Lett. 2003, 44:4057–4061.
55. Alcaide B, Almendros P, Alonso JM: Chem Eur J. 2003, 9:5793–5799.
56. Kotha S, Mandal K: Tetrahedron Lett. 2004, 45:1391–1394.
57. Schmidt B: J Org Chem. 2004, 69:7672–7687.
58. Hong SH, Sanders DP, Lee CW, Grubbs RH: J Am Chem Soc. 2005, 127:17160–1716.
59. Bennasar ML, Roca T, Monerris M, Garcia-Diaz D: J Org Chem. 2006, 71:7028–7034.
60. Hanessian S, Giroux S, Larsson A: Org Lett. 2006, 8:5481–5484.
61. Courchay FC, Sworen JC, Ghiviriga I, Abboud KA, Wagener KB: Organometallics. 2006, 25:6074–6086.
62. Raju R, Allen LJ, Le T, Taylor CD, Howell AR: Org Lett. 2007, 9:1699–1701.
63. Moiese J, Arseniyadis S, Cossy J: Org Lett. 2007, 9:1695–1698.
64. Gurjar MK, Yakambram P: Tetrahedron Lett. 2001, 42:3633–3636.
65. Greenwood ES, Parsons PJ, Young MJ: Synth Commun. 2003, 33:223–228.
66. Werner H, Grunwald C, Stuer W, Wolf J: Organometallics. 2003, 22:1558–1560.
67. Edlin CD, Faulkner J, Fengas D, Knight CK, Parker J, Preece I, Quayle P, Richards SN: Synlett. 2005, 572–576.
68. Finnegan D, Seigal BA, Snapper ML: Org Lett. 2006, 8:2603–2606.

Application of Stir Bar Sorptive Extraction to Analysis of Volatile and Semivolatile Organic Chemicals of Potential Concern in Solids and Aqueous Samples from the Hanford Site

J. M. Frye and J. M. Kunkel

Introduction

Stir bar sorptive extraction (SBSE) was applied to aqueous and solid samples for the extraction and analysis of organic compounds from the Hanford Site chemicals of potential concern (COPC) list, as identified in the vapor data quality objectives. The 2224 Laboratory analyzed these compounds from vapor samples on thermal desorption tubes (TDU) as part of the Hanford Site industrial hygiene vapor sampling effort. For this reason, these compounds were chosen for testing SBSE (interoffice memorandum7S110-JMF-07-132, "Development of Stir Bar Sorptive Extraction as Applied to Hanford Tank Waste Liquids and Solids or Soils").

The extractions were performed by either immersing the SBSE in the aqueous sample for a set period of time and a set revolutions per minute, or by suspending it above the surface of the sample in a sealed headspace vial for a set period of time and temperature. Unlike classical extraction techniques in which the compounds of interest are extracted from a complicated matrix using liquid-liquid extraction, in SBSE the compound of interest is extracted into the solid sorbent phase, which is coated onto the surface of a glass stir bar. The SBSEs used for this test were GERSTEL-Twister© (Twister) bars. Twisters are glass-covered magnetic stir bars that have a thick film (0.5 mm) of polydimethylsiloxane (PDMS). The PDMS is used to extract nonpolar compounds from polar matrices. It is possible to obtain an indication of how well a particular compound will extract into the Twister using the octanol/water coefficient (KOw). Compounds with extremely low KOw are less likely to be absorbed into the PDMS phase. SBSE technique was applied successfully to the less polar volatile organic analysis (VOA) compounds on the COPC list on both solids and liquids, and to semivolatile organic analysis (SVOA) compounds on the list from liquids.

Procedures ATS-LT-523-161, "222-S Laboratory Analysis of Semivolatile Organic Compounds Collected by Twister Stir-bar Sorptive Extraction by Gas Chromatography / Mass Spectrometry," and ATS-LT-523- 162, "Analysis of Volatile Organic Compounds Collected on Twister Stir Bar Sorptive Extractors by Gas Chromatography/Mass Spectrometry," were developed as a result of this research. Experimental notes and results are documented in controlled notebook HNF-N-422, 2, Analytical Development Projects, pages 35 to 63 and 66 to 93. The compounds in Table 1 were successfully analyzed using this technique. The SVOA liquid method detection limits (MDL) were determined on two different gas chromatography/mass spectrometry (GC/MS) instruments, identified in the laboratory database as Spock and Sulu. The SVOA results obtained on these two different instruments were comparable. The MDL values determined for the same compounds from TDU tubes using a modified U. S. Environmental Protection Agency method are listed for comparison (Compendium Method TO-17, Determination of Volatile Organic Compounds in Ambient Air Using Active Sampling onto Sorbent Tubes). The TDU VOA MDLs were determined from CarbotrapTM-300 tubes, and the TDU SVOA MDLs were determined from Carbotrap 150 tubes in 2007 from GC/MS Scotty and GC/MS Sulu. These MDLs are recalculated periodically and change.

Table 1. Method Detection Limit Comparison between Twister and Thermal Desorption Tubes Methods

Compound	VOA from Solid, MDL (ng)	VOA from Liquid, MDL (ng)	SVOA from Liquid, MDL (ng)	MDLs Using TDU Tubes (ng)
Instruments	Sulu	Sulu	--	Scotty
2-Methylfuran	10.78	20.57	NA	5.95
2,5-Dihydrofuran	10.75	16.00	NA	6.68
Butanal	34.43	31.92	NA	10.53
Benzene	6.98	9.99	NA	6.91
Butanenitrile	7.08	14.59	NA	20.32
3-Butene-2-one	11.32	NA	NA	6.49
1-Butanol	15.29	NA	NA	7.41
Pyridine	9.26	14.75	NA	16.83
2-Hexanone	9.66	11.74	NA	5.26
Pentanenitrile	14.48	12.09	NA	5.11
2,4-Dimethylpyridine	9.56	10.08	NA	12.60
Hexanenitrile	12.64	8.05	NA	3.82
Instruments	--	--	Spock, Sulu	Sulu
Biphenyl	NA	NA	3.73, 2.77	1.49
Diethylphthalate	NA	NA	6.20, 3.25	6.56
Dibutylbutylphosphonate	NA	NA	4.05, 2.39	1.03
Tributylphosphate	NA	NA	4.10, 4.06	1.26

Some VOA compounds like 3-Butene-2-one and 1-Butanol were too polar to be extracted from the aqueous phase with good repeatability into the PDMS phase of the Twister. Note that butanal, which is also a small polar compound, had an elevated MDL when compared to the other compounds. The SVOA compounds could not reliably be extracted from solids. The MDL values obtained for the less polar, longer chain VOA compounds, and the SVOA compounds are in the same range as those determined off of TDU tubes.

Experimental Conditions

Semivolatile Organic Extraction and Analysis

Semivolatile Organic Analysis from Liquids Extraction

The SVOA from liquids were extracted by placing 10 mL of water in a 10-mL glass headspace vial. The water was adjusted to pH 10 with five drops of 0.125M NaOH.Calibration standards were added to the vials along with a Twister stir bar extractor. The vials were capped and stirred at 1000 rpm for 60 minutes. After this extraction time, the Twisters were removed from the vial and placed in an empty glass TDU tube and desorbed on the GERSTEL TDS 33 at 280 "C. The analysis was run on an Agilent4 6890GC/Agilent 5973MSD equipped with a GERSTEL TDS 3, using the conditions listed in the following:

Initial column temperature:	40°C for 1 minute
Column temperature program:	40-300°C at 15 °/minute, 300-315°C at 6 °/minute, hold 6 minutes
Final column temperature:	315°C for 6 minutes
Column:	Agilent 122-5536 DB-5 ms, 30 m x 0.25mm x 0.5 pm, (or equivalent) max temperature 325°C installed in back inlet.
Transfer line temperature:	300°C
Back inlet, GERSTEL CIS 3:	Back inlet mode: solvent vent
Initial temp:	-30°C
Equilibration time:	0.20 minutes
Initial time:	0.20 minutes
Ramp rate:	12 °C/s
Final temperature:	300°C
Hold time:	3.00 minutes
Carrier gas:	Helium at flow recommended by manufacturer.
GC run time:	29.83 minutes

The following conditions were used on the Gerstel TDS3:

TDS 3 parameters:

- Sample mode: Sample remove
- Flow mode: Splitless
- Initial temp: 30°C
- Initial time: 0.00 minutes
- Delay time: 0.50 minutes
- Purge time: 0.00 minutes
- Transfer temp: 300°C
- 1st Rate: 60 Wminutes
- 1st Final temp: 280 "C

- 1st Final time: 8.00 minutes Znd Rate: 0 “C
- 2nd Final temp: 0 “C
- 2nd Final time: 0.00 minutes

Run parameters:

- Run time: 15.00 minutes

Solutions were adjusted to pH 10 to mimic what might be encountered from Hanford radioactive tank waste samples, which are strongly basic. The pH 10 solutions did increase the number of siloxane type compounds present in the chromatograms. If the samples to be analyzed are not basic, then the calibration and samples could be run without adjusting the pH to 10. Tank simulant samples were extracted at pH values above pH 10, but fewer siloxane peaks were seen and better recovery oftarget compounds were obtained when the pH was adjusted back down to pH 10 with dilute nitric acid. For this comparison, an aliquot (1 mL) of tank simulant was added to 10 mL water in the Twister vial and spiked with a laboratory control standard (LCS). Another vial was prepared in the same way, but the pH was adjusted with 0.5 N nitric acid. The amount of nitric acid needed to bring the solution back down to pH 10 nearly doubled the volume in the vial. Extractions and analysis were performed in the same manner.

Semivolatile Organic Analysis from Solids Extraction

Efforts were made to determine if SVOA compounds could be extracted from a solid by suspending the Twister above the solid in a sealed headspace vial. Glass beads, 100 to 120 mesh size, were used to simulate a clean solid. These efforts were unsuccessful as the SVOA compounds had insufficient vapor pressure to be determined from a solid with the necessary accuracy, even when heated up to 200°C dry or to 135°C with a few drops of water added.

Volatile Organic Extraction and Analysis

Volatile Organic Analysis from Liquids Extraction Conditions

Liquids were extracted by placing 10 mL of water adjusted to pH 10 using five drops of 0.125M NaOH in a 10-mL Twister vial. Approximately 1 g of sodium sulfate, measured with a scoop, was added to the vials. The vials were spiked with 0.5 pL of calibration standards and 0.05 pL of internal standard solution. A Twister was added, and the vial was capped immediately. The vials were allowed to stir at room temperature at 1000 rpm for 60 minutes.

Volatile Organic Analysis from Solids Extraction Conditions

Approximately 1 g, measured volumetrically with a marked scoop, of 100- to 200-mesh glass beads were placed in a 20-mL headspace vial. The glass beads were spiked with calibration standards, a Twister was suspended above the glass beads, and the vial was sealed. The vials were heated at 65°C in a block heater for 2 hours. The Twisters were then removed and placed in an empty glass TDU tube and desorbed at 250°C. Initially in these studies the Twisters were heated to 75°C for the extraction and 280°C for the desorption. While this worked and produced usable calibration curves, it caused larger siloxane-type artifact peaks than using the lower temperature.

Volatile Organic Analysis Conditions

Gas Chromatography Conditions

Initial column temperature:

ï 40°C for 6 minutes

Column:

ï Restek Rtx-VMS,

ï 60M x 0.25 mm with a

ï 1.4-micron film thickness.

Column temperature program:

ï 40-80°C at 6 degrees/minute,

ï hold for 5 minutes, 8°C/minute

ï to 95°C, hold for 11 minutes,

ï 12°C/minutes, to 197°C

Final column temperature:

ï 197°C for 5 minutes

Run time:

ï 44.04 minutes

Transfer line temperature:

ï 300°C

Carrier gas:

ï Helium at flow recommended by manufacturer.

Mass Spectrometrv Conditions

Electron energy:

- 70 V (nominal)

Mass range:

- 30-300 amu

Scan time:

- Not to exceed 1 second/scan.

Results

Initial evaluation of the K_{Ow} values for the compounds on the COPC list is shown in Table 2.

In "New Concepts in Sorption Based Sample Preparation for Chromatography," Baltussen (2000) notes that the KOw value for a compound can be used to predict the Kpdms value for a compound. In other words, if the KOw is low, the Kpdms predicted to be low as well. Based on this prediction, compounds with very low KOw values like acetonitrile, N-nitrosopyrollidine, N-Nitrosodiethylamine, N-Nitrosomethlyethylamine, and N-Nitrosodimethylamine would not be expected to work.

Table 2. Chemicals of Potential Concern Compound &Values

Semivolatile Compound	CAS Number	Log KoW*
Semivolatile Compound	**CAS Number**	**Log KoW***
Biphenyl	92-52-4	3.76
Diethylphthalate	84-66-2	2.65
Dibutylbutylphosphonate	78-46-6	3.83
Tributylphosphate	126-73-8	3.82
N-Nitrosodimethylamine	62-75-9	-0.64
2,6-Di-tetra-butyl-p-cresol	128-37-0	5.03
Volatile compounds		
Acetonitrile	75-05-8	-0.15
Benzene	71-43-2	1.99
Butanal	123-72-8	0.82
1-Butanol	71-36-3	0.84
3-Buten-2-one	78-94-4	0.41
Butanenitrile	109-74-0	0.84
2,4-Dimethylpyridine	108-47-4	1.90
2,5-Dihydrofuran	1708-29-8	0.72
Furan	110-00-9	1.36
2-Hexanone	591-78-6	1.24
Hexanenitrile	628-73-9	1.82
2-Methylfuran	534-22-5	1.91
Pentanenitrile	110-59-8	1.33
Propionitrile	107-12-0	0.35
Pyridine	110-86-1	0.80
N-nitrosodimethylamine	62-75-9	-0.64
N-nitrosomethylethylamine	10595-95-6	-0.15
Methylene Chloride	75-09-2	1.34
N-Nitrosodiphenylamine	86-30-6	3.16
N-Nitrosodi-n-propylamine	621-64-7	1.33
N-Nitrosodiethylamine	55-18-5	0.34
N-Nitrosodi-n-butylamine	924-16-3	2.31
N-Nitrosopyrollidine	930-55-2	0.23
N-Nitrosopiperidine	100-75-4	0.72
SVOA Internal Standards		
1,4-Dichlorobenzene-d4	3855-82-1	NA
Naphthalene-d8	1146-65-2	3.17
Acenaphthalene-d10	15067-26-2	NA
Phenanthrene-d10	1517-22-2	NA
Chrysene-d12	1719-03-5	NA
Perlene-d12	1520-96-3	NA
VOA Internal Standards		
Hexafluorobenzene	392-56-3	3.20
Trifluoromethylbenzene	98-08-8	2.96
1-Bromo-4-fluorobenzene	460-00-4	3.08
Bromobenzene-d5	4165-57-5	NA

* KowWin Program, Interactive LogKow (KowWin) Demo, Syracuse Research Corporation, Copyright 1999, http://www.svres.com/eSc/est_kowdemo.htm

In this work the chemicals marked in bold in Table 2 were evaluated for analysis using Twisters to extract them from liquids or solids. Some of the other compounds listed on the COPC list were not available as standards. N-nitrosodimethylamine and Nitrosomethylethylamine were tried even though the KOw values indicated that they might not work and as expected, they did not work. However, KOw values are not a foolproof predictor for how well a compound will work. The 2,6-ditetra-butyl-p-cresol, and furan have high enough KOw values that one would expect them to extract with Twister, but the laboratory was unable to get either of them to repeatedly extract from liquid or solid matrices. When examining a homologous series such as acetonitrile through hexanenitrile, it is seen that the KOw values increase from acetonitrile, which has a KOw of -0.15, to hexanenitrile, which has a KOw of 1.82. The MDLs for these nitriles get progressively lower as the compounds get larger with higher KOw values, indicating increased sensitivity for each compound in the series as the carbon chain gets progressively longer. Acetonitrile is recommended by the manufacturer of the Twister as a solvent of choice for cleaning the Twisters because of the low KOw value. Propionitrile could not be extracted from liquids or solids using Twister, but butanenitrile, pentanenitrile, and hexanenitrile were successfully extracted and determined by the method.

Figure 1 is an example of a chromatogram of an LCS for the VOAs.

Figure 2 shows an example of a chromatogram of a LCS standard for SVOAs extracted and analyzed using Twister.

Conclusions

Repeatability studies were performed for VOA and SVOA compounds by analyzing replicate extractions of a mid-range standard from a source independent from the calibration standard, (LCS). Results of these studies are shown in Tables 3 and 4. All results are given in percent recovery.

The results of a SVOA LCS determination in diluted tank simulant are shown in Table 5. Terminal Liquor Standard (Terliq), a nearly saturated, extremely basic solution was used as a tank simulant. The Terliq standard is composed of a high concentration of aluminum, sodium nitrate, and sodium nitrite in 50% sodium hydroxide with lower quantities of sodium chloride, sodium fluoride, sodium sulfate, and sodium phosphate, tribasic decahydrate (LR-332-111, "Terminal Liquor Standards").

The results in the sample, adjusted to pH 10, were within the administrative limits for standard percent recovery of 65 to135Y0.

The method was applied to analysis of tank simulants containing oil reported in interoffice memorandum WRPS-0800035, "Report for Hydraulic Fluid Interaction Study." The samples were successfully analyzed for SVOA compounds. Analysis of SVOA standards in pH 10 water and standards spiked into basic solutions of tank simulants can be successfully carried out using Twister stir bar sorptive extraction providing the pH of the solution is kept to about pH 10 or less. Volatile organic analysis compounds can be analyzed from solids or from pH 10 liquids.

Figure 1. Volatile Organic Analysis Chromatogram Example

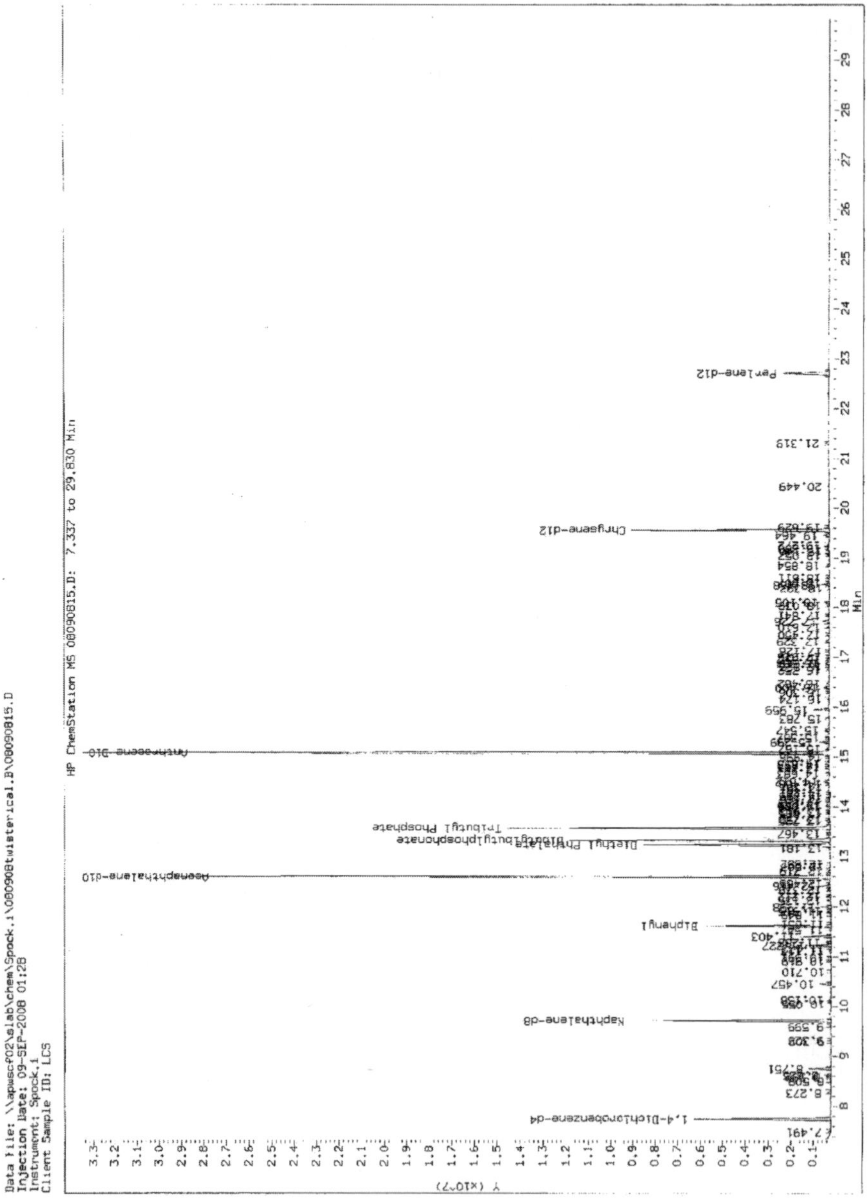

Figure 2. Semivolatile Organic Analysis Chromatogram Example

Table 3. Volatile Organic Analysis Repeatability

Compound from Solids	Number of Replicates	Average % Recovery	Standard Deviation	-1 σ	+1 σ
2-Methylfuran	7	104.41	27.61	76.80	132.02
2,5-Dihydrofuran	7	103.55	10.13	93.42	113.69
Benzene	7	108.23	10.57	97.67	118.80
Pyridine	7	99.25	8.19	91.07	107.44
2-Hexanone	7	102.65	5.99	96.65	108.64
Pentanenitrile	7	120.51	7.10	113.41	127.61
Hexanenitrile	7	111.36	6.09	105.27	117.45
1-Butanol	7	111.90	16.54	95.36	128.44
Butanenitrile	7	97.23	4.56	92.67	101.80
2,4-Dimethylpyridine	7	101.58	6.03	95.55	107.61
Butanal	7	107.21	8.02	99.19	115.23
3-Buten-2-one	7	115.07	13.48	101.59	128.55
Compounds from Liquid					
2-Methylfuran	7	130.68	30.92	99.77	161.60
2,5-Dihydrofuran	7	89.54	17.92	71.61	107.46
Benzene	7	120.29	9.87	110.42	130.17
Pyridine	7	100.92	17.55	83.37	118.47
2-Hexanone	7	98.57	16.80	81.77	115.37
Pentanenitrile	7	111.69	12.48	99.20	124.17
Hexanenitrile	7	105.82	10.78	95.05	116.60
Butanenitrile	7	84.72	12.72	72.00	97.43
2,4-Dimethylpyridine	7	90.44	10.82	79.62	101.27
Butanal	7	91.54	11.70	79.84	103.24

Table 4. Semevolatile Organic Analysis Repeatibility

Compounds from Liquid	Number of replicates	Average % Recovery	Standard Deviation	-1 σ	+1 σ
Biphenyl	21	115.55	15.80	99.75	131.36
Diethylphthalate	21	92.03	26.75	65.28	118.78
Dibutylbutylphosphonate	21	98.77	35.22	63.55	133.99
Tributylphosphate	21	100.96	27.80	73.16	128.76

Table 5. Percent Recovery of LCS Standards in Terliq with Twister

Compound	1mL Terliq + LCS pH>10	1mL Terliq + LCS pH =10
Biphenyl	95.6	102.3
Diethyl phthalate	61.9	98.6
Dibutylbutyl phosphonate	69.0	74.8
Tributyl phosphate	63.9	82.5

References

1. 7S110-JMF-07- 132, 2007, "Development of Stir Bar Sorptive Extraction as Applied to Hanford Tank Waste Liquids and Solids or Soils" (interoffice memorandum from J. M. Frye to K. M. Hall, dated October 9), CH2M HILL Hanford Group, Inc., Richland, Washington.
2. Baltussen H. A,, "New Concepts in Sorption Based Sample Preparation for Chromatography," E. G. Gerstel, H. Gerstel, R. Bremer, GERSTEL GmbH & Co, KG, Technische Universiteit Eindhoven, 2000, Proefschrift, ISBN 90-386-2971-0, NUGI 813.
3. Compendium Method TO-17, 1999, Determination of Volatile Organic Compounds in Ambient Air Using Active Sampling onto Sorbent Tubes, Second Edition, U. S. Environmental Protection Agency, Cincinnati, Ohio.
4. KowWin Program, Interactive LogKow (KowWin) Demo, Syracuse Research Corporation, Copyright 1999, http://www.swes.condeSc/est kowdemo.htm
5. LR-332-111, Rev. 1-0, "Terminal Liquor Standards," Advanced Technologies and Laboratories International, Inc., Richland, Washington.
6. WRPS-0800035, 2008, interoffice memorandum, "Report for Hydraulic Fluid Interaction Study" (interoffice memorandum from H. J. Huber to W. G. Barton, dated October 14), Washington River Protection Solutions LLC, Richland, Washington.

Analogues of Amphibian Alkaloids: Total Synthesis of (5R,8S,8as)-(-)-8-Methyl-5-Pentyloctahydroindolizine (8-Epi-Indolizidine 209B) and [(1S,4R,9as)-(-)-4-Pentyloctahydro-2H-Quinolizin-1-Yl]Methanol

Joseph P. Michael, Claudia Accone, Charles B. de Koning and Christiaan W. van der Westhuyzen

ABSTRACT

Background

Prior work from these laboratories has centred on the development of enaminones as versatile intermediates for the synthesis of alkaloids and other nitrogen-containing heterocycles. In this paper we describe the enantioselective synthesis of indolizidine and quinolizidine analogues of bicyclic amphibian alkaloids via pyrrolidinylidene- and piperidinylidene-containing enaminones.

Results

Our previously reported synthesis of racemic 8-epi-indolizidine 209B has been extended to the laevorotatory enantiomer, (-)-9. Attempts to adapt the synthetic route in order to obtain quinolizidine analogues revealed that a key piperidinylidene-containing enaminone intermediate (+)-28 was less tractable than its pyrrolidinylidene counterpart, thereby necessitating modifications that included timing changes and additional protection-deprotection steps. A successful synthesis of [(1S,4R,9aS)-4-pentyloctahydro-2H-quinolizin-1-yl]methanol (-)-41 from the chiral amine tert-butyl (3R)-3-{benzyl [(1R)-1--phenylethyl]amino}octanoate (+)-14 was achieved in 14 steps and an overall yield of 20.4%.

Conclusion

The methodology reported in this article was successfully applied to the enantioselective synthesis of the title compounds. It paves the way for the total synthesis of a range of cis-5,8-disubstituted indolizidines and cis-1,4-disubstituted quinolizidines, as well as the naturally occurring trans-disubstituted alkaloids.

Background

The astonishingly diverse range of alkaloids isolated from the skins of amphibians includes numerous 1-azabicyclic systems belonging to the indolizidine (1-azabicyclo [4.3.0]nonane), quinolizidine (1-azabicyclo [4.4.0]decane) and lehmizidine (1-azabicyclo [5.3.0]decane) classes.[1,2] The first of these classes is by far the most populous, and has commanded enormous attention from organic chemists stimulated by the challenges of designing novel total syntheses.[3] The more recently discovered amphibian quinolizidines constitute a smaller group of alkaloids; they embrace homopumiliotoxins (e.g. (+)-homopumiliotoxin 223G 1; Figure 1) and related systems, 4,6-disubstituted quinolizidines (e.g. rel-quinolizidine 195C 2) and 1,4-disubstituted quinolizidines (e.g. (-)-quinolizidine 217A 3). In the latter group, it appears that most of the well-characterised alkaloids have a 1,4-trans disposition of the substituents; the only alkaloid in which the substituents are unambiguously cis is (-)-quinolizidine 207I 4. Comparatively few syntheses of quinolizidine 207I, 217A and related compounds have been reported. [4-9]

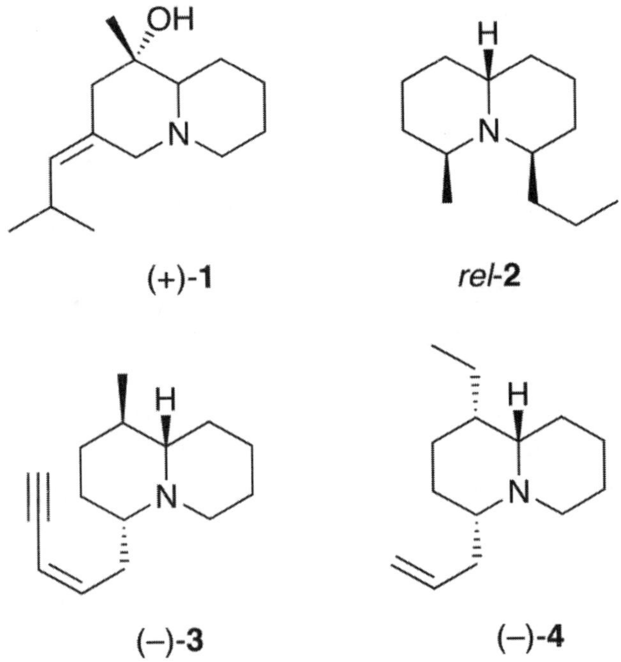

Figure 1. Representative quinolizidine alkaloids from amphibians.

As part of a long-standing investigation into the utility of pyrrolidinylidene- and piperidinylidene-containing enaminones (vinylogous urethanes) 5 and 6 as key intermediates in the synthesis of alkaloids and other nitrogen-containing heterocycles,[10] we previously reported total syntheses of (-)-indolizidine 167B 7,[11,12] the 5,8-disubstituted indolizidine (-)-209B 8 and its racemic diastereomer (±)-9,[13] and the 5,6,8-trisubstituted indolizidines (+)-10 and (+)-11,[14] among other similar compounds (Figure 2). While our attempts to prepare quinolizidines have been less successful, we have synthesised two simple lupin alkaloids, lupinine 12 and epilupinine 13, in racemic form.[15] Although it might seem that reactions of the enaminones 5 and 6 should be directly comparable, we [15,16] and others [17,18] have previously found unexpected differences in the preparation and reactions of cyclic enaminones of different ring sizes. In this article we report our progress in preparing 1,4-disubstituted quinolizidine analogues of amphibian alkaloids by an extension of our approach to the synthesis of 5,8-disubstituted indolizidine alkaloids. [19]

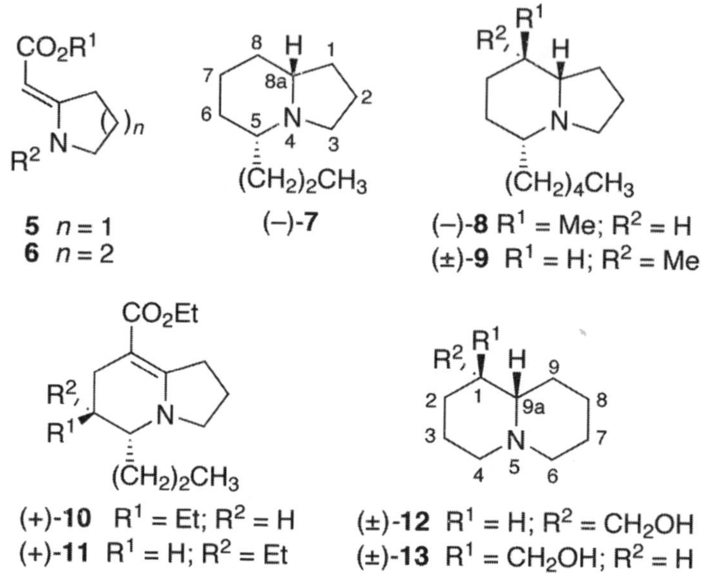

Figure 2. Indolizidines and quinolizidines prepared from enaminone precursors 5 and 6. The conventional numbering scheme for both bicyclic systems is also shown.

Results and Discussion

Steps in our reported total synthesis of (-)-indolizidine (-)-209B 8 [13] are shown in Scheme 1. Absolute stereocontrol resulted from use of the Davies protocol,[20,21] whereby the homochiral amine (+)-14 prepared from tert-butyl (E)-oct-2-enoate and (R)-N-benzyl-1-phenylethylamine, was converted into the primary amine (-)-15 and thence in several steps into the thiolactam (+)-16. Eschenmoser sulfide contraction [22,23] with ethyl bromoacetate yielded the key enaminone intermediate (+)-17, chemoselective reduction of the saturated ester of which produced the alcohol (-)-18. The bicyclic core of the alkaloid was then constructed by a cycloalkylation that took advantage of the nucleophilic reactivity of the enaminone, following which a chemoselective and reasonably diastereoselective (88:12) reduction of the alkene bond of the bicyclic enaminone (+)-19 set up the desired stereochemistry at C-8 and C-8a. Epimerisation of the ester in the reduced compound (-)-20 produced (-)-21, reduction of which gave the alcohol (-)-22. Reduction of the corresponding methanesulfonate with lithium triethylborohydride, as described by Holmes et al.,[24] completed the total synthesis of (-)-indolizidine 209B 8.

Scheme 1. Reagents: (i) H2 (7 atm), 10% Pd/C, AcOH, rt; (ii) Cl(CH2)3COCl, NaHCO3, CHCl3, reflux; (iii) KOBut, ButOH, rt; (iv) Lawesson's reagent, PhMe, reflux; (v) BrCH2CO2Et, MeCN, rt; (vi) Ph3P, Et3N, MeCN, rt; (vii) LiAlH4, THF, rt; (viii) I2, imidazole, Ph3P, PhMe, 110°C; (ix) H2 (1 atm), PtO2, AcOH, rt; (x) NaOEt (cat.), EtOH, reflux; (xi) LiAlH4, THF, 0°C to rt; (xii) MeSO2Cl, NEt3, CH2Cl2, 0°C to rt; (xiii) LiEt3BH, THF, 0°C.

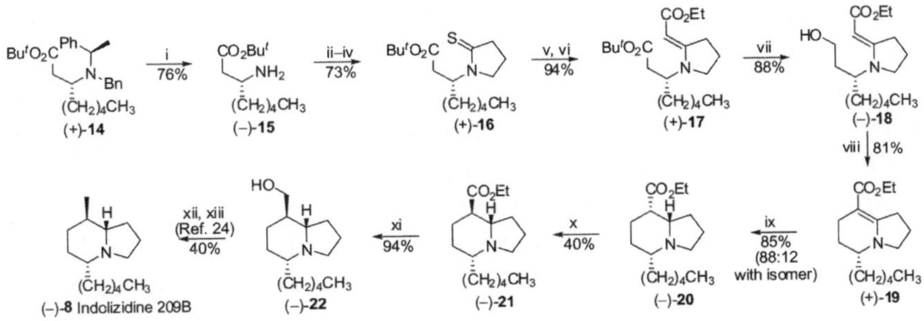

As a postscript to the above synthesis, we have now completed an enantioselective synthesis (Scheme 2) of the indolizidine analogue of 8, viz. (5R,8S,8aS)-8-epi-indolizidine 209B (-)-9, which we had previously made as a racemate.[13] Intermediate (-)-20 was reduced with lithium aluminium hydride in diethyl ether to give the alcohol (-)-23 in 97% yield. The corresponding methanesulfonate (-)-24 (66%) was then defunctionalised by an improved procedure, which entailed treatment with freshly prepared Raney nickel [25] in boiling ethanol to give (-)-(5R,8S,8aS)-8-methyl-5-pentyloctahydroindolizine (8-epi-indolizidine 209B) 9 in 74% yield. The spectroscopic data for this product agreed with those reported for the racemate. Support for the cis-relationship of the hydrogen atoms at C-5 and C-8a in all of these compounds was provided by Bohlmann bands [26] at ca. 2790 cm-1 in the FTIR spectra, a feature that also implies a trans-disposition of the lone pair and 8a-H across the ring junction.

Scheme 2. Reagents: (i) LiAlH4, THF, 0°C to rt; (ii) MeSO2Cl, NEt3, CH2Cl2, 0°C to rt; (iii) Raney Ni, EtOH, reflux.

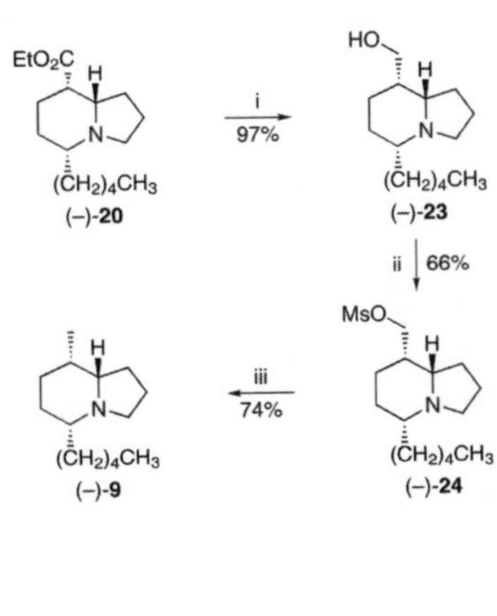

Extending the route illustrated in Scheme 1 to the synthesis of quinolizidine analogues required initial acylation of the chiral amine (-)-15, prepared as described in our prior work,[13] with 5-bromopentanoyl chloride (obtained in two steps from δ-valerolactone).[27,28] This afforded tert-butyl (3R)-[(5-bromopentanoyl)amino]octanoate (+)-25 in 98% yield (Scheme 3). However, subsequent cyclisation to the lactam (+)-26 was troublesome, giving at best a yield of 52% when performed with sodium hydride and tetrabutylammonium iodide in N,N-dimethylformamide. An effortless thionation of 26 with Lawesson's reagent in boiling toluene produced the thiolactam (+)-27 in 92% yield. Eschenmoser sulfide contraction was then effected by first treating the thiolactam with ethyl bromoacetate, after which reaction of the resulting S-alkylated intermediate with triethyl phosphite and triethylamine in acetonitrile gave the vinylogous urethane (+)-28 in 75% yield.

Scheme 3. Reagents: (i) Br(CH2)4COCl, NaHCO3, ClCH2CH2Cl, rt; (ii) NaH, Bu4NI, DMF, rt; (iii) Lawesson's reagent, PhMe, reflux; (iv) BrCH2CO2Et, MeCN, rt; (v) P(OEt)3, Et3N, MeCN, rt; (vii) LiAlH4, THF, rt.

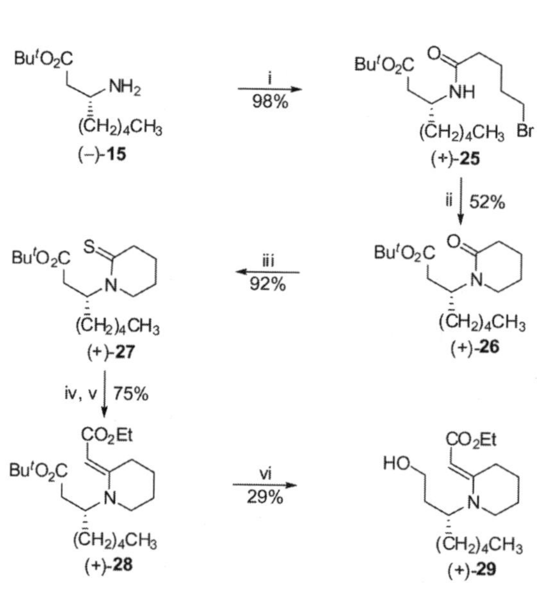

At this stage, however, our fears of the discrepant behaviour of five- and six-membered enaminones proved to be all too well founded. In the indolizidine series, the robust enaminone 17 survived reduction with lithium aluminium hydride, leaving only the saturated ester to be reduced. With the six-membered analogue 28, the enaminone unit was far more susceptible to reduction, and despite many attempts to modify conditions, over-reduction led to a plethora of basic products that could neither be separated nor properly characterised. Although the desired alcohol (+)-29 containing an intact enaminone system could be isolated on occasion, the best yield obtained was 29% when the reaction was not allowed to go to completion. Thus a change of strategy was required to produce 29, the pivotal intermediate from which the quinolizidine nucleus needs to be constructed.

The reduction of the tert-butyl ester clearly needed to be performed at an early stage of the synthesis before the introduction of other incompatible functional groups (lactam, thiolactam, enaminone). The only feasible option was to go back to the chiral amine (+)-14, reduction of which with lithium aluminium hydride gave the unstable amino alcohol (+)-30 in 97% yield as long as the amine was added slowly to a stirred suspension of the hydride in diethyl ether (Scheme 4). If the order of addition were reversed, the best yield obtained was 48%. The amino alcohol was protected as its tert-butyl(dimethyl)silyl ether (-)-31 (99%) before hydrogenolysis of the benzyl groups over Pearlman's catalyst in glacial acetic acid gave the free amine (-)-32 in quantitative yield. Treatment with 5-bromopentanoyl chloride as described above afforded the unstable bromoamide 33 as an orange oil in 89% yield. In this case, cyclisation of the crude intermediate to the lactam (+)-341 was most successfully effected by adding potassium tert-butoxide to a solution of the bromoamide in dry tetrahydrofuran at room temperature, a yield of 81% being obtained by keeping the reaction time short (25 min). To our dismay, however, the attempted thionation of 34 with Lawesson's reagent under a variety of conditions was uniformly unsuccessful, apparently because the silyl ether failed to survive the reaction conditions.

Scheme 4. Reagents: (i) LiAlH4, Et2O, 0°C, then add (+)-2 in Et2O, rt; (ii) TBDMSCl, imidazole, DMF, rt; (iii) H2 (5 atm), 20% Pd(OH)2/C, AcOH, rt; (iv) Br(CH2)4COCl, NaHCO3, ClCH2CH2Cl, rt; (v) t-BuOK, THF, rt, 25 min.

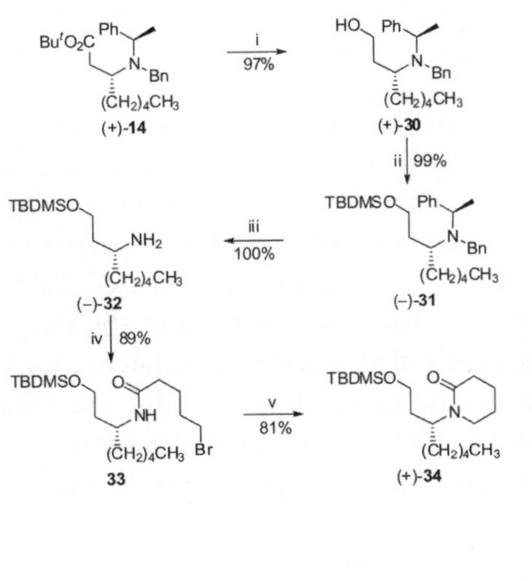

Inelegant though it was, we were forced at this stage to change protecting groups on the alcohol. Fortunately, the drop in yield was not too serious when desilylation of 34 with aqueous hydrofluoric acid to give the free alcohol (+)-35 was followed by acetylation with acetic anhydride in pyridine (Scheme 5). The lactam (+)-36, obtained in an overall yield of 89%, was then successfully thionated with Lawesson's reagent in boiling toluene to give the thiolactam (+)-37 in 94% yield. Finally, reaction with ethyl bromoacetate followed by treatment with triphenylphosphine and triethylamine in acetonitrile give the vinylogous urethane (+)-38 in 80% yield. Hydrolysis of the acetate with potassium carbonate in methanol then afforded the pivotal alcohol (+)-29 (70%). The scene was now set for cyclisation to the quinolizidine system. Immediate conversion of the unstable free alcohol into the corresponding iodide with iodine, triphenylphosphine and imidazole in a mixture of toluene and acetonitrile [29] and heating the reaction mixture under reflux gave the desired 3,4,6,7,8,9-hexahydro-2H-quinolizine-1-carboxylate (-)-39 in 70% yield.

Scheme 5. Reagents: (i) aq. HF (40%), MeOH, rt; (ii) Ac2O, pyridine, 0°C to rt; (iii) Lawesson's reagent, PhMe, reflux; (iv) BrCH2CO2Et, MeCN, rt; (v) Ph3P, Et3N, MeCN, rt; (vi) K2CO3, MeOH, rt; (vii) I2, PPh3, imidazole, MeCN-PhMe (2:1), reflux; (viii) H2 (1 atm), PtO2, AcOH, rt; (ix) LiAlH4, THF, 0°C to rt; (x) MeSO2Cl, NEt3, CH2Cl2, 0°C to rt; (xi) Raney Ni, EtOH, reflux.

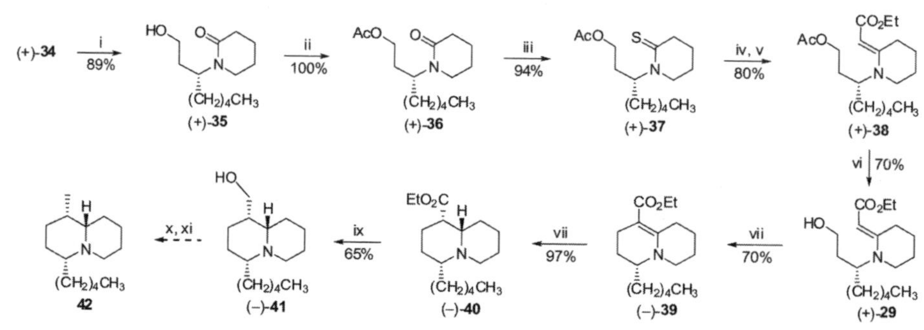

In order to introduce the remaining stereogenic centres of the target system, the alkene bond of the bicyclic vinylogous urethane 39 needs to be reduced stereoselectively. Based on our previous success with the indolizidine analogue 19, we opted for catalytic hydrogenation, which is expected to produce not only a cis-relationship between C-1 and C-9a, but also a cis-relationship between C-4 and C-9a. The developing chair conformation of the six-membered ring in the transition state should result in an equatorial preference for the pentyl side chain, which in turn should bias the approach of the reductant towards the more remote face of the double bond. Gratifyingly, hydrogenation of intermediate 39 over platinum oxide catalyst in ethanol at a pressure of five atmospheres produced the quinolizidine (-)-40 as a single diastereomer in 97% yield. The diastereoselectivity is manifestly better than in the indolizidine case. Support for the cis-relationship of the hydrogen atoms at positions C-4 and C-9a and the trans-ring junction in the product was once again provided by Bohlmann bands in the FTIR spectrum at ca. 2790 cm-1. However, further confirmation of the relative stereochemistry by consideration of the 1H NMR spectrum was not feasible because overlap of signals prevented the extraction of coupling constants for 1-H and 9a-H.

Finally, reduction of the ester to the primary alcohol (-)-41 was accomplished in moderate yield (65%) with lithium aluminium hydride. Again, coupling constants could not be determined for 1-H and 9a-H. In this case, however, there is good precedent for assigning the relative stereochemistry of the hydroxymethyl substituent at C-1 on the basis of 13C chemical shifts. For example, the chemical shift of C-1 in lupinine 12, which possesses an axial hydroxymethyl substituent, is 38.8 ppm; whereas the corresponding chemical shift in epilupinine 13, the equatorial hydroxymethyl epimer, is 43.8 ppm.[30] The chemical shift difference of about 5 ppm between the C-1 equatorial and axial hydroxymethyl epimers appears to be general for quinolizidines.[31] A similar effect has been reported for 8-hydroxymethylindolizidine epimers, for which the chemical shift difference is even larger (ca 10 ppm).[24] In the present case, the observed chemical shift of 38.4 ppm for 41 is consistent with an axial disposition of the C-1 substituent, and thus with the expected cis-hydrogenation of 39.

While it would have been desirable to conclude this investigation by preparing (1S,4R,9aS)-4-pentyloctahydro-2H-quinolizine 42, the ring homologue of 8-epi-indolizidine 209B, this target eluded us. Attempts to reduce the corresponding methanesulfonate of 41 with Raney nickel in boiling ethanol gave ambiguous results no matter how we modified the reaction conditions.

Conclusion

Few approaches to 1,4-cis-disubstituted quinolizidines and 5,8-cis-disubstituted indolizidines of amphibian origin have been reported in the literature. Because the route we have devised proceeds through bicyclic enaminone intermediates in which the alkene bond is located between the bridgehead position and the adjacent site, we have a convenient and dependable method for introducing the correct relative stereochemistry at these two sites by means of catalytic hydrogenation. However, the differences in behaviour of pyrrolidinylidene- and piperidinylidene-containing enaminones that we have come to expect [15,16] was again apparent, necessitating several protection-deprotection steps that lengthened the route to the quinolizidine system. Nevertheless, our success in preparing the chiral alcohol 41 opens up a route to quinolizidine alkaloids containing C-1 methyl substituents (provided, of course, that we can find a better method for deoxygenation, probably by radical-mediated reaction). In addition, alkyl homologues at C-1 should be accessible; one could, for example, replace the alcohol by a leaving group that can be displaced by organometallic reagents (e.g. cuprates) of appropriate chain length. Substituents at C-4 can also be varied by choosing appropriate analogues of the chiral amine 14, which should also be available in both enantiomeric forms by the Davies procedure.[32] Finally, since the pendent substituents in the indolizidine series can be induced to adopt a trans-orientation by base-catalysed epimerisation of a carbonyl substituent adjacent to the bridgehead position (cf Scheme 1), it should in principle be possible to effect a similar epimerisation in the quinolizidine series, thereby providing a route to most of the known 1,4-disubstituted amphibian quinolizidine alkaloids.

Acknowledgements

This work was supported by grants from the National Research Foundation, Pretoria (grant number 2053652) and the University of the Witwatersrand. We are grateful to Dr Tom Spande (NIH-NIDDK) for GC-FTIR and chiral GC analysis of (-)-8-epi-indolizidine 209B.

References

1. Daly JW, Garraffo HM, Spande TF: Alkaloids from amphibian skins. In Alkaloids: Chemical and Biological Perspectives. Edited by: Pelletier SW. Amsterdam: Pergamon Press; 1999:1–161. [Alkaloids: Chemical and Biological Perspectives, vol 13.]
2. Daly JW, Spande TF, Garraffo HM: J Nat Prod. 2005, 68:1556–1575.
3. Michael JP: Nat Prod Rep. 2007, 24:191–222 and previous reviews in the series.
4. Toyooka N, Tanaka K, Momose T, Daly JW, Garraffo HM: Tetrahedron. 1997, 53:9553–9574.
5. Pearson WH, Suga H: J Org Chem. 1998, 63:9910–9918.
6. Michel P, Rassat A, Daly JW, Spande TF: J Org Chem. 2000, 65:8908–8918.
7. Huang H, Spande TF, Panek JS: J Am Chem Soc. 2003, 125:626–627.
8. Kinderman SS, de Gelder R, van Maarseveen JH, Schoemaker HE, Hiemstra H, Rutjes FPJT: J Am Chem Soc. 2004, 126:4100–4101.
9. Maloney KM, Dannheiser RL: Org Lett. 2005, 7:3115–3118.

10. Michael JP, de Koning CB, Gravestock D, Hosken GD, Howard AS, Jungmann CM, Krause RWM, Parsons AS, Pelly SC, Stanbury TV: Pure Appl Chem. 1999, 71:979–988.
11. Michael JP, Gravestock D: Eur J Org Chem. 1998, 865–870.
12. Michael JP, Gravestock D: S Afr J Chem. 1998, 51:146–157.
13. Michael JP, Gravestock D: J Chem Soc, Perkin Trans 1. 2000, 1919–1928.
14. Michael JP, de Koning CB, van der Westhuyzen CW: Org Biomol Chem. 2005, 3:836–847.
15. Michael JP, de Koning CB, San Fat C, Nattrass GL: Arkivoc. 2002, ix:62–77.
16. Michael JP, de Koning CB, van der Westhuyzen CW: J Chem Soc, Perkin Trans 1. 2001, 2055–2062.
17. David O, Fargeau-Bellassoued M-C, Lhommet G: Tetrahedron Lett. 2002, 43:3471–3474.
18. Russowsky D, da Silvera Neto BA: Tetrahedron Lett. 2004, 45:1437–1440.
19. Cai G, Zhu W, Ma D: Tetrahedron. 2006, 62:5697–5708. For another approach to the synthesis of quinolizidines and related systems via enaminones, and references cited therein.
20. Davies SG, Ichihara O: Tetrahedron: Asymmetry. 1991, 2:183–186.
21. Costello JF, Davies SG, Ichihara O: Tetrahedron: Asymmetry. 1994, 5:1999–2008.
22. Roth M, Dubs P, Götschi E, Eschenmoser A: Helv Chim Acta. 1971, 54:710–734.
23. Shiosaki K: The Eschenmoser coupling reaction. In Comprehensive Organic Synthesis. Volume 2. Edited by: Trost BM. Oxford: Pergamon Press; 1991:865–892.
24. Holmes AB, Smith AL, Williams SF, Hughes LR, Lidert Z, Swithenbank C: J Org Chem. 1991, 56:1393–1405.
25. Covert LW, Adkins H: J Am Chem Soc. 1932, 54:4116–4117.
26. Bohlmann F: Chem Ber. 1985, 91:2157–2167.
27. Bagnall WH, Goodings EP, Wilson CL: J Am Chem Soc. 1951, 73:4794–4798.
28. Collman JP, Groh SE: J Am Chem Soc. 1982, 104:1391–1403.
29. Garegg PJ, Samuelsson B: J Chem Soc, Perkin Trans 1. 1980, 2866–2869.
30. Wenkert E, Chauncy B, Dave KG, Jeffcoat AR, Schell FM, Schenk HP: J Am Chem Soc. 1973, 95:8427–8436.
31. Tourwé D, van Binst G: Heterocycles. 1978, 9:507–533.
32. Davies SG, Smith AD, Price PD: Tetrahedron: Asymmetry. 2006, 16:2833–2891.

An Efficient Preparation, Spectroscopic Properties, and Crystal Structure of 1,1-Bis(4-[2-(dimethylamino)ethoxy]phenyl)-2-(3-guaiazulenyl)ethylene

Shin-ichi Takekuma, Seiki Hori, Toshie Minematsu and Hideko Takekuma

ABSTRACT

Reaction of 2-(3-guaiazulenyl)-1,1-bis(4-hydroxyphenyl)ethylene with 2-chloroethyldimethylammonium chloride in acetone in the presence of K2CO3 at reflux temperature for 24 hours gives a new title compound in 89% yield. Spectroscopic properties and crystal structure of the target molecule are reported.

Introduction

It is well known that chlorotrianisene [2-chloro-1,1,2-tris(4-methoxyphenyl)ethylene] (1) [1] exhibits significant estrogenic activity, while tamoxifen [(Z)-1-[4-(2-dimethylaminoethoxy)phenyl]-1,2-diphenyl-1-butene] (2) [1–4] exhibitssignificant antiestrogenic activity, owing to the difference between the substituents of 1 and those of 2 (see Figure 1). Furthermore, tamoxifen, which serves as a nonsteroidal antiestrogen [3], has become a widely used drug for a first-line endocrine therapy for all stages of breast cancer in pre- and postmenopausal women [5]. On the other hand, azulene, possessing a large dipole moment, is regarded as one of the representative examples of non-benzenoid aromatic hydrocarbons, and naturally occurring guaiazulene (7-isopropyl-1,4-dimethylazulene) (see Figure 1) [6] has been widely used clinically as an anti-inflammatory and antiulcer agent. However, none have really been used as other industrial materials. In relation to azulene chemistry, in 2004, we reported that the reaction of guaiazulene with 1,2-bis(4-methoxyphenyl)-1,2-ethanediol in methanol in the presence of hydrochloric acid at 60°C for 3 hours gave 2-(3-guaiazulenyl)-1,1-bis(4-methoxyphenyl) ethylene (5) (see Figure 2), possessing a similar structure to 1, in 97% yield [7]. Similarly, the reaction of guaiazulene with 1,2-bis(4-hydroxyphenyl)-1,2-ethanediol under the same reaction conditions as the above afforded 2-(3-guaiazulenyl)-1,1-bis(4-hydroxyphenyl)ethylene (3) (see Figure 3) in 73% yield [7]. As a series of basic studies on our azulene chemistry [7–17], our interest has quite recently been focused on an efficient preparation and properties of new tamoxifen analogues with a 3-guaiazulenyl (7-isopropyl-1,4-dimethylazulen-3-yl) group, for example, the title compound 4. We now wish to report a facile preparation as well as the spectroscopic properties and the crystal structure of the target compound 4 (see Figure 3).

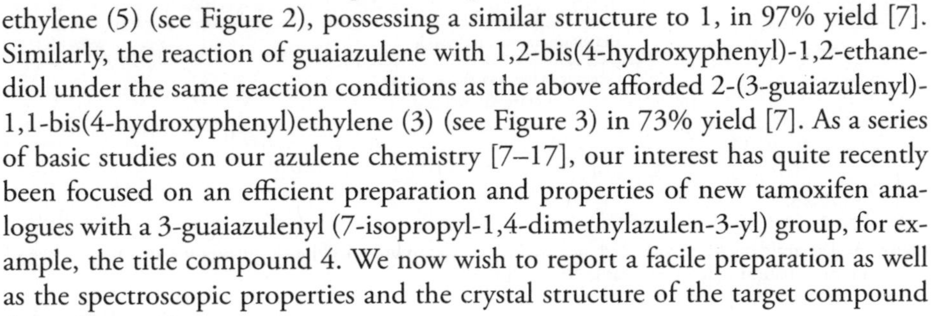

1 2 Guaiazulene

Figure 1

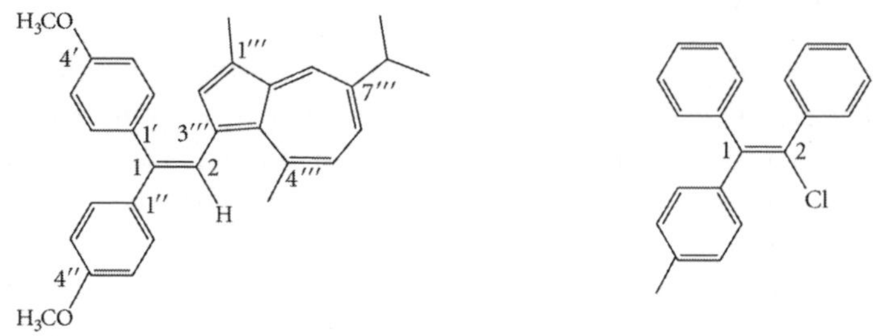

5 6

Figure 2

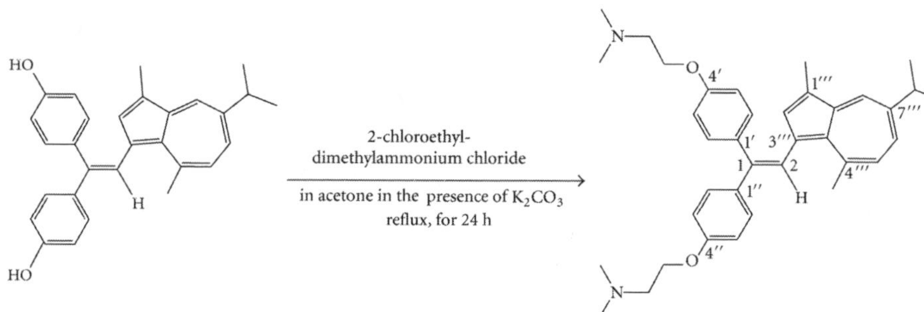

3 4

Figure 3. The reaction of 3 with 2-chloroethyldimethylammonium chloride in acetone in the presence of potassium carbonate at reflux temperature for 24 hours.

Results and Discussions

Compound 4 (89% yield) was prepared in acetone, according to the procedure shown in Figure 3, the molecular structure of which was established on the basis of elemental analysis and spectroscopic data [UV-vis, IR, exact FAB-MS, and 1H and 13C NMR including NOE and 2D NMR (i.e., H–H COSY, HMQC, and HMBC)]. 1H NMR signals (δ and J values) were carefully assigned using computer-assisted simulation based on first-order analysis (1H NMR signals were assigned using computer-assisted simulation (software: gNMR developed by Adept Scientific plc) on a DELL Dimension XPS T500 personal-computer with a Pentium III processor.).

The UV-vis spectrum of 4, possessing a 1,1-bis(4-[2-(dimethylamino)ethoxy] phenyl)ethylene part, compared with that of structurally related π-electron system 5 [7] showed that the spectral pattern of 4 resembled that of 5; however, the longest absorption wavelength of 4 (λmax 632 nm, log ε=2.72) revealed a slight hyperchromic effect (Δlog ε=0.18) in comparison with that of 5 (λmax 635 nm, log ε=2.54). The IR spectrum showed specific bands based on the C–O, C–N, aromatic C=C, and C–H bonds. The protonated molecular formula C37H47O2N2 was determined by exact FAB-MS spectrum. The elemental analysis confirmed the molecular formula C37H46O2N2. The 1H NMR spectrum showed signals based on a 3-guaiazulenyl group, signals based on two nonequivalent 4-[2-(dimethylamino)ethoxy]phenyl groups, and a signal based on an ethylene part (>C=CH–). The 13C NMR spectrum exhibited 29 carbon signals, which could be assigned using HMQC and HMBC techniques. Thus, the elemental analysis and the spectroscopic data for 4 led to the title molecular structure.

The crystal structure of 4 was then determined by means of X-ray diffraction. Two molecules of 4 were found to exist in the unit cell of the crystal. The ORTEP drawing of 4 (one of the two molecules) with a numbering scheme is shown in Figure 4(a). The structural parameters of 4 revealed that the C1–C1′, C1–C1″, C2–C3‴, and C1=C2 bond lengths of 4 coincided with those of 5 [7]. Similarly, as in the case of 5, the aromatic rings of the two 4′- and 4″-[2-(dimethylamino) ethoxy]phenyl and 3‴-guaiazulenyl groups of 4 twisted by 47.9°, 135.0°, and 47.4° from the plane of the >C=CH– part, owing to the influence of large steric hindrance and repulsion between the aromatic rings. The average C–C bond lengths for the seven- and five-membered rings of the 3‴-guaiazulenyl group of 4 (1.412 and 1.422 Å) coincided with those of 5 (1.405 and 1.427 Å) and the average C–C bond lengths for the benzene rings of the two 4′- and 4″-[2-(dimethylamino)ethoxy]phenyl groups of 4 (1.386 and 1.385 Å) also coincided with those of the two 4′- and 4″-methoxyphenyl groups of 5 (1.382 and 1.380 Å).

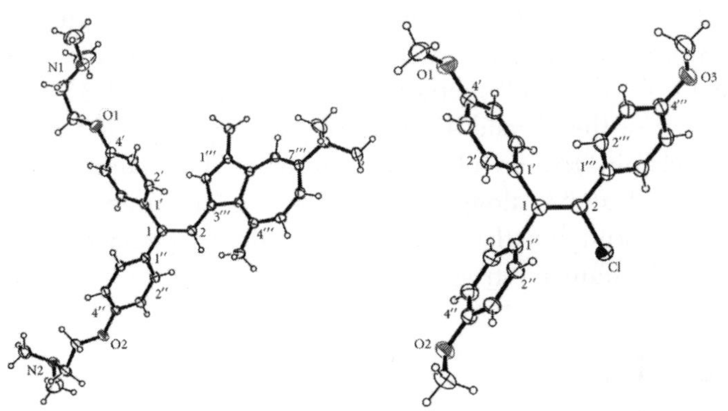

4

(a)

1

(b)

Figure 4. The ORTEP drawings with the numbering scheme (30% probability thermal ellipsoids) of 4 and 1.

Comparing the bond lengths observed for 4 to those of 2 [4], it was found that the C1=C2 bond length of 4 coincided with that of 2 (1.34 Å) and the C1–C1′ and C1–C1″ bond lengths of 4 also coincided with those of 2 (1.50 Å each). The average C–C bond lengths observed for the 1′- and 1″-phenyl groups of 4 coincided with those of 2 (1.39 Å each), and the bond lengths observed for the two (dimethylamino)ethoxy groups of 4 also coincided with those of 2 (C4′–O:1.38, O–CH2:1.44, CH2–CH2:1.51, CH2–N:1.50, N–CH3:1.34, and N–CH3:1.26 Å). The (dimethylamino)ethoxy side chain of 2 is essential to the antiestrogenic action [1]. From the accurate structural parameters of 4, a study on the antiestrogenic activity of 4, compared with that of 2, is noteworthy.

The crystal structure of the estrogenic drug 1 was not documented. For comparative purposes, the crystal structure of 1 (Commercially available 1 was recrystallized from diethyl ether (several times) to provide stable single crystals (mp 114°C).) has been determined by means of X-ray diffraction (Crystallographic data for 1: C23H21O3Cl (FW = 380.87), colorless prism (the crystal size, 0.50×0.40×0.40 mm3), monoclinic, C2/c (#15), a=18.577(5) Å, b=9.787(5) Å, c=23.738(4)Å, β=115.82(1)°, V=3885(2) Å3, Z=8, Dcalcd=1.302g/cm3, μ(Mo K α)=2.17 cm−1, scan width=(1.26 + 0.30 tan θ)°, scan mode=ω−2θ, scan rate=8.0°/min, measured reflections = 4882, observed reflections = 4461, No. of parameters = 244, R1=0.041, wR2=0.167, and goodness of fit indicator = 1.35. Deposition no. CCDC-252847 for compound no. 1.). The ORTEP drawing of 1 with a numbering scheme is shown in Figure 4(b). The structural parameters of 1 revealed that the C1=C2 bond length of 1 coincided with those of (Z)-2-chloro-1-(4-methylphenyl)-1,2-diphenylethylene (6) [1.340(3) Å] (see Figure 2) [18], 5 [7], and 4. The C1–C1′, C1–C1″, and C2–C1‴ bond lengths of 1 also coincided with those of 4–6. The benzene rings of the three 4′-, 4″-, and 4‴-methoxyphenyl groups of 1 twisted by 64.4°, 103.5°, and 41.3° from the plane of the >C=CCl– part, owing to the influence of large steric hindrance and repulsion between the benzene rings. The average C–C bond lengths for the benzene rings of the 4′-, 4″-, and 4‴-methoxyphenyl groups of 1 (1.384 Å each) coincided with those of the 4′- and 4″-methoxyphenyl groups of 5 and the 4′- and 4″-[2-(dimethylamino)ethoxy]phenyl groups of 4. The accurate crystal structure determination of 1 and 4, along with that of 5, is important from a viewpoint of computational study on the compounds' potential interactions with specific residues within the human estrogen receptor.

Experimental

Preparation of 4

Compound 4 was prepared according to the following procedure. To a solid of anhydrous K2CO3 (17 mg, 123 mmol) was added a solution of 3 (50 mg, 123 mmol) in acetone (10 mL). The mixture was stirred at 25°C for 12 hours, and then 2-chloroethyldimethylammonium chloride (36 mg, 250 mmol) was added to the mixture. The mixture was stirred at reflux temperature for 24 hours. After the reaction, distilled water was added to the mixture, and then the resulting product was extracted with diethyl ether (10 mL × 3). The extract was washed with distilled water, dried ($MgSO_4$), and evaporated in vacuo. The residue thus obtained was carefully separated by silica-gel column chromatography with methanol-dichloromethane (3:7, vol/vol) as an eluant. The crude product was recrystallized from petroleum ether to provide pure 4 (60 mg, 109 mmol, 89% yield) as stable crystals.

Compound 4: dark-green needles [Rf = 0.10 on silica-gel TLC (solv. methanol:dichloromethane = 1:9, vol/vol)]; mp 100°C [determined by thermal analysis (TGA and DTA)]; found: C, 80.69; H, 8.41; N, 5.08%; calcd for C37H46O2N2:C, 80.69; H, 8.42; N, 5.09%; UV-vis λmax (CH3CN) nm (log ε), 271 (4.52), 329 (4.40), 357sh (4.30), 409 (4.27), and 632 (2.72); IR υmax (KBr) cm−1, 2955–2766 (C–H), 1605, 1508 (C=C), 1285 (C–N), and 1242, 1042 (C–O); exact FAB-MS (3-nitorobenzylalcohol matrix), found: m/z 551.3664; calcd for C37H47O2N2: [M + H]+, m/z 551.3638; 500 MHz H NMR: (CD3CN), signals based on the 3-guaiazulenyl group: δ1.28 (6H, d, J = 7.0 Hz, (CH3)2CH-7‴), 2.38 (3H, s, CH3-1‴), 2.94 (3H, s, CH3-4‴), 2.97 (1H, sept, J = 7.0 Hz, (CH3)2C H-7‴), 6.81 (1H, d, J = 10.5 Hz, H-5‴), 6.89 (1H, s, H-2‴), 7.24 (1H, dd, J = 10.5, 2.5 Hz, H-6‴), and 7.92 (1H, d, J = 2.5 Hz, H-8‴); signals based on the (Z)-4-[2-(dimethylamino)ethoxy]phenyl group: δ2.22 (6H, s, (CH3)2N), 2.60 (2H, t, J = 6.0 Hz, CH2–N), 3.98(2H, t, J = 6.0 z, CH2–O), 6.74 (2H, dd, J = 8.5, 2.5 Hz, H-3′,5′), and 7.04 (2H, dd, J = 8.5, 2.5, Hz H-2′,6′); and signals of based on the (E)-4-[2-(dimethylamino)ethoxy]phenyl group: δ2.24 (6H, s, (CH3)2N), 2.63 (2H, t, J = 6.0 Hz, CH2–N), 4.03 (2H, t, J = 6.0 Hz, CH2–O), 6.84 (2H, dd, J = 8.5, 2.5 Hz, H-3″,5″), and 7.22 (2H, dd, J = 8.5, 2.5 Hz, H-2″,6″); and a signal based on the >C=CH– part: δ7.52 (1H, s, H-2); 125 MHz 13C NMR (CD3CN), δ159.3 (C-4″), 158.9 (C-4′), 147.1 (C-4‴), 141.5 (C-7‴), 140.2 (C-2‴), 139.5 (C-8a‴), 138.9 (C-1″), 137.5 (C-1′), 135.7 (C-6‴), 135.4 (C-3a‴), 134.4 (C-1), 134.1 (C-8‴), 133.4 (C-2′,6′), 129.6 (C-2″,6″), 127.2 (C-5‴), 127.1 (C-3‴), 125.8 (C-2), 125.4 (C-1‴), 115.1 (C-3″,5″), 115.0 (C-3′,5′), 67.2 (CH2–O-4″), 67.1 (CH2–O-4′), 59.0 (CH2–N″), 58.9 (CH2–N′), 46.1 (Me2N × 2), 38.4 ((CH3)2CH-7‴), 27.9 (Me-4‴), 24.6 ((CH3)2CH-7‴), and 12.9 (Me-1‴).

Crystallographic Data for 4

$C_{37}H_{46}O_2N_2$ (FW = 550.78), dark-green prism (the crystal size, 0.50 × 0.20 × 0.50 mm^3), triclinic, P-1 (#2), a = 16.904(3) Å, b = 20.322(4) Å, c = 10.372(1) Å,

α = 103.00(1)°, β = 96.90(1)°, γ = 108.53(1)°, V = 3220.4(9) Å3, Z = 4, Dcalcd = 1.136 g/cm^3, μ(Mo Kα = 0.69 cm^{-1}, scan width = 1.26+0.30 tan θ)°, scan mode = ω–2θ, scan rate = 16.0°/min, measured reflections = 15303, observed reflections = 11731, no. of parameters = 739, R1 = 0.063, wR2 = 0.177, and goodness of fit indicator = 1.48. Crystallographic data of 4 have been deposited with Cambridge Crystallographic Data Center, Deposition no. CCDC-635049 for compound no. 4. Copies of the data can be obtained free of charge via http://www.ccdc.cam.ac.uk/ conts/retrieving.html (or from the Cambridge Crystallographic Data Center, 12, Union Road, Cambridge CB2 1EZ, UK; Fax: +44 1223 336033; e-mail: deposit@ccdc.cam.ac.uk).

Conclusions

We have reported the following three interesting points in this paper, namely, (i) the reaction of 2-(3-guaiazulenyl)-1,1-bis(4-hydroxyphenyl)ethylene (3) with 2-chloroethyldimethylammonium chloride in acetone in the presence of K_2CO_3 at reflux temperature for 24 hours gave a new title compound 4 in 89% yield; (ii) the elemental analysis and the spectroscopic data for 4 led to the target molecular structure; (iii) the crystal structure of 4 compared with those of structurally related compounds 2-(3-guaiazulenyl)-1,1-bis(4-methoxyphenyl)ethylene (5), tamoxifen (2), and chlorotrianisene (1) was documented. From the accurate structural parameters of 4, a comparative study on the antiestrogenic activity of 4 with that of 2 is noteworthy.

Acknowledgement

This work was partially supported by a Grant-in-Aid for Scientific Research from the Ministry of Education, Culture, Sports, Science, and Technology, Japan.

References

1. V. C. Jordan, J. M. Schafer, A. S. Levenson, et al., "Molecular classification of estrogens," Cancer Research, vol. 61, no. 18, pp. 6619–6623, 2001.
2. V. Lubczyk, H. Bachmann, and R. Gust, "Investigations on estrogen receptor binding. The estrogenic, antiestrogenic, and cytotoxic properties of C2-alkyl-substituted 1,1-bis(4-hydroxyphenyl)-2-phenylethenes," Journal of Medicinal Chemistry, vol. 45, no. 24, pp. 5358–5364, 2002.

3. M. J. Meegan, R. B. Hughes, D. G. Lloyd, D. C. Williams, and D. M. Zisterer, "Flexible estrogen receptor modulators: design, synthesis, and antagonistic effects in human MCF-7 breast cancer cells," Journal of Medicinal Chemistry, vol. 44, no. 7, pp. 1072–1084, 2001.

4. G. Precigoux, C. Courseille, S. Geoffre, and M. Hospital, "[p-(diméthylamino-2 éthoxy)phényl]-1 trans-diphényl-1,2 butène-1 (tamoxfène) (ICI-46474)," Acta Crystallographica Section B, vol. 35, part 12, pp. 3070–3072, 1979.

5. L. J. Lerner and V. C. Jordan, "Development of antiestrogens and their use in breast cancer: eighth cain memorial award lecture," Cancer Research, vol. 50, no. 14, pp. 4177–4189, 1990.

6. Y. Matsubara, H. Yamamoto, and T. Nozoe, "Oxidation products of guaiazulene and other azulenic hydrocarbons," in Studies in Natural Products Chemistry, Atta-Ur-Rahman, Ed., vol. 14 of Stereoselective Synthesis (Part I), pp. 313–354, Elsevier, Amsterdam, The Netherlands, 1994.

7. M. Nakatsuji, Y. Hata, T. Fujihara, et al., "Reactions of azulenes with 1,2-diaryl-1,2-ethanediols in methanol in the presence of hydrochloric acid: comparative studies on products, crystal structures, and spectroscopic and electrochemical properties," Tetrahedron, vol. 60, no. 28, pp. 5983–6000, 2004.

8. S. Takekuma, Y. Hata, T. Nishimoto, et al., "Reactions of guaiazulene with methyl terephthalaldehydate and 2-hydroxy- and 4-hydroxybenzaldehydes in methanol in the presence of hexafluorophosphoric acid: comparative studies on molecular structures and spectroscopic, chemical and electrochemical properties of monocarbocations stabilized by 3-guaiazulenyl and phenyl groups," Tetrahedron, vol. 61, no. 28, pp. 6892–6907, 2005.

9. S. Takekuma, K. Takahashi, A. Sakaguchi, et al., "Reactions of guaiazulene with 2-furaldehyde, thiophene-2-carbaldehyde and pyrrole-2-carbaldehyde in methanol in the presence of hexafluorophosphoric acid: comparative studies on molecular structures and spectroscopic, chemical and electrochemical properties of monocarbocations stabilized by 3-guaiazulenyl and 2-furyl (or 2-thienyl or 2-pyrrolyl) groups," Tetrahedron, vol. 61, no. 43, pp. 10349–10362, 2005.

10. S. Takekuma, K. Takahashi, A. Sakaguchi, M. Sasaki, T. Minematsu, and H. Takekuma, "Reactions of (1R,2S)-1,2-di(2-furyl)-1,2-di(3-guaiazulenyl)ethane and (1R,2S)-1,2-di(3-guaiazulenyl)-1,2-di(2-thienyl)ethane with tetracyanoethylene (TCNE) in benzene: comparative studies on the products and their spectroscopic properties," Tetrahedron, vol. 62, no. 7, pp. 1520–1526, 2006.

11. S. Takekuma, M. Hirosawa, S. Morishita, M. Sasaki, T. Minematsu, and H. Takekuma, "Reactions of (E)-1,2-di(3-guaiazulenyl)ethylene and

2-(3-guaiazulenyl)-1,1-bis(4-methoxyphenyl)ethylene with tetracyanoethylene (TCNE) in benzene: comparative studies on the products and their spectroscopic properties," Tetrahedron, vol. 62, no. 15, pp. 3732–3738, 2006.

12. S. Takekuma, K. Sonoda, C. Fukuhara, and T. Minematsu, "Reaction of guaiazulene with o-formylbenzoic acid in diethyl ether (or methanol) in the presence of hexafluorophosphoric acid: comparative studies on 1H and 13C NMR spectral properties of 3-guaiazulenylmethylium- and 3-guaiazulenium-ion structures," Tetrahedron, vol. 63, no. 11, pp. 2472–2481, 2007.
13. S. Takekuma, K. Tone, M. Sasaki, T. Minematsu, and H. Takekuma, "Reactions of guaiazulene with thiophene-2,5-dicarbaldehyde and furan-2,5-dicarbaldehyde in methanol in the presence of hexafluorophosphoric acid: a facile preparation and properties of delocalized dicarbenium-ion compounds stabilized by two 3-guaiazulenyl groups and a thiophene (or furan) ring," Tetrahedron, vol. 63, no. 11, pp. 2490–2502, 2007.
14. S. Takekuma, K. Mizutani, K. Inoue, et al., "Reactions of azulene and guaiazulene with all-trans-retinal and trans-cinnamaldehyde: comparative studies on spectroscopic, chemical, and electrochemical properties of monocarbenium-ions stabilized by expanded π-electron systems with an azulenyl or 3-guaiazulenyl group," Tetrahedron, vol. 63, no. 18, pp. 3882–3893, 2007.
15. S. Takekuma, M. Tamura, T. Minematsu, and H. Takekuma, "An efficient preparation, crystal structures, and properties of monocarbenium-ion compounds stabilized by 3-guaiazulenyl and anisyl groups," Tetrahedron, vol. 63, no. 48, pp. 12058–12070, 2007.
16. S. Takekuma, K. Sonoda, T. Minematsu, and H. Takekuma, "An efficient preparation and spectroscopic and electrochemical properties of quinodimethane derivatives with four 3-(methoxycarbonyl)azulen-1-yl groups," Tetrahedron, vol. 64, no. 17, pp. 3802–3812, 2008.
17. S. Takekuma, S. Hori, T. Minematsu, and H. Takekuma, "Preparation, crystal structures, and properties of new conjugated π-electron systems with 3-guaiazulenyl and 4-(dimethylamino)phenyl groups," Bulletin of the Chemical Society of Japan, vol. 81, no. 11, pp. 1472–1484, 2008.
18. C. Muthiah, K. P. Kumar, S. Kumaraswamy, and K. C. Kumara Swamy, "An easy access to trisubstituted vinyl chlorides and improved synthesis of chloro/bromostilbenes," Tetrahedron, vol. 54, no. 47, pp. 14315–14326, 1998.

Characterizing the Formation of Secondary Organic Aerosols

Melissa Lunden, Douglas Black and Nancy Brown

Executive Summary

Organic aerosol is an important fraction of the fine particulate matter present in the atmosphere. This organic aerosol comes from a variety of sources; primary organic aerosol emitted directly from combustion process, and secondary aerosol formed in the atmosphere from condensable vapors. This secondary organic aerosol (SOA) can result from both anthropogenic and biogenic sources. In rural areas of the United States, organic aerosols can be a significant part of the aerosol load in the atmosphere. However, the extent to which gas-phase biogenic emissions contribute to this organic load is poorly understood. Such an understanding is crucial to properly apportion the effect of anthropogenic emissions in these rural areas that are sometimes dominated by biogenic sources.

To help gain insight on the effect of biogenic emissions on particle concentrations in rural areas, we have been conducting a field measurement program at the University of California Blodgett Forest Research Facility. The field location includes has been used to acquire an extensive suite of measurements resulting in a rich data set, containing a combination of aerosol, organic, and nitrogenous species concentration and meteorological data with a long time record. The field location was established in 1997 by Allen Goldstein, a professor in the Department of Environmental Science, Policy and Management at the University of California at Berkeley to study interactions between the biosphere and the atmosphere. The Goldstein group focuses on measurements of concentrations and whole ecosystem biosphere-atmosphere fluxes for volatile organic compounds (VOC's), oxygenated volatile organic compounds (OVOC's), ozone, carbon dioxide, water vapor, and energy. Another important collaborator at the Blodgett field location is Ronald Cohen, a professor in the Chemistry Department at the University of California at Berkeley. At the Blodgett field location, his group his group performs measurements of the concentrations of important gas phase nitrogen compounds.

Experiments have been ongoing at the Blodgett field site since the fall of 2000, and have included portions of the summer and fall of 2001, 2002, and 2003. Analysis of both the gas and particle phase data from the year 2000 show that the particle loading at the site correlates with both biogenic precursors emitted in the forest and anthropogenic precursors advected to the site from Sacramento and the Central Valley of California. Thus the particles at the site are affected by biogenic processing of anthropogenic emissions.

Size distribution measurements show that the aerosol at the site has a geometric median diameter of approximately 100 nm. On many days, in the early afternoon, growth of nuclei mode particles (<20 nm) is also observed. These growth events tend to occur on days with lower average temperatures, but are observed throughout the summer. Analysis of the size resolved data for these growth events, combined with typical measured terpene emissions, show that the particle mass measured in these nuclei mode particles could come from oxidation products of biogenic emissions, and can serve as a significant route for SOA partitioning into the particle phase.

During periods of each year, the effect of emissions for forest fires can be detected at the Blodgett field location. During the summer of 2002 emissions from the Biscuit fire, a large fire located in Southwest Oregon, was detected in the aerosol data. The results show that increases in particle scattering can be directly related to increased black carbon concentration and an appearance of a larger mode in the aerosol size distribution. These results show that emissions from fires can have significant impact on visibility over large distances. The results also reinforce the view that forest fires can be a significant source of black carbon in the atmosphere, which has important climate and visibility.

Continuing work with the 2002 data set, particularly the combination of the aerosol and gas phase data, will continue to provide important information on the extent to which biogenic emissions contribute to secondary organic aerosol and may elucidate important interactions between anthropogenic and biogenic sources. The results of these studies, performed in the field, will contribute to the growing effort to produce robust models for particulate formation that are necessary for air quality planning and source apportionment.

Introduction

Fine particulate matter comes from a number of sources, and has received considerable attention lately due to aerosol influences on human health, global climate, and visibility. An important fraction of particulate matter is organic aerosol. For example, organic compounds contribute from 10 to 70 % of dry fine particle mass in the atmosphere. (U.S. EPA, Air Quality Criteria for Particulate Matter (Fourth External Review Draft), 2003.) This organic aerosol consists of primary aerosol, emitted directly from the sources as an aerosol, and secondary, formed in-situ in the atmosphere from condensable vapors. Primary organic aerosols often result from combustion processes. (Rogge et al., 1993, Schauer et al. 1999a,b, 2001) Secondary organic aerosol (SOA) is formed from the atmospheric oxidation of volatile organic compounds to produce condensable species. Secondary organic aerosols form when organic vapors emitted into the atmosphere are oxidized to produce less volatile products. These products then partition between the gas and particle phase. SOA production from a given VOC depends on: (1) the abundance and reactivity of the given VOC compound; (2) the abundance of radicals in the atmosphere; (3) the nature of its reaction pathways; (4) the volatility and gas-to-particle partitioning properties of its products; (5) the ambient aerosol mass concentration; and (6) temperature.

SOA can form from both anthropogenic and biogenic precursors. The contribution of biogenic precursors to SOA can be significant. Terrestrial vegetation releases a number of reactive organic compounds into the atmosphere, including isoprene, monoterpenes, sesquiterpenes, and others. The total annual global biogenic organic emissions are estimated to be 1150 Tg yr-1, while the estimate for total anthropogenic emissions is 142 Tg yr-1 (Seinfeld and Pandis, 1998)

Our knowledge of the chemistry of SOA formation is not well understood, particularly in rural environments where the role of biogenic SOA formation is of greater importance. Much of the research on biogenic emissions has been performed in the Eastern and Southern United States. There is limited information on the emissions from the pine forest that dominate much of the Western United States. Research concerning SOA formation from biogenic organic emissions is sparse, and primarily consists of controlled experiments in laboratory smog chambers. There are very few field measurements of speciated biogenic gas phase measurements coupled with aerosol characterization measurements. These would serve as a comparison to SOA products and yields measured in laboratory experiments. An understanding of the reaction pathways, kinetics, and aerosol forming behavior of biogenic emission sources is important to understanding regional air quality issues including particle loading, ozone formation, and visibility.

The objective of this study is to investigate the biosphere-atmosphere exchange of VOCs and their effects on aerosol formation and processing. This is being achieved by conducting aerosol characterization measurements in concert with measurements of the gas phase concentrations of biogenic and anthropogenic VOCs and meteorological variables over the period of several months at an established sampling site located at the University of California Blodgett Forest Research Station. The field site is located on the western slope of the Sierra Nevada Mountains, approximately 75 km northeast of Sacramento at the site of an evenly aged ponderosa pine plantation. This project provides crucial data needed to characterize the formation of biogenic SOA under actual conditions in the rural atmosphere. The field program offers the advantage of having all atmospheric reactants present at atmospheric concentrations, something that is not usually achieved with smog chamber experiments. Data are collected over an extended period of time, providing data during both the height of summer and cooler periods in the spring and fall. The results from the study contribute to the understanding of the chemistry active in the formation of SOA, and provide critical information necessary for the modeling of SOA.

This report provides a brief review of the theory behind SOA formation, relevant experimental results, and approaches to modeling SOA behavior. The report will also provide details on the experimental design of the Blodgett forest research site. Finally, the report will present initial results from the field site.

Secondary Organic Aerosol

The interactions between organic gases and aerosols in the atmosphere are complex. Some organic aerosol can be directly emitted into the atmosphere as a result of fossil fuel combustion or biomass burning; these organic aerosols are termed primary organic aerosol (POA). A large fraction of organic gases are volatile, and are therefore emitted into the atmosphere in the gas phase. These organic gases are generally quite reactive and are thus termed reactive organic gases (ROG). The oxidation of ROGs is a key step in the formation of photochemical smog. In addition, the oxidation products of some ROGs have sufficiently low vapor pressure that they partition into the aerosol phase. The aerosol that is comprised of these condensed products of ROG oxidation is called secondary organic aerosol. SOA is important because it can form a significant fraction of the fine particle burden in both rural and urban regions. In addition, there is evidence to suggest that the organic fraction is suspected to contribute to the adverse health effects of particulate matter. (Dreher 2000, Mauderly 2003)

Theory of Secondary Organic Aerosol Formation

The atmospheric chemical reaction pathways of ROGs are complex, leading to a large number of oxidation products, all of which have different propensities to condense and form SOAs. These chemical pathways can be depicted in general terms as follows. Consider the production of semivolatile organic gases, S1, S2, … from the gas-phase reaction of the parent hydrocarbon, HC, with the OH radical. (Griffen 1999, Pankow 1994a,b)

$$HC + OH \xrightarrow{k_{OH}} \ldots + \alpha_1 S_1 + \alpha_2 S_2 + \ldots \quad (1)$$

where k_{OH} is the OH reaction rate constant, and α1, α2, … are the product stoichiometric coefficients. It is also possible for many hydrocarbons to react with O_3 and NO_3 radicals, providing additional pathways for semivolatile product formation,

$$HC + O_3 \xrightarrow{k_{O_3}} \ldots + \alpha_{1,0_3} S_{1,0_3} + \alpha_{2,0_3} S_{2,0_3} + \ldots \quad (2)$$

$$HC + NO_3 \xrightarrow{k_{NO_3}} \ldots + \alpha_{1,NO_3} S_{1,NO_3} + \alpha_{2,NO_3} S_{2,NO_3} + \ldots \quad (3)$$

The first generation products S1, S2 … may subsequently undergo gas-phase reactions themselves, creating another generation of condensable products, S1a, S1b, … and S2a, S2b, … etc.

$$S_1 + OH \xrightarrow{k_{OH,S_1}} \ldots + \alpha_{1a} S_{1a} + \alpha_{1b} S_{1b} + \ldots \quad (4)$$

$$S_2 + OH \xrightarrow{k_{OH,S_2}} \ldots + \alpha_{2a} S_{2a} + \alpha_{2b} S_{2b} + \ldots \quad (5)$$

Initially, it was believed that each parent ROG possessed a unique value for the amount of SOA formed from the reaction of each ROG. (Grosjean 1989, Pandis et al., 1992, 1993) These theories contrasted with measured yields for individual organic gases that showed a degree of variation that cannot be reconciled by such a single value theory. An improvement over this approach was presented in Pandis et al. (1992), where each secondary organic reaction product was assigned a volatility based upon its saturation vapor pressure. Under conditions where the gas phase concentration exceeds its saturation vapor pressure, the excess was assumed to condense into the aerosol phase. If the atmospheric concentration was reduced below this saturation concentration, a portion of the condensed product was assumed to return to the gas phase, thus the model allowed for the secondary reaction products to partition reversibly between the aerosol and gas phases depending upon atmospheric conditions and concentrations.

The authors recognized that this modeling approach likely underestimated the total amount of gas-to-particle conversion that occurs, due to the fact that both adsorption of the secondary product onto available particle surfaces and/or absorption of the gas phase compound into pre-existing organic aerosol phase allows for gas-to-particle conversion below the saturation vapor pressure.

Pankow (1994a) developed a combination adsorption/absorption model in order to describe SOA formation at gas phase concentrations below the saturation vapor pressure. As a starting point, the equation used to parameterize gas/particle partitioning is

$$K_p = \frac{A_i / TSP}{G_i} \tag{6}$$

where K_p ($m^3\ \mu g^{-1}$) is a temperature-dependent partitioning constant, TSP ($\mu g\ m^{-3}$) is the concentration of total suspended particulate matter, Ai ($\mu g\ m^{-3}$) is the fraction of the organic compound i in the particle phase and Gi ($\mu g\ m^{-3}$) is the fraction in the gas phase. This equation does not assume anything about the nature of the partitioning.

If partitioning is simple physical adsorption, then the partitioning coefficient can be calculated as

$$K_p = \frac{N_s a_{tsp} T e^{(Q_1 - Q_v)/RT}}{1600 p_L^{\circ}} \tag{7}$$

where N_s (sites cm^{-2}) is the surface concentration of sorption sites, a_{tsp} ($m^2 g^{-1}$) is the specific surface area of the aerosol, Q_1 ($kJ\ mol^{-1}$) is the enthalpy of desorption from the surface, Q_v ($kJ\ mol^{-1}$) is the enthalpy of vaporization of the liquid (sub-cooled if necessary), p_L^0 is the vapor pressure (torr) of the compound of interest (sub-cooled if necessary), R is the gas constant, and T is the temperature (K).

The partitioning coefficient for absorption can be defined as

$$K_{p,i} = \frac{760 RT f_{om}}{MW_{om} 10^6 \zeta_i p_{L,i}^{\circ}} \tag{8}$$

where f_{om} is the mass fraction of the aerosol that is the absorbing organic matter, MW_{om} (g mol^{-1}) is the mean molecular weight of the absorbing organic matter, ζ_i is the activity coefficient of compound i in the organic matter phase, and pL,i^0 is the vapor pressure of the absorbing compound as a liquid (sub-cooled if necessary.) The factor of 10^6 accomplishes the appropriate unit conversions. A similar partitioning coefficient can be defined for the total amount of condensed organic material rather than the total aerosol mass as

$$K_{om,i} = \frac{A_i / M_O}{G_i} = \frac{K_{p,i}}{f_{am}} \tag{9}$$

where Mo is total amount of organic in the aerosol. $K_{om,i}$ is constant at a particular temperature, therefore Eq. 8 implies that as the total organic aerosol fraction increases, the fraction of species i that will partition into the aerosol phase will also increase.

SOA Yields

The most accepted approach to describe the amount of SOA formed from a specific parent reactive organic gas is the fraction mass yield, Y. (Pandis et al. 1991, Odum et al. 1996, Hoffman et al. 1997, Griffin et al. 1999) The yield for each product i is the mass concentration of SOA produced from the oxidation of ROG and it is defined as

$$Y_i = \frac{A_i}{\Delta ROG} \tag{10}$$

where ΔROG is the amount of the ROG that has reacted. The total SOA yield for product i will be the sum of the individual product yields

$$Y = \sum_{i=1}^{N} Y_i \tag{11}$$

where N is the total number of condensable products.

The total concentration of each oxidation product i, Ci, is proportional to the amount of parent hydrocarbon that reacts

$$\alpha_i \Delta ROG = C_i \tag{12}$$

(It is important to note that in Eq. 11, αi is a mass-based stoichiometric coefficient, which can be related to the more traditional molar stoichiometric coefficient by the ratio of the molecular weights of the product to the precursor hydrocarbon, $\alpha_{i,mass} = \alpha_{i,mol} MW_i / MW_{ROG}$.) Ci is also equal to the sum of gas (Gi) and aerosol (Ai) phase concentrations of i,

$$C_i = G_i + A_i \tag{13}$$

Combining Eqs. 8 through 12 provides an expression for the fractional mass yield as a function of the organic partitioning coefficient, $K_{om,i}$ and the organic aerosol mass concentration, M_0,

$$Y = M_0 \sum_{i=1}^{N} \frac{\alpha_i K_{om,i}}{1 + K_{om,i} M_0} \tag{14}$$

It is interesting to examine the limits of the yield formula. When the concentration of organic mass is small, or the volatility of the products is high (decreasing the value of $K_{om,i}$), the yield is directly proportional to M_0,

$$Y: \; M_o \sum_{i=1}^{N} \alpha_i K_{om,i} \tag{15}$$

In the limit of large M0 or products of low volatility the yield is independent of M0,

$$Y: \; \sum_{i=1}^{N} \alpha_i \tag{16}$$

and is simply the sum of the mass-based stoichiometric coefficients of the products.

The theory detailed above has been successfully employed to describe SOA yield data from smog chamber experiments. (Odum et al., 1996,1997, Griffen et al., 1999) These experiments resulted in a series of yield data for different amounts of reacted parent hydrocarbon, and were fitted using Eq. 13. The results of these experiments, conducted for a variety of parent hydrocarbons, were adequately described by two hypothetical products resulting in four fitted parameters, $\alpha 1$, Kom,1, $\alpha 2$, and Kom,2. These two products can be roughly thought of as the SOA yields for an average low and high vapor pressure product, respectively.

While the finding that the two-product model successfully fits laboratory data was an important result, the model is semi-empirical, and as such is not a suitable method for predictively modeling the more complex systems found in the atmosphere. Such a predictive approach is difficult to implement due to an incomplete knowledge of the important reaction products and stoichiometric coefficients for each ROG, and the unavailability of thermodynamic data necessary to provide a values of the vapor pressure pL0 and the activity coefficients ζi for each reaction product. Note that ζi is a function of the SOA mixture composition.

Experimental Investigations of SOA Formation

Experimental studies of secondary organic aerosol formations have largely employed laboratory chambers. These chamber studies provide a means to investigate mechanistic understanding of individual chemical and physical processes that would be difficult to study in the atmosphere due to the complexities inherent in atmospheric flow and mixing process. Chamber studies allow for detailed, lab-based characterization of gas-phase species and species concentrations, aerosol size distribution, and aerosol chemical characterization. The chambers usually are at least a few cubic meters in volume in order to minimize the losses to the chamber walls, and are primarily constructed of Teflon as a non-reactive surface. Both natural and artificial sunlight are used to drive the chemical processes in the chambers. Outdoor chambers use the available natural light to perform experiments, however experiment reproduction can be difficult due to weather and atmospheric variability. Indoor chambers use artificial lights to simulate sunlight, and allow for more precise control of light, temperature, and relative humidity, allowing for better reproducibility between experiments. There is some concern that the artificial light does not simulate the solar spectrum as closely as desired.

Chamber experiments have been performed for a number of different compounds in the last decade, including aromatic hydrocarbons that are prevalent in gasoline (Odum et al. 1996, 1997, Jang and Kamens 2001, Kleindienst et al. 1999) and biogenic hydrocarbons (Hoffman et al.1997, Jang and Kamens 1999, Griffen et al. 1999, Yu et al. 1999a, Cocker et al. 2001, Kamens and Jaoui 2001). The laboratory studies have utilized as an oxidant O_3 reactions, NOx photooxidation, and OH reactions. The results have shown that formation of SOA from terpenes is significant, exceeding the formation rates measured for typical aromatic compounds found in gasoline. Of all the VOC's studies, sesquiterpenes show the highest SOA formation potential.

Several recent studies have provided direct measurements of the condensation of the oxidation products of biogenic hydrocarbons in forest atmospheres. (Kavouras et al. 1998, Leaitch et al. 1999, Yu et al. 1999, O'Dowd et al. 2002) Yu et al. (1999b,c) used laboratory filter-based organic compound speciation techniques to identify monoterpene oxidation components of organic aerosol from forests in both California and Nova Scotia.

Modeling SOA Formation

SOA formation is a highly complex process that depends on a thorough understanding of the emissions of the parent hydrocarbons, their gas phase oxidation pathways and reaction rates, the stoichiometry of the oxidation products, and the thermodynamics of these products in the condensed phase. This section briefly focuses on the modeling efforts of the research community to date, and the strengths and limitations of each approach.

The most basic of the approaches to SOA modeling has been to use measured experimental yield parameters in oxidation models for the smog chamber. (Hoffman et al. 1997, Bowman et al. 1997, Bathelemie and Pryor 1999) This approach employs a box model of the chamber, with known initial conditions. When coupled with appropriate time resolved gas phase data, this approach enables detailed analysis of smog chamber results with a goal of understanding the evolution of the system over time. The results can provide information about the importance of different oxidation pathways and detail of the microphysical processes that provide routes to aerosol formation. This modeling approach is limited when applied to the atmosphere because the atmosphere is not a constant and/or controlled environment. Its behavior is not deterministic but stochastic, and can be quite variable.

Recently, there have been attempts to model SOA yields of specific organic precursors, such as alpha-pinene, using a relatively complete detailed description of the photochemical oxidation products. (Pankow et al., 2001, Seinfeld et al., 2001) These models have been used to calculate yield parameters from know oxidation products and theoretically derived thermodynamics. A few laboratory investigations exist that were able to identify and quantify appreciable fractions of the organic compounds contributing to SOA for a few parent hydrocarbons. (Yu et al. 1999, Kalberer et al. 2000) The values of the stoichiometric coefficients, α_i, were determined experimentally. Values for pL^0 and ζ_i were estimated using UNIFAC (Universal Quasi-Chemical Functional Group Activity Coefficients) principles. (The UNIFAC method uses the idea of a solution-of-groups model utilizing existing phase equilibrium data to predict phase equilibria for systems where no experimental data exist. The method represents the molecules in a mixture by a system of structural subgroups. The properties of the subgroups and interactions between the subgroups are used to calculate activity coefficients.) The model was able to successfully predict aerosol yield values for five different hydrocarbons using ozone as an oxidant, thus providing good evidence that SOA formation can be modeled by using a multicomponent absorbtive partitioning process. This type of modeling also offers possible insight on the usefulness of a proposed chemical mechanism for the oxidation of a specific parent hydrocarbon, or a corroboration of experimentally determined yield data.

Some recent efforts have addressed the representation of the formation of secondary organic aerosol in atmospheric models. (Griffin et al. 2002a,b, 2003, Pun et al. 2002) These models used the Statewide Air Pollution Research Center (SAPRC) mechanism, modified to represent both ozone chemistry as well as the formation of individual organic oxidation products that are capable of forming SOA. In the mechanism, individual parent organics were aggregated into lumped surrogate structures. A thermodynamic equilibrium partitioning model was developed that considered both hydrophobic and hydrophilic constituents. As in the previously discussed models, individual compound activities and the liquid vapor pressure were calculated using UNIFAC. The gas phase mechanism and thermodynamics were combined into a 3-D chemical transport model and used to simulate gas and aerosol concentrations during a 1993 smog episode in Southern California. The results of the models show that the SOA formation in urban areas is dominated by the partitioning of hydrophobic organic compounds formed from the oxidation of anthropogenic organics. In addition, they found that the biogenic contribution to total SOA increased in rural areas as did the fraction of hydrophilic SOA. This type of modeling is clearly the most detailed, and will be necessary for modeling areas for compliance purposes. However, this model requires detailed information on meteorology and emissions, as well as a data set to be compared to, which are not available in many areas. Biogenic emissions inventories are particularly problematic – emissions from many important species in the Sierra Nevadas are not well characterized.

Figure 1. Picture of the tower at the Blodgett Forest Field Site. The boxes mounted on the tower contain the aerosol equipment.

Field Studies

Site Description

The field site is near the Blodgett Forest Research Station (38°53'42.9"N, 120°37'57.9"W, 1315 m) on the western slope of the Sierra Nevada Mountains in California, approximately 75 km northeast of Sacramento. Measurements at the site were established in 1997 in order to investigate the transport processes and photochemistry that occurs as air moves upslope carrying air pollution to our site in a regular diurnal pattern. (Dillon et al. 2002, Dreyfus et al. 2002) The site is characterized by a Mediterranean climate, with warm dry summers and rainfall between September and May. The site is an evenly aged ponderosa pine plantation owned by Sierra Pacific Industries planted in 1990. Measurements were conducted using a 12-m walk-up tower, shown in Figure 1.

The meteorology at the site is fairly consistent. The predominant daytime airmass trajectory is upslope from the Sacramento Valley in the southwest direction, carrying anthropogenic pollutants from Sacramento and the agricultural Central Valley of California. During the evening, a drainage flow moves cleaner air downslope from the Sierra Nevada Mountains in the North East direction.

The site is powered by a diesel generator located ~130 m to the northwest of the tower. During the day, the flow from the west is strong enough that plumes from the generator are rarely observed. At night the winds are weaker, and an occasional plume of short duration is observed. These data points can be identified by brief increases in total particle concentration and black carbon concentration.

Aerosol Measurements

The physical properties of the aerosols at the site were measured using a variety of instruments. The optical properties of the aerosol were measured using an aethalometer (Model. AE2, McGee Scientific, Berkeley, CA) and an integrating nephelometer (NGN-2, OPTEC, Inc, Michigan). An integrated measure of the number concentration of aerosols was provided by a condensation particle counter (CPC, TSI 3022A). Particle size distributions at the site were measured using an optical particle counter (Lasair 1003, Particle Measurement Systems) providing size distributions from 0.1 to 10 μm, and a scanning electrical mobility analyzer (SEMS) providing size distributions between 10 and 400 nm. The SEMS system utilized a differential mobility analyzer (TSI 3071A) coupled with a CPC (TSI 3760) as a detector. All instruments except the OPC were mounted on the tower. The OPC was located in an adjacent instrument container. A schematic of the instrument placement on the tower is shown in Figure 2. The air was sampled at 16.7 L min-1 through a PM10 inlet followed by a PM2.5 sharp cut cyclone

(Rupprecht & Patashnick Co. models 57-000596 and 57-005896, respectively). The inlet was mounted above the top of the tower at a height of approximately 13 m. Sample flows for the aethalometer, CPC, SEMS, and OPC were isokinetically sampled from this main flow. Measurements were recorded from 9 May to 14 November in 2002. Measurements from the CPC, SEMS, OPC, and nephelometer were recorded every 2 minutes, while those from the aethalometer were recorded every 5 minutes.

The aethalometer measures light absorption of the aerosol, babs, using a filter-based light transmission technique. (Hansen 1984) The instrument measures decreases in transmitted light through an increasingly particle-laden filter. Black carbon (BC) is the insoluble, most resistant to heating, graphitic, and strongly light absorbing component of the aerosol mass. The aethalometer used for this study measured light absorption at two wavelengths, 880 nm (near IR) and 350 nm (near UV). BC is believed to be the only aerosol component that absorbs in the near infrared spectral region, so measured transmission of 880 nm wavelength light is used to determine BC concentration. An empirically derived calibration equation is used to estimate the mass concentration of BC in air (in μg m-3) from the measured decrease in light transmission. (Gundel et al. 1984) Blue and near ultraviolet light transmission measurements are used to indicate the presence of aerosol materials that absorb in these spectral regions, such as mineral dust (d'Almeida, 1987) and some types of organic compounds. (Kirchstetter, 2004; Bond, 2001) The flow rate through the instrument was 4 L min-1.

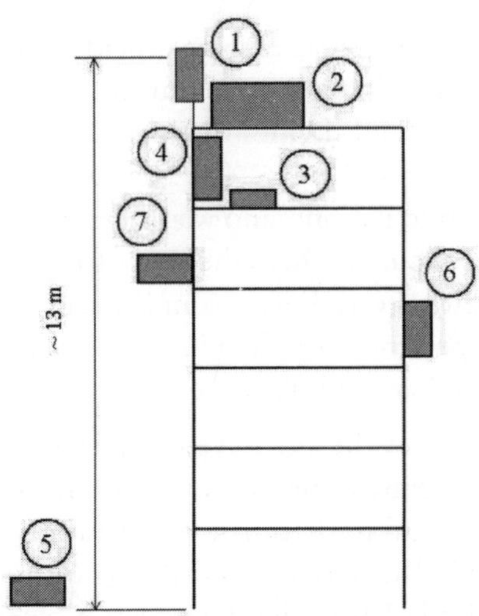

Figure 2. Schematic of the aerosol instrument on the tower. The instruments are as indicated: (1) 2.5 mm cyclone inlet, (2) aethalometer, (3) condensation particle counter, (4) scanning mobility particle scanner, (5) optical particle sizer, (6) nephelometer, and (7) filter samplers.

The nephelometer measures the light scattering coefficient, bscat, at an effective wavelength of 550 nm. The instrument, described in more detail by Molenar et al. (1989), has an open air design, allowing the air to pass through a large opening in the side of the instrument. This design minimizes the changes in relative humidity and temperature that has lead to underestimated scattering measurements in other nephelometers. Because the instrument is open, it allows a wide range of particle sizes to pass through it, including particles > 2.5 µm. The cutpoint of the instrument has not been characterized.

The SEMS system provides aerosol size distributions using a mobility classifier. Particles are classified by their electrical mobility while moving across an electric potential. Only particles of a certain mass to charge ratio can exit the instrument. Particles exiting the instrument are counted using the condensation particle counter. By ramping over the voltage range of the instrument exponentially, it is possible to obtain size distributions relatively rapidly. (Wang and Flagan 1990) The SEMS system at the Blodgett field locations sampled aerosol at a flow rate of 0.7 L min-1. Utilizing a sheath flow rate in the mobility classifier of 7 L min-1, a full voltage scan of the instrument resulted in size distributions from 10 nm to 400 nm. The SEMS system was controlled using Labview software and National Instruments control boards. One complete size distributions was obtained in a little over 1 minute using software developed by Donald Collins of Texas A&M University and Patrick Chuang of the Univeristy of California at Santa Cruz. The inversion program used to process the raw data was based on that presented by Collins et al. (2002).

Size distributions of larger particles were provided by an optical particle counter. The instrument sized particles by measuring the amount of light that each individual particle scatters, and divides the measured size distribution up into 8 size bins between 0.1 µm and 10 µm. The particular instrument employed was sensitive to temperature fluctuations, and was therefore housed in the temperature controlled instrument container where the other gas phase measurement systems were conducted. The aerosol was drawn from the sampling system down into the container by a 1/8" copper line. Sampling line losses were characterized on site for each size bin and averaged around 50 percent.

Aerosol measurements at the site have been ongoing since the fall of 2000, with the instruments operating primarily from late spring into late fall. The instruments vary a great deal in their ease of operation and degree of sensitivity to operating in the field. This has affected the quantity and availability of data for each instrument. Table I lists the dates that each instrument operated for each year at Blodgett.

Table 1. Dates of operation for each aerosol instrument used at the Blodgett Field Site.

	2000	2001	2002	2003
CPC	10/5 – 11/21	7/12 – 10/20	5/9 – 11/13	8/6 – present
Nephelometer	10/5 – 11/11	7/12 – 11/5	5/9 – 11/13	8/6 – present
Aethalometer	9/28 – 11/21	7/12 – 10/24	5/9 – 11/12	6/23 - present
SEMS	N/A	7/12 – 10/24*	5/29 – 11/13	7/31 – 11/19
Lasair	N/A**	N/A	5/9 – 11/13	N/A

*The SEMS was not functioning correctly for much of the 2001 measurement season due to a noise problem in the signal lines. This problem was not identified and corrected until September, 2001.
** The Lasair proved to be extremely sensitive to temperature, and measurements were not successful until we installed it in the temperature controlled container.

During the late summer and fall of 2003, a semi-regular set of filter samples was collected to provide measures of PM2.5 mass as well as carbon, nitrate, sulfate, and ammonium through chemical analysis. These filter measurements provide important data concerning the chemical composition of the aerosol, including what fraction is organic, and important quantity that affects the partitioning coefficient (Eq. 8). Each measurement period consisted of four sets of samples collected over 48 hours, with the filter samples running from 9:30 AM to 7:30 PM, and from 8 PM to 9 AM. The times during which the filters were changed correspond to periods when the wind was shifting - downslope to upslope in the morning and upslope to downslope in the evening. This results in particle chemistry at the site during periods of primarily anthropogenic emissions and primarily biogenic. The first sample line consisted of a NaCl coated honeycomb denuder followed by a 25 mm Teflon filter (Gelman), and then one 47 mm Nylon filter (Gelman) and a cellulose fiber filters (Whatman) in a Teflon filter holder (Savillex). The nylon filter was analyzed for gaseous nitric acid. The cellulose filter was impregnated with citric acid to allow for the analysis of gaseous ammonia. (Anlauf et al. 1988) The second sampling line used dual 25 mm diameter quartz filters operated in series for total and organic carbon concentrations. The third sampling line was similar to the second, consisting of a 25 mm Teflon filter (Gelman) followed by a 25 mm quartz filter. This third line was used to quantify the organic sampling artifact (cite – Kirchstetter et al. 2001, Turpin et al. 1994). All three lines drew a sample through a different inlet than that used by the aerosol instrumentation. Each line employed a 2.5 μm cyclone precut (John and Reischl 1980) and was sampled at a flow rate of approximately 25 L/min.

PM2.5 mass was gravimetrically determined using a Cahn Model 21 electrobalance. Before weighing, the Teflon filters were equilibrated for at least 4 hours at 20 °C and a relative humidity of 35%. For chemical analysis, the filters were extracted by sonication in distilled deionized water. Extracts from both the Teflon filters were analyzed for nitrate and sulfate by ion chromatography and for ammonium by an ion-specific (ammonia gas-sensing) electrode. The nylon filters were analyzed by ion chromatography. The citric acid impregnated cellulose fiber filters were analyzed for ammonium ion by an ion-specific electrode. The quartz filters were assayed for carbon using the evolved gas analysis method of Novakov et al. (1982, 1983). The ambient organic carbon particulate was calculated as the difference between the values measured on the front and back filter.

A similar set of filter measurements was also performed in the late fall of 2002. During this period, a series of controlled burns was performed at other locations on the Blodgett property. These prescribed burns were performed by Scott Stephens and his group from the Department of Environmental Science, Policy, and Management at the University of California at Berkeley as part of a project to study fire and fire surrogates for forest restoration. Forest fires can be a significant source of organic aerosol, and can serve to reduce visibility over large regions. The air masses at the measurement tower were heavily impacted by the prescribed burns, resulting in an opportunity to sample and analyze the aerosol resulting from the fires.

VOC Measurements

Continuous in-situ hydrocarbon measurements were taken at 1 hour intervals using a fully automated gas chromatograph dual flame ionization detector (GC-FID) system (Lamanna and Goldstein 1999). Air was sampled at 8 L min-1 through 2 μm PTFE particulate filters and Teflon tubing and then subsampled at 20 mL min-1 onto a graphitized carbon black and carbon molecular sieve preconcentration system packed in series and mounted in a cold block. After preconcentration, the samples were thermally desorbed (~250°C for 1.2 min) onto two Rtx-WAX columns for separation and quantification by the two FIDs. The system was fully automated (HP Chemstation and Campbell Scientific data logger) and processed as a 20 min average sample every hour. The experimental detection limit depended upon the species, but averaged 10-20 parts per trillion. Mixing ratios above 0.2 ppb are considered accurate to 15%.

While gas chromatography has been a useful instrument to measure atmospheric concentrations of VOCs, the long time resolution of the technique limits its usefulness in the analysis of the dynamics of processes that occur on faster time scales. To enable gas phase measurements with higher time resolution, a proton-transfer-reaction mass spectrometer (PTR-MS) was employed at the site beginning July 2002. The PTR-MS utilizes chemical ionization for the analysis of trace constituents, with a VOC detection limit at the parts per trillion level and a time response of ~0.2s. Because the measurements of the PTR-MS are based upon mass, the molecular weight of the target analyte must be unique. Thus the system can is used to measure the total concentration of monoterpenes because all monoterpene species share the same molecular weight. Similarly, the concentration of different groups of monoterpene oxidation products will be detected because many of the oxidation products of different monoterpenes have common molecular weights as well.

Table 2. VOC species measured at the Blodgett forest field location by year. During 2002 and 2003 the species are grouped by instrument.

2000	2001	2002	2003
alphapinene	Isoprene	*GC:*	*GC:*
betapinene	Acetaldehyde	alphapinene	alphapinene
Toluene	Acetone	betapinene	betapinene
Isoprene	Butanol	3-carene	3-carene
Methylvinyl ketone (MVK)	Methacrolein (MACR)	Limonene + b-phellendrene	Limonene + b-phellendrene
methacrolein	Methyl furan	alphaterpinene	alphaterpinene
3-carene	Methanol	Camphene	Camphene
	Methylethyl ketone (MEK)	Myrcene	Myrcene
	Ethanol		
	Benzene	*PTR-MS:*	*PTR-MS:*
	MVK	Total monoterpene	Acetonitrile
	Pentanal	Acetaldehyde	Isoprene + MBO
	alphapinene	Acetone	Total monoterpene
	Toluene	MACR + MVK	Benzene
	2-methyl-3-buten-2-ol (MBO)	Toluene	Toluene
	Camphene	Isoprene + MBO	Methyl chavicol
	Hexenal	3-methylfuran	MACR + MVK
	betapinene		Methanol
	3-Carene		Acetone
	Myrcene		Nopinone
	o-xylene		Pinonaldehyde
	limonene		

The VOC measurements at the site have been ongoing since 1997, and have been performed by Allen Goldstein and his group in the Department of Environmental Science, Policy and Management at the University of California at Berkeley. Table II lists the individual VOC species that have been measured in each year that aerosol data is available. Their research has been supported by the California Air Resources Board (CARB) and the National Science Foundation (NSF).

NO_2 Measuremements

A system utilizing thermal dissociation at several specific temperatures and catalytic oxidation coupled with laser induced fluorescence of NO_2 was used for in situ measurements of the important nitrogen containing trace gas systems in the atmosphere. The gases measured included NO, NO_2, the sum total of peroxy nitrates, the sum total of alkyl nitrates and hydroxyalkyl nitrates, and nitric acid. The instrument is capable of measuring these compounds with a detection limit of 90 ppt 10 s^{-1}, with an accuracy of 15%. Ronald Cohen and his group from the Chemistry Department at the University of California at Berkeley conducted the measurements.

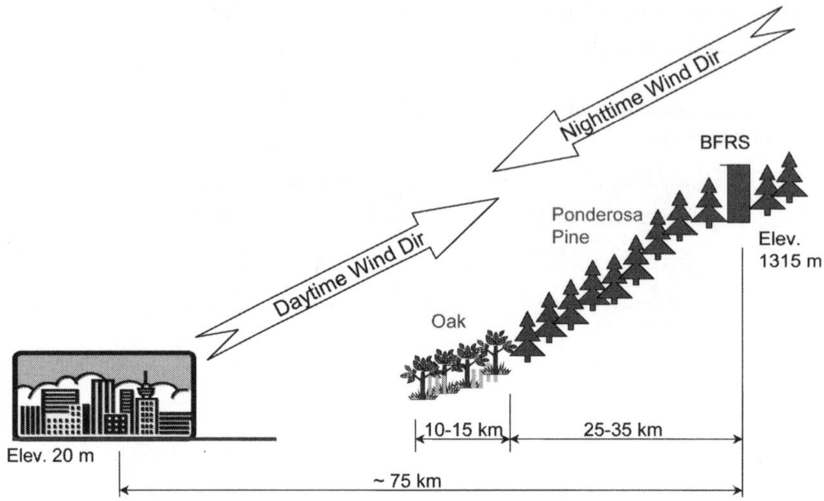

Figure 3. A diagram of the wind patterns experienced by the Blodgett Forest Research Tower.

Meteorological Measurements

Meteorological parameters such as air temperature, humidity, wind speed and direction, and net and photosynthetically active radiation (PAR) were measured continuously and stored as 30-minute averages. Meteorological data and trace gas mixing ratios and fluxes (CO_2, H_20, O_3, and hydrocarbons) were measured approximately 5-6 m above the average tree height.

Results and Discussion

Aerosol data exist for the Blodgett field location for the years 2000, 2001, and 2002. Support from LBNL-Laboratory Directed Research and Development Funds was used to initiate the particle measurements in 2000, and was continued with support from the California Air Resources Board during 2001 and 2002. The data from 2000 was obtained for a limited number of weeks, and contains no size resolved data. However, VOC measurements are available from the Goldstein group for 2000, and therefore we performed some preliminary correlations between the aerosol and VOC measurements with this data. The aerosol measurements were expanded in 2001, however it appears that it may be some time before the VOC data from this year will be available. Therefore, much of the results that we will present in this report are from the 2002 data set. This data set is the most complete, with integrated and size resolved data as well as measured optical properties for a number of months. This data has been analyzed to identify any inconsistencies, biases, or artifacts, and periods of faulty instrument performance were removed to bring it to Level I readiness. The VOC data from 2002 is currently being processed and some subsets of the data will be available soon.

An important final task necessary for more sophisticated data analysis to proceed is to remove the signal resulting from generator plumes from the data. As noted previously, during the evening when the winds are weak, plumes of particulate emissions from the generator can be observed in the particulate data. Diesel generators often emit large concentrations of small (< 100 nm) carbonaceous particles. These emissions are reflected in the data by brief increases in total particle and black carbon concentrations, as well as the appearance of a smaller mode in the size distributions. We are currently pursuing several routes to remove this diesel signal from the data. We have obtained data from the sonic anemometer that provides data on wind speed and wind direction at one-minute time intervals, and are testing to see if the spike occurs when the wind direction is from the same direction as the generator. By combining this data set with black carbon and size resolved measurements, we can more precisely identify times that are affected by the generator. Lastly, we are developing data rejection routines that can test for quick, short, large-scale departures from any individual data set that are due to the generator. By using these techniques alone or in combination, we will have a good data rejection criterion to bring the particle data to Level II readiness.

Meteorology, Pollutant Formation and Transport

The predominant daytime airmass trajectory at the site is southwest, directed upslope from the Sacramento Valley. At night, air flows northeast, downslope from the Sierra Nevada Mountains. Figure 3 shows this diurnal wind pattern graphically. During the daytime, the upslope flow brings pollution to the site that is a mixture of urban outflow from the Bay Area and Sacramento and agricultural outflow upwind of Sacramento. During the evening the downslope drainage flow from the mountains brings cleaner air containing primarily biogenic emissions.

The effect of the diurnal wind patterns at the measurement tower on transport of anthropogenic pollutants to the site are shown in Figure 4. The top graph shows how the wind direction shifts from downslope to upslope flow between 7 and 9 AM, and shifts back again between 6 and 8 PM. The bottom plot shows the typical pattern in total particle concentration. The concentration begins to rise in the afternoon as pollutants are transported to the tower location. The concentration decreases after the wind shifts to cleaner downslope flow from the Sierras. Figure 4 also shows the effect of the generator that provides power to the site on the particle measurements.

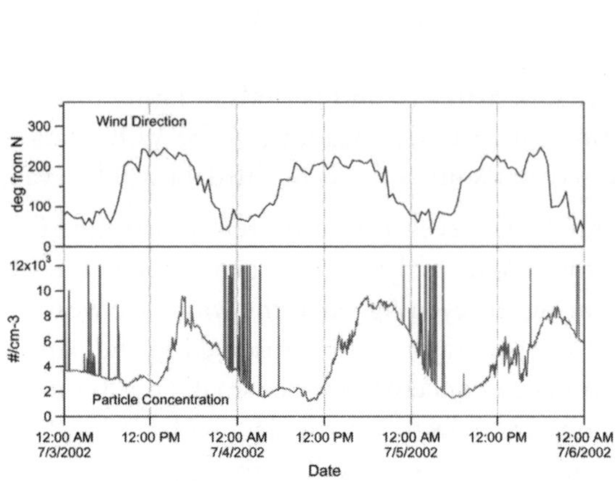

Figure 4. Wind direction and particle concentration data measured at the Blodgett tower for a three day period in July of 2002.

Typical patterns in the VOC concentrations measured at the tower during the year 2000 are shown in Figure 5. Isoprene, a reactive VOC emitted from oaks, has maximum concentrations in the afternoon and minimum in the early morning. Toluene, an anthropogenic compound, increases rapidly around noon after the wind shifts direction and increases in speed, and remains high until cleaner air descends from above. Terpenes, represented by alpha-pinene, have a diurnal pattern with highest concentrations at night when vertical mixing is weakest and low concentrations during the day when mixing is more vigorous.

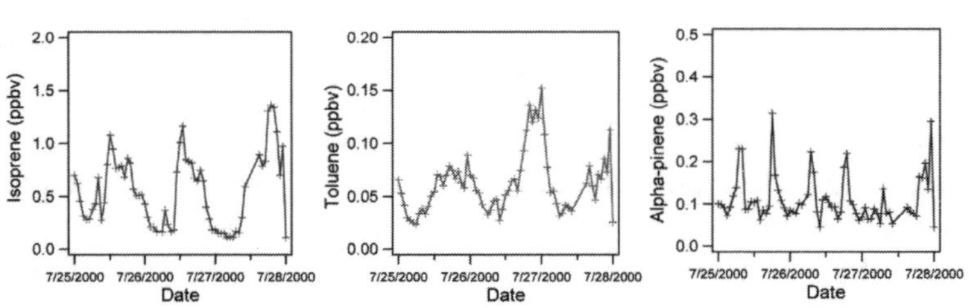

Figure 5. Typical diurnal patterns of the three important VOC species measured during 2000: (green) isoprene, (red) toluene, and (blue) alpha-pinene.

Half hour averages of total particle concentration measured with the CPC and black carbon concentrations measured using the aethalometer were analyzed with the VOC data from the year 2000 to look for correlations between different types of organic precursors and particle loading. The results of the correlations between particle and VOC concentrations are shown in Table 3. The results show correlation with both biogenic and anthropogenic (toluene) precursors, implying that aerosol at the site result from a mix of sources. The correlation with the biogenic VOCs a-pinene and b-pinene that are emitted by the pines are relatively weak. The correlation with isoprene, emitted by oak trees, is stronger and is accompanied by strong correlations with the oxidation products of isoprene (MVK and methacrolein), implying that oxidation of oak emissions upstream are contributing to the particle load at the tower.

Table 3. Results of multivariate correlations between total particle concentration and VOC concentration. Note that these correlations are for 422 hours of complete data from the fall of 2000.

Gas Phase Compound	Correlation
α-pinene	0.14
β-pinene	0.15
Toluene	0.56
Isoprene	0.38
Methyl vinly ketone	0.45
methacrolein	0.39

Black carbon and carbon monoxide both serve as excellent tracers of anthropogenic emissions as they primarily result from combustion. The correlation between particle concentration and black carbon was 0.69, and the correlation with carbon monoxide was 0.73. These correlations, combined with the correlation with toluene, imply that the total particle loading at the site is associated with anthropogenic emissions transported to the site from the Sacramento Valley. The results, however, do not definitively demonstrate that the aerosol at the site results from anthropogenic sources; the urban outflow may simply be arriving at the site on the same time frame that other local biogenic processes are simultaneously contributing to the particle load.

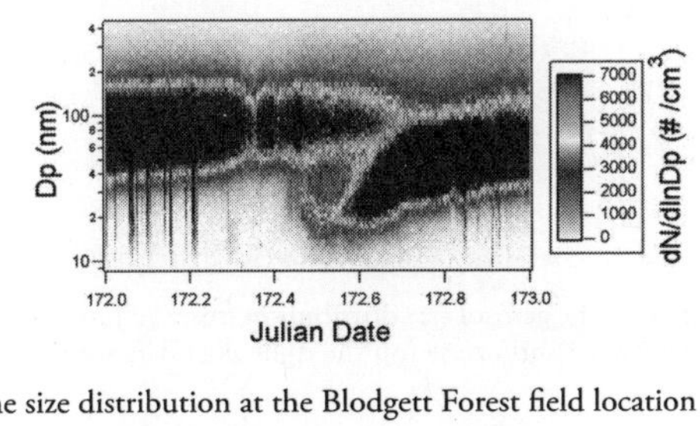

Figure 6a. A time line of the size distribution at the Blodgett Forest field location for 21 June, 2002.

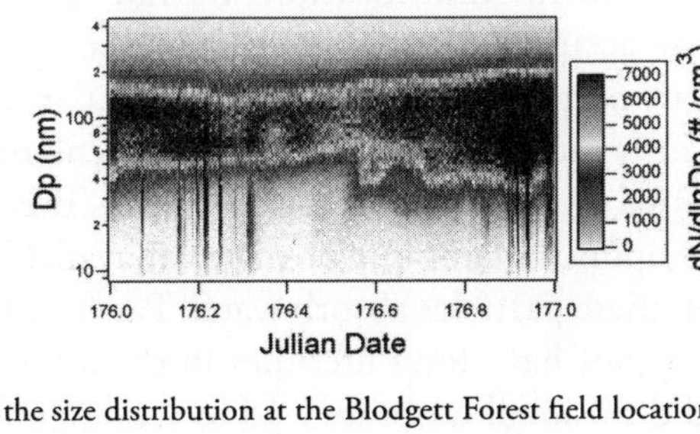

Figure 6b. A time line of the size distribution at the Blodgett Forest field location for 25 June, 2002.

Particle Formation and Growth

The size distribution measurements performed at the tower site during 2002 provide important additional information about the nature of the aerosol that cannot be ascertained from measures of total concentration alone. An important feature of atmospheric aerosol distributions is the fact that they often exhibit multiple modes. Particles of size with an aerodyamic diameter, D_p, greater than approximately 1 μm are often referred to as the coarse mode. These particles are generated by mechanical processes, and consist of soil dust, sea salt, tire and break wear particles, etc. Fine mode is the term for particles of size less than 1 μm; fine mode particles contain particles from combustion sources, and secondary aerosols formed by chemical reactions that result in gas-to-particle conversion. The fine mode can be further divided into an accumulation mode (0.1 μm < D_p< 1.0 μm) and a nuclei mode (D_p< 0.1 μm). The primary mechanism for movement of particles from the nuclei to the accumulation mode is by coagulation and growth by condensation of vapors onto existing particles. Coagulation among accumulation mode particles is slow and does not result in growth into the coarse mode. It should be noted that the size boundaries between these different sections are not precise, and depend on the type of region (urban, remote, marine etc.) in which the aerosol exists.

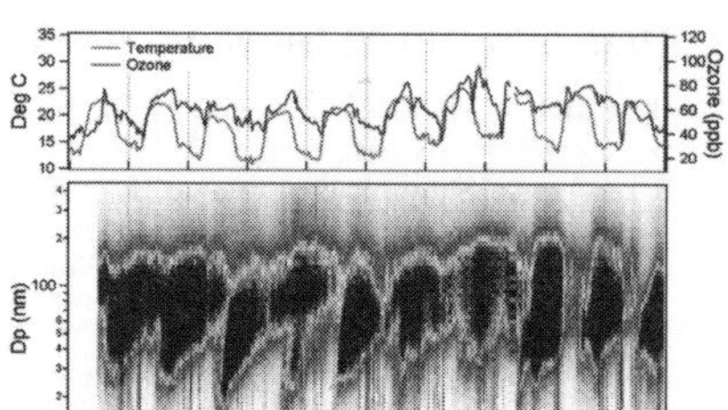

Figure 7. Time resolved evolution of the aerosol size distribution from 19 June to 29 June 2002. The upper plot shows the temperature (on the left axis) and ozone (on the right axis) data for the time period.

Aerosol size distributions in rural locations do not always contain particles in the nuclei mode. The accumulation mode provides a significant surface area for condensable gases to partition onto. Nuclei mode particles will form only when there is a sufficiently large concentration of condensable gases for it to be thermodynamically favorable for new particles to form. The size to which these particles grow depends upon the amount of material available for condensation and the degree to which these particles absorb water. Particles that are quite small, 50 nm or less in size, do not have long lifetimes in the atmosphere. Their small size and thus enhanced diffusional mobility lead to scavenging by larger particles and deposition to surfaces. Therefore, when such small particles exist, it can be concluded that they were formed from gas phase precursors relatively close to the measurement location.

A time history of the particle size distribution of two separate days at the Blodgett forest location are shown in Figure 6. The figure shows each two-minute size distribution vertically, with color representing the number concentration at each measured diameter. From midnight until midday, the peak of both size distributions is around 100 nm. The particles in this mode are likely a mixture of sulfates, nitrates, and black carbon formed by atmospheric processing of pollutants transported from upstream and local emissions sources. Note that in Figure 6a, a smaller size mode of aerosol begins to appear just after noon. These nuclei mode particles, initially below 20 nm have a limited lifetime in the atmosphere (~1hr), suggesting local formation mechanisms. Figure 6b shows a day when such a growth event does not occur.

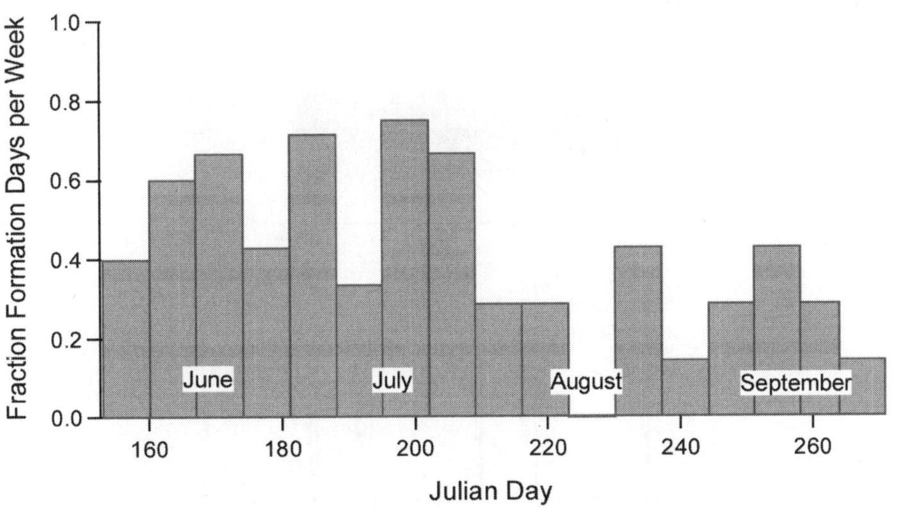

Figure 8. The fraction of days per week that a growth event occurred for the months of June through September.

Figure 7 shows a time line of 10 days from 19 June to 29 June, 2002. During this week, it appears that the fine particle growth events happen on 7 of the 10 days. The lifetime of 20 nm sized particles in the atmosphere is approximately 1 hour. Combining typical wind speeds and the size of the forest results in transport times within the forest that are greater that this lifetime, implying that the particle growth events occur over the forest. These particles are formed either from a combination of biogenic processing of the anthropogenic air mass or solely by biogenic reactions. Figure 8 shows the fraction of days each week that these growth events occurred for the months of June through September. Most weeks had a significant fraction of days with these formation events. Observations by other investigators at other locations have suggested that particle formation events are preceded by a period of relatively particle-free air. In contrast, we have observed formation both with and without the presence of a larger aerosol mode.

A detailed look at the aerosol size distributions before, during, and after the appearance of the nuclei mode aeroso l is shown in Figure 9 for July 24, 2002. At 11:45, the size distribution is peaked at approximately 80 nm. The appearance of the nuclei mode can be clearly seen in Figure 9b at approximately 25 nm, with the primary accumulation mode peak still at 80 nm. By the early evening, the nuclei mode has continued to grow by condensation of more gas phase product, and the particles in the smaller mode have been coagulating with those in the larger mode. This results in the broader distribution, seen in Figure 9c, where the nuclei and accumulation modes have almost merged, although the nuclei mode peak can still be discerned at approximately 45 nm.

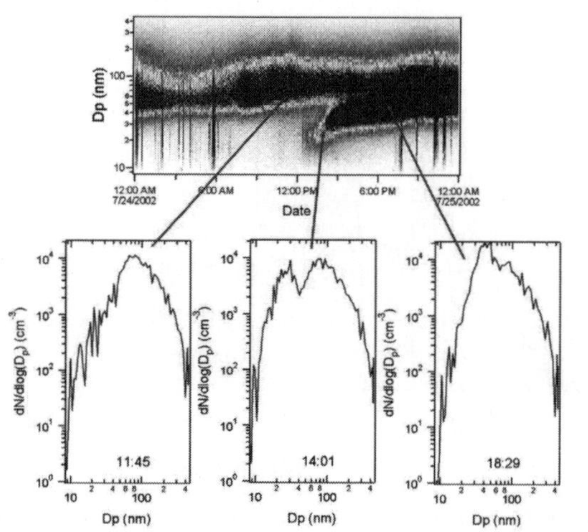

Figure 9. Aerosol size distributions (a) before, (b) during, and (c) after the particle growth event measured on July 24, 2002.

The growth events seen in the size distribution data provide evidence for the physical processes that lead to particle formation and growth. As discussed previously, when the concentration of condensable gases increase, they have two routes to the particle phase; they may nucleate to form new particles or condense onto existing particle surface area. The route to condensation depends upon the thermodynamics of both the gas and particle phases. At the Blodgett site, the oxidation products will most often condense onto accumulation mode particles. Arguments using thermodynamics suggest that low volatility products resulting from the oxidation of biogenic VOC's are not likely to nucleate. It does appear that new particles are being formed during these particle growth events. It has been suggested that these nuclei mode particles are instead growing by condensation onto pre-existing, very small "seed" particles (< 3nm) that likely consist of sulfuric acid. (Czoschke et al. 2003) There is, however, reason to believe that the majority of the mass addition in these growth mode particles is due to condensed organic oxidation products.

There is some experimental evidence available to provide an explanation for what conditions lead to the occurrence of the growth events. Among the factors that affect when and to what extent the organic oxidation products partition to the particle phase are the gas phase concentration of condensable products, the rate at which the gas accumulates, the concentration of seed particles, and the temperature and relative humidity of the ambient environment. An important variable is temperature. Recall that in Eq. 7 that the partitioning coefficient, Kp,i, is inversely proportional to the vapor pressure of the absorbing compound. As the temperature decreases, the vapor pressure also decreases, thereby increasing the partitioning coefficient. When the temperature is reduced, the oxidation products are more likely to partition to the aerosol phase, however it is also possible that fewer oxidation products form. These two competing processes, the increase in partitioning and decrease in reaction rates, suggest that optimum conditions for the temperature at which growth events occur are rather sharply defined.

The effect of temperature can be observed in Figure 7, which shows time resolved temperature and ozone measurements recorded for the same time period at the aerosol size distributions. The figure shows that when the temperature is higher, the particle growth events are less likely to occur. This is evidenced in the difference between day 172, one of the colder days during this 10-day period and with a pronounced growth event versus day 176, the warmest day of this period with no discernable growth in the nuclei mode. A summary of this phenomenon for the summer months can be seen in Table 4, which gives the daily maximum temperature for the days with and without growth events. The days with growth events had a maximum temperature approximately 4 °C lower than days with no events.

Table 4. Average maximum daily temperature for days in 2002 on which growth events occurred and those with no event.

	June	July	Aug	Sept
Growth Event	21.1 °C	24.3 °C	23.0 °C	21.7 °C
No Event	25.0 °C	28.0 °C	26.5 °C	24.8 °C

Figure 10 shows the same series of size distribution data shown in Figure 7, as well as several integrated quantities calculated from the size distribution: total number concentration, total particle volume, and particle volume less than 50 nm. It is not shown here, but the total particle concentration calculated from the size distribution data matches well with that measured by the CPC. The total calculated particle volume is between 1 and 10 μm-3 cm-3, and does not vary over the same dynamic range that total particle number does. This is due to the fact that particle volume, being the cube of particle diameter, is more sensitive to larger particles. The behavior of the particle volume calculated for particles smaller than 50 nm looks very similar to the time resolved behavior of total particle concentration. Thus, total particle number is often dominated by smaller sizes, particularly on days with particle growth events, when large numbers of small particles exist.

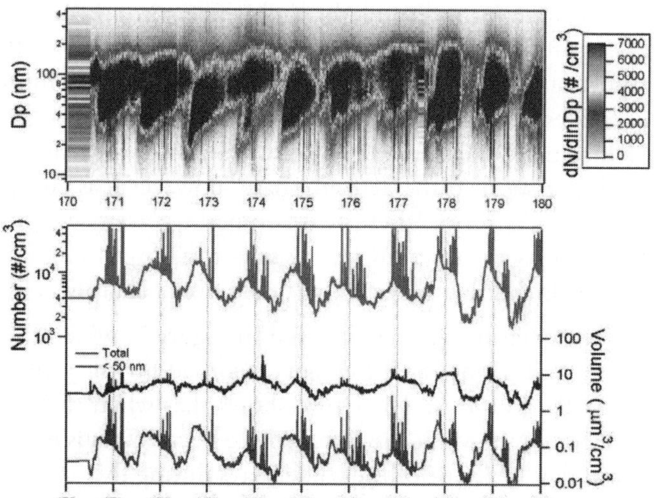

Figure 10. The upper plot shows the time resolved evolution of the aerosol size distribution from 19 June to 29 June 2002. The lower plot shows the total particle number, particle volume, and particle volume for those particles less than 50 nm calculated from the size distribution data.

The percentage of total particle volume that is contributed by particles less than 50 nm in size is shown in Figure 11 calculated from the same size distribution data shown in Figure 10. In general, the percentage of total particle volume contributed by particles less than 50 nm is quite small, around 1 percent. However, during the growth events in the nuclei mode, the fraction of total particle volume from this size range can increase to as high as 10 percent. This is a significant fraction, particularly if it consists primarily of biogenic organic oxidation products. The fate for the majority of these nucleation mode particles is coagulation with the larger particle mode. Consequently, the processing of these small particles via coagulation with larger ones can be a significant pathway of SOA into the ambient aerosol.

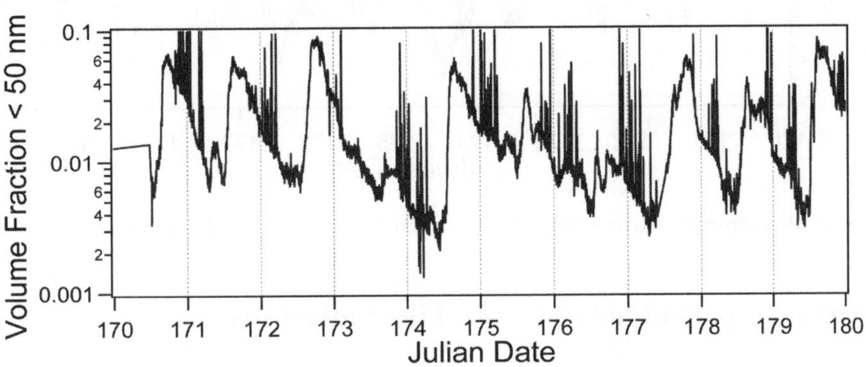

Figure 11. Fraction of total particle volume for those particles less than 50 nm calculated from the size distribution shown in Figure 10.

To further explore the link between secondary oxidation products and condensation onto the nucleation mode particles, we next focus on the total particle mass in the nuclei mode to ascertain whether this mass is of the same order of magnitude as the measured concentration of oxidation products. Figure 12 shows the mass calculated for particles less than 50 nm. This calculation assumes that the particles are entirely organic with a molecular weight of 136, equivalent to that of a monoterpene. The units have been converted into ppb for ease of comparison with gas phase data. The results show that the calculated mass is between 0.005 and 0.1 ppb. The concentrations of the monoterpene alpha-pinene, shown in Figure 5, are between 0.1 and 0.5 ppb. Other biogenic precursors, not shown in this report, also show similar concentrations in the atmosphere. Based upon these measured values, it is possible that the mass of the nucleation mode products could consist entirely of the oxidation products of measured biogenic precursors. It should also be noted that an estimate of 136 as the mass of aerosol organic phase, is probably low; the lower volatility oxidation products often have higher molecular weights than the parent compound.

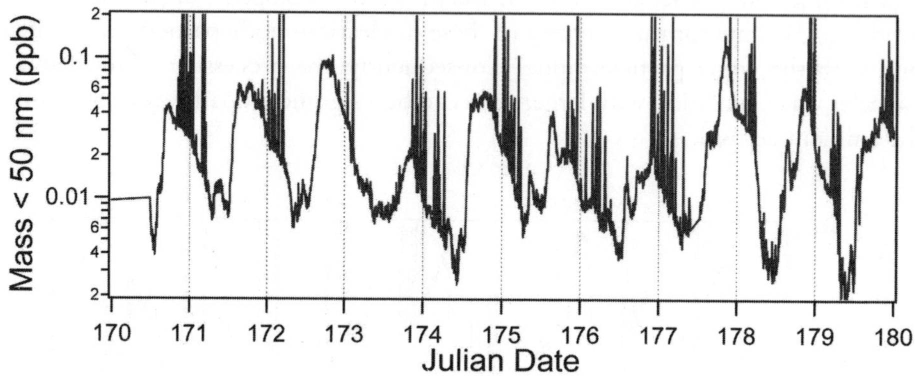

Figure 12. Mass of particles less than 50 nm calculated from the size distribution shown in Figure 10.

Aerosol Impacts Due to Forest Fire Emissions

During the 2002 measurement season, the Blodgett location was affected by a number of regional forest fires. The largest fire signal in the data was due to the Biscuit fire in Southeast Oregon. This fire, started by a lightning strike, lasted from July 13th to November 9th, eventually burning approximately 500,000 acres. The Blodgett field site was strongly affected by the fire from approximately August 15th through August 22. Figure 13 shows plots of total particle scattering, black carbon concentration, and particle size distribution for the period of time affected by the Biscuit Fire. The period of time most affected by the fire can be detected by an increase in the black carbon data, denoted by the area in the red squares in Figure 13. The particle size distributions show the presence of a larger size mode during the period most affected by the fires, as emissions from the fires age and grow in the atmosphere as they are transported to the site. The presence of a larger mode in the size distribution has been observed for fire events from other periods in the years 2002 and 2001. Figure 13 also shows an increase in particle scattering during the same period. Indeed, the particle scattering signal tracks the black carbon signal very well. Scattering is proportional to particle diameter by a power law, therefore the larger particles in the smoke affected periods lead to increased scattering. The correlation between the black carbon and scattering signals implies that the majority of the large particles are largely carbonaceous. This data show how forest fires can dramatically affect visibility, even several hundred miles from the source.

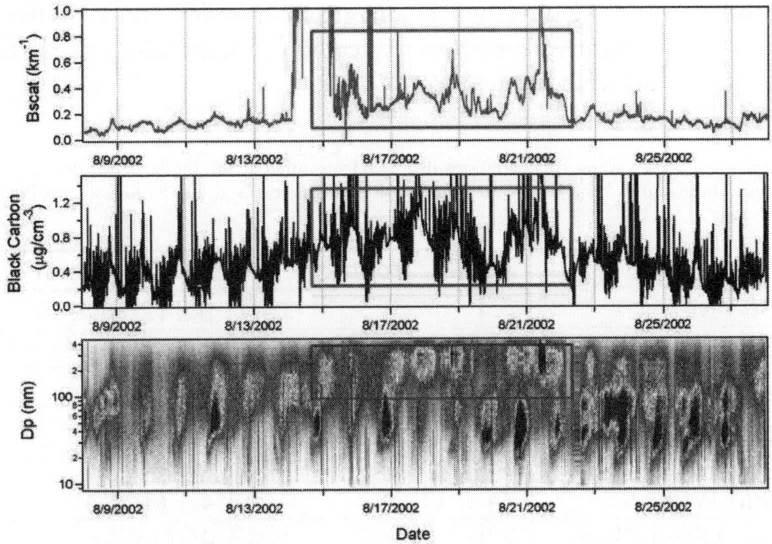

Figure 13. Graph of total particle scattering (km-1), black carbon concentration (µg/cm3), and particle size distribution for the period of time affected by the Biscuit fire.

Table 5 shows the results from filter measurements performed during the prescribed burns that occurred during late October 2002. The fires were controlled and small in scope, no more than a few acres each. Due to dry conditions, the fires were only set during the evening. The AM filter samples were run from 9:30 AM to 9:30 PM, with the PM sample running from 9:30 PM to 9:30 AM. The data in the table show the total carbon measured on the filter as well as the amount and percentage (in parentheses) that was black carbon. The sample measured on 10/30 during the evening was the only sample of the set where fire emissions were noticeably present at the tower location, which is reflected by the larger concentrations of carbon for that period. The table also includes carbon measurements acquired at different locations: urban (Fresno, CA), remote (also measured in the Sierra Nevada mountains), large fire plume in South Africa. The results show that the carbon levels measured during the controlled burns are similar to those measured in the urban location during the fall, but the percentage of the total carbon that was black carbon was smaller. This result implies that the carbonaceous aerosols emitted from the fire were composed primarily of organic carbon. The total carbon concentrations measured in the Sierra during routine monitoring were not significantly different from many of the controlled burn samples. This is most probably due to the small scope of the burns having a limited effect at the tower site. The results listed for Fresno during the winter show high carbon concentrations, a result largely influenced by the large number of fireplace emissions from homes. The concentrations measured in the South African Plume show how large carbon emissions can be directly over a large uncontrolled fire. Both the Fresno and South Africa measurements show that carbon concentrations due to fires can be large, even in urban locations.

Table 5. Total and black carbon concentrations in mg/m3 measured at the Blodgett field location as well as for locations indicative of urban (Fresno), remote (US Remote Sierra), and heavily fire affected (South Africa Plume) locations.

	Total Carbon	Black Carbon
10/30 PM	7.4	0.2 (2%)
10/31 AM	2.4	0.2 (7%)
10/31 PM	2.2	0.13 (6%)
11/1 AM	2.0	0.15 (8%)
Fresno Fall	7 ± 3	1.4 (20%)
Fresno Winter	34 ± 11	4.1 (12%)
South Africa Plume	220	45 (18%)
US Remote Sierra	2.35	0.25 (11%)

Acknowledgements

This research was supported by the Assistant Secretary for Fossil Energy, Office of Natural Gas and Petroleum Technology, through the National Petroleum Technology Office under U.S. Department of Energy Contract No. DE-AC03-765F00098, and the Independent Petroleum Association of Mountain States. The authors are grateful for the efforts of other colleagues who have provided technical assistance; Dennis Debartalomeo, Ken Hom, Kenneth Revzan, Erin McNamara and Toshi Hotchi of LBNL Megan McKay of UC Berkeley, and Sheryl and Dave Rambeau at the Blodgett Forest Research Center.

References

1. Anlauf, K.G., MacTavish, D.C., Wiebe, H.A., Schiff, H.I., Mackay, G.I. (1988) Atmos. Environ. 22(8), 1589.
2. Barthelmie, R.J., and Pryor, S.C. (1999) J. of Geophys. Res. 104(D19), 23657–23669.
3. Bond, TC. (2001) Geophys. Res. Letters, 28, 4075–4078.
4. Bowman, F.M., Odum, J.R., Seinfeld, J.H., and Pandis, S.N. (1997) Atmospheric Environment 31(23), 3921–3931.

5. Cocker, D.R. III, Clegg, S.L., Flagan, R.C., and Seinfeld, J.H. (2001) Atmospheric Environment 35, 6049–6072.
6. Collins, D.R., Flagan, R.C., and Seinfeld, J.H. (2002) Aerosol Science and Technology, 36(1), 1–9.
7. Czoschke, N.M, Jang, M., and Kamens, R.C. (2003) Atmospheric Environment 37(30), 4287–4299.
8. Dillon, M.B, Lamanna, M.S., Schade, G.W., Goldstein, A.H., and Cohen, R.C. (2002) J. of Geophys. Res. 107(D5) 10.1029/2001JD000969.
9. Dreher, K.L. (2000) Inhalation Toxicol. 12(suppl. 3), 45–57.
10. Dreyfus, G.B., Schade, G.W., and Goldstein, A.H. (2002) J. Geophys. Res. 107(D19) 4365, 10.1029/2001JD001490.
11. Griffin, R.J., Cocker, D.R., and Seinfeld, J.H. (1999) Environ. Sci. Technol. 33, 2403–2408.
12. Griffin, R.J., Dabdub, D., and Seinfeld, J.H. (2002a) J. Geophys. Res. 107(D17), 4332.
13. Griffin, R.J., Dabdub, D., Kleeman, M.J., Fraser, M.P., Cass, G.R., and Seinfeld, J.H. (2002b) J. Geophys. Res. 107(D17), 4334.
14. Griffin, R.J., Nguyen, K., Dabdub, D., and Seinfeld, J.H. (2003) Journal of Atmospheric Chemistry 44, 171–190.
15. Griffin, R.J., Cocker III, D.R., Flagan, R.C., and Seinfeld, J.H. (1999) J. Geophys. Res. 104(D3), 3555–3567.
16. Grosjean, D, and Seinfeld, J.H. (1987) Atmospheric Environment 23, 1733.
17. Gundel, L.A., Dod, R.J., Rosen, H., and Novakov, T. (1984) Sci. Tot. Environ. 36, 197.
18. d'Almeida, G.A. (1987) J. Geophys. Res. 92, 3017–3026.
19. Hansen, ADA; Rosen, H; Novakov, T. (1984) Sci. Tot. Environ. 36, 191–196, 1984.
20. Hoffmann, T., Odum, J.R., Bowman, F., Collins, D., Klockow, D., Flagan, R.C., and Seinfeld, J.H. (1997) Journal of Atmospheric Chemistry 26, 189–222.
21. Jang, M., and Kamens, R.M. (1999) Atmospheric Environment 33, 459–474.
22. Jang, M., and Kamens, R.M. (2001) Environ. Sci. Technol. 35, 4758–4766.
23. John W. and Reischl G. (1980) J. Air Pollution Control Assoc. 30, 872.
24. Kalbere, M., Yu, J., Cocker, D.R., Flagan, R.C., and Seinfeld, J.H. (2000) Environ. Sci. Technol. 34, 4894–4901.

25. Kamens, R.M. and Jaoui, M. (2001) Environ. Sci. Technol. 35, 1394–1405.

26. Kavouras, I.G., Mihalopoulos, N., Lemonakis, A., and Stehanou, E.G. (1998) Nature 395, 683–686.

27. Kirchstetter, T.W.; Corrigan, C.E.; Novakov, T. (2001) Atmos. Environ., 35, 1663.

28. Kirchstetter, TW; Novakov, T. On the wavelength dependence of light absorption by atmospheric aerosol, in preparation for submission to Geophys. Res. Letters, 2004.

29. Kleindienst, T.E., Smith, D.F., Li. W., Edney, E.O., Driscoll, D.J. et al. (1999) Atmospheric Environment 33,3669–3681.

30. Lamanna, M., and Goldstein, A.G. (1999) J. Geophys. Res. 104(D17), 21247–21262.

31. Leaitch, W.R., Bottenheim, J.W., Biesenthal, T.A., Li, S.M., Liu, S.K., Asalian, K., Dryfhout-Clark, H., and Hopper, F. (1999) J. Geophys. Res. 104(D7), 8095–8111.

32. Mathai, Ed., 374–383. Air and Waste Management Association, Pittsburgh.

33. Mauderly, J (2003) Presentation at the fourth colloquium on PM and Human Health, Pittsburgh, PA, April 1, 2003.

34. Molenar, J.F., Dietrich, D.L., and Tree, R.M. (1989) In Visibility and Fine Particles, C.V. Novakov, T. (1982) In: Particulate Carbon: Atmospheric Life Cycle, Wolff, G.T., Klimish, R.L. (Eds.). Plenum, New York, pp. 19–41.

35. Novakov, T. (1981) In: Malissa, H., Grasserbaure, M., Belcher, R. (Eds.), Nature, Aim and Methods of Microchemistry. Springer-Verlag, New York, pp. 141–165.

36. O'Dowd, C.D., Aalto, P., Hämeri, K, Kulmala, M., Hoffmann, T. 2002 Nature 416, 497.

37. Odum, J.R., Hoffmann, T., Bowman, F., Collins, D., Flagan, R.C., and Seinfeld, J.H. (1996) Environ. Sci. Technol. 30(6), 2580–2585.

38. Odum, J.R, Jungkamp, T.P.W., Griffen, R.J., Forster, H.J.L., Flagan, R.C., and Seinfeld, J.H. (1997) Science 276, 96–99.

39. Pandis, S.N., Paulsen, S.E., Seinfeld, J.H., and Flagan, R.C. (1991) Atmospheric Environment 25, 997.

40. Pandis, S.N., Harley, R.A., Cass, G.R., and Seinfeld, J.H. (1992) Atmospheric Environment 26A, 2269–2282.

41. Pandis, S.N., Wexler, A.S., and Seinfeld, J.H. (1993) Atmospheric Environment 27, 2403.

42. Pankow, J.F. (1994) Atmospheric Environment 28(2), 185–188.

43. Pankow, J.F. (1994) Atmospheric Environment 28(2), 189–193.

44. Pankow, J.F., Seinfeld, J.H, Asher, W.E., and Erdakos, G.B. (2001) Environ. Sci. Technol. 35, 1164–1172.

45. Pun, B.K., Griffin, R.J., Seigneur, C., and Seinfeld, J.H. (2002) J. Geophys. Res. 107(D17), 4333.

46. Rogge, W.F., Mazurek, M.A., Hildemann, L.M., and Cass, G.R. (1993) Atmospheric Environment 27, 1309–1330.

47. Schauer, J.J., Kleeman, M.J., Cass, G.R., and Simoneit, B.R.T. (1999a) Environ. Sci. Technol. 33, 1566–1577.

48. Schauer, J.J., Kleeman, M.J., Cass, G.R., and Simoneit, B.R.T. (1999b) Environ. Sci. Technol. 33, 1578–1587.

49. Schauer, J.J., Kleeman, M.J., Cass, G.R., and Simoneit, B.R.T. (2001) Environ. Sci. Technol. 35, 1716–1728.

50. Seinfeld, J. and Pandis, S. (1998) Atmospheric Chemistry and Physics John Wiley & Sons, New York.

51. Seinfeld, J.H., Erdakos, G.B., Asher, W.E., and J.F. Pankow (2001) Environ. Sci. Technol. 35, 1806–1817.

52. Turpin, B.J.; Huntzicker, J.J.; and Hering, S.V. (1994) Atmos. Environ., 28, 3061.

53. U.S. EPA (2003) Air Quality Criteria for Particulate Matter (Fourth External Review Draft). U.S. Environmental Protection Agency, Office of Research and Development, National Center For Environmental Assessment, Research Triangle Park Office, Research Triangle Park, NC, EPA/600/P-99/002aD and bD.

54. Yu, J., Cocker, D.R., Griffin, R.J., Flagan, R.C., and Seinfeld, J.H. (1999a) Journal of Atmospheric Chemistry 34, 207–258.

55. Yu, J., Griffin, R.J., Cocker, D.R. III, Flagan, R.C., Seinfeld, J.H., and Blanchard, P. (1999b) Geophys. Res. Lett. 26, 1145–1148.

56. Yu, J., Flagan, R.C., and Seinfeld, J.H. (1999c) Environ. Sci. Technol. 32, 2357–2370.

57. Wang, S.C., and Flagan, R.C. (1990) Aerosol Sci. and Technol. 13, 230.

Combining Two-Directional Synthesis and Tandem Reactions, Part 11: Second Generation Syntheses of (±)-Hippodamine and (±)-Epi-Hippodamine

Annabella F. Newton, Martin Rejzek, Marie-Lyne Alcaraz and Robert A. Stockman

ABSTRACT

Background

Hippodamine is a volatile defence alkaloid isolated from ladybird beetles which holds potential as an agrochemical agent and was the subject of a synthesis by our group in 2005.

Results

Two enhancements to our previous syntheses of (±)-hippodamine and (±)-epi-hippodamine are presented which are able to shorten the syntheses by up to two steps.

Conclusion

Key advances include a two-directional homologation by cross metathesis and a new tandem reductive amination/double intramolecular Michael addition which generates 6 new bonds, 2 stereogenic centres and two rings, giving a single diastereomer in 74% yield.

Background

Ladybird beetles (Coleoptera: Coccinellidae) are important predators contributing to the natural control of pest aphid populations and are therefore of considerable commercial interest. However, ladybirds themselves are attacked by a range of natural enemies. General predation on ladybirds by vertebrates such as birds is largely prevented by highly toxic defence alkaloids contained in a reflex bleed released when the ladybird is attacked. To date, eight alkaloids of this type have been isolated from coccinellid beetles,[1] all of them being formally derivatives of perhydro-9b-azaphenalene (Figure 1). Another group of natural enemies, parasitic insects, can cause substantial reductions in populations of ladybird species. Recent research [2] has shown that the parasites locate the ladybirds through perception of certain defence alkaloids that they emit. If ladybirds are to be used effectively in insect pest control then their parasites must be controlled as well. The significant attraction of parasitic insects to the ladybird alkaloids suggests that there is potential for development of control strategies for this particular natural enemy. To further test this theory significant amounts of the defensive alkaloids will be needed. Coccinelid beetles seem to be the sole source of the defence alkaloids. Consequently much attention has been paid to developing syntheses of these compounds.

Hippodamine (1) is a naturally occurring alkaloid isolated from a ladybird beetle Hippodamia convergens by Tursch and co-workers in 1972. [3] The structure of hippodamine (1) was established two years later by the same group[4] on the basis of a single-crystal X-ray diffraction experiment (Figure 1). Epi-hippodamine (2) is its unnatural isomer with an axial C-5 methyl group. Both hippodamine (1) [5-7] and epi-hippodamine (2) [8] have been synthesized previously, and we reported syntheses of these two compounds using a two-directional synthesis/tandem, reaction approach in 2005. [9] Scheme 1 details the key aspects to our earlier work. [10] Herein, we report two refinements to our earlier work which have allowed even more concise routes to azaphenalene alkaloids 1 and 2.

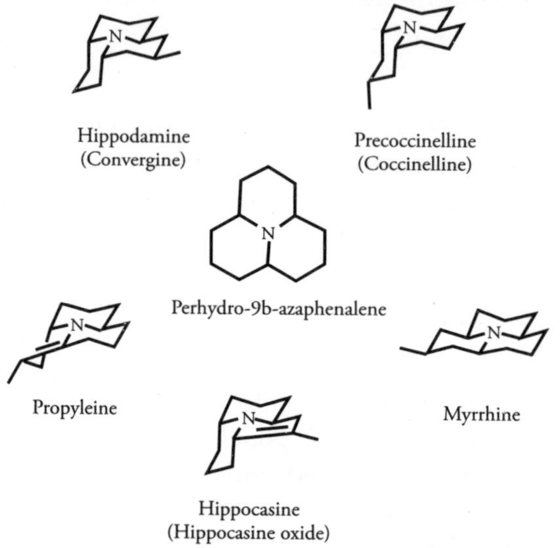

Figure 1. Structures of Coccinellid Alkaloids (N-Oxides and their names in brackets).

Scheme 1. Summary of our previous syntheses of hippodamine (1) and epi-hippodamine (2).

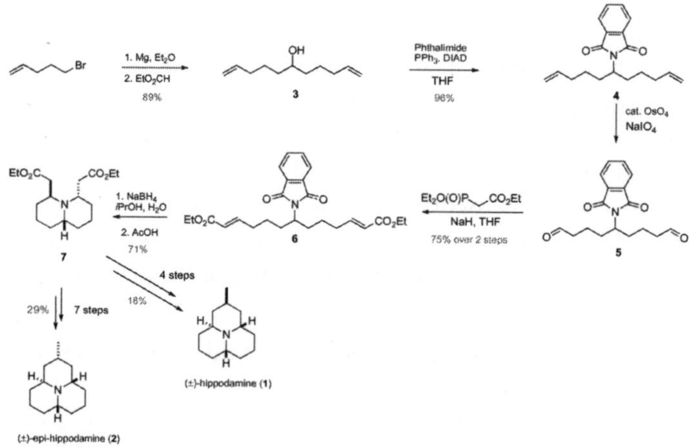

Results and Discussion

When we decided to take a second look at the syntheses of hippodamine and epi-hippodamine, we decided to focus on the synthesis of the key common intermediate 7 and try to realise an improvement over our earlier work. This paper discloses two such improvements. The first of these is the conversion of dialkene 4 into the diacrylate derivative 6. Originally this was achieved by oxidative cleavage of the two alkene moieties of 4 to form the rather sensitive dialdehyde 5. Whilst we were able to purify compound 5, this resulted in a significant loss of material through degradation of the dialdehyde on the purification media, be it silica gel or neutral alumina. Thus, we found that use of the crude dialdehyde in the subsequent Horner-Wadsworth-Emmons reaction was preferable, and gave a good 75% yield of the doubly homologated compound 6 after purification by column chromatography. Whilst this process did allow us to produce multigram quantities of 6, this sequence of reactions had the drawbacks that the olefination reaction needed to be carried out immediately after isolation of the dialdehyde 5 (this was found to decompose upon storage, even at low temperature), and also the oxidative cleavage reaction produced large amounts of toxic osmium waste. During our recent synthesis of histrionicotoxin,[11] we found that two-directional homologation of a symmetrical dialkene similar to 4 using cross-metathesis with acrylonitrile was possible using the Hoveyda modification of Grubbs second generation catalyst[12] (Scheme 2). Thus, it seemed a logical extension of this thinking to see if we could carry out a direct double homologation of dialkene 4 with ethyl acrylate as the cross-metathesis coupling partner. In fact, due to the non-co-ordinating nature of ethyl acrylate (in comparison to acrylonitrile) and the inert phthalimide group, this reaction proved to be an outstanding success, delivering diester 6 in 88% yield over one step after a five day reaction in dichloromethane. This step therefore reduces the overall number of steps for the synthesis of hippodamine to eight, and increases the overall yield from 8 to 10%. Similarly it reduces our synthesis of epi-hippodamine to eleven steps and increases the yield from 13 to 16% overall. The two-directional cross-metathesis reaction is shown in Scheme 2 below.

Scheme 2. Improved synthesis of tandem reaction precursor 6.

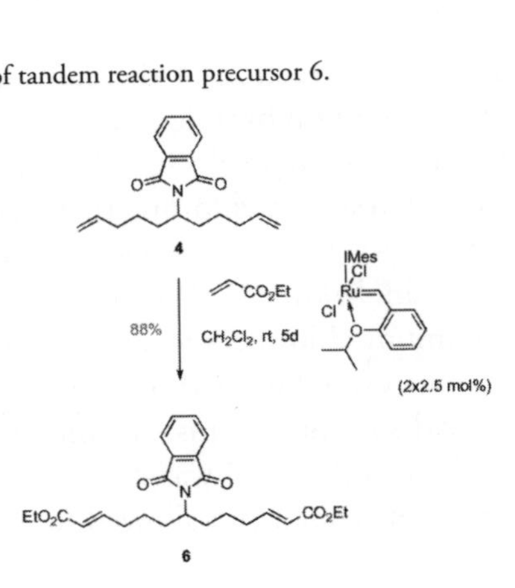

Having refined our synthesis of diester 7 by providing a shortened synthesis of its precursor, we decided to see if we could attain a synthesis of this common intermediate for the synthesis of both hippodamine and epi-hippodamine without the use of the phthalimide protecting group. This would render the entire synthesis of hippodamine free of protecting group chemistry – a distinct driving force for a compound which may find use as an agrochemical. Thus we postulated whether we would be able to transform keto-diester 9 into quinolizidine 7 by a tandem reductive amination/double intramolecular Michael addition. Our results are shown in scheme 3 below.

Scheme 3. Tandem reductive amination/double intramolecular Michael addition.

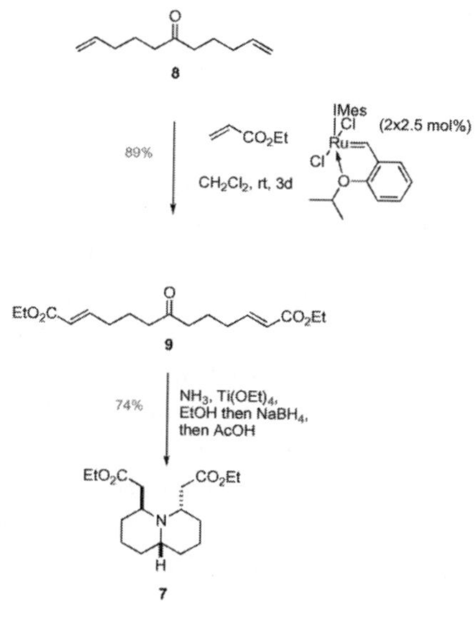

Thus ketone 8 was formed by reaction of the commercially available hex-5-enylnitrile with 4-pentenylmagnesium bromide in 70% yield. [13] Double cross-metathesis was found to proceed smoothly in 89% yield using the Hoveyda-Grubbs second generation catalyst in dichloromethane at room temperature for 3 days, giving keto diester 9. [14] We tried a range of reductive amination conditions for the formation of quinolizidine 7. The ammonia equivalents tried were ammonium acetate, ammonium chloride and ammonium formate, along with sodium borohydride, sodium cyanoborohydride and Hantzsch ester in either ethanol or ethanol/acetic acid solvent systems. A summary of conditions tried is shown in Table 1 below. The ketodiester 9 was dissolved in ethanol and the ammonia source and desiccant were added and allowed to stir overnight to form the iminium species, before the hydride source was added and the reaction allowed to proceed for a further 24 h. The hydride source was quenched with acetone before excess glacial acetic acid was added and the reaction mixture was heated for the given time before quenching with brine. Entries 1–4 and entry 6 showed reduction of the ketone to the alcohol (by 1H NMR), and entry 5 using the Hantzsch ester showed no reaction at all.

Table 1. Summary of our efforts to effect quinolizidine formation in a one-pot reaction.

Entry	Amine Source	Desiccant	Hydride reagent	Temp (°C)	Time (h)	% Yield
1	NH_4Cl and NEt_3	-	$NaBH_4$	75	48	-
2	NH_4Cl and NEt_3	-	$Na(BH_3CN)$	80	48	-
3	NH_3 in EtOH	4Å Sieves	$Na(BH_3CN)$	rt	96	-
4	NH_4OAc and NEt_3	4Å Sieves	$Na(BH_3CN)$	60	24	-
5	NH_4OAc and NEt_3	4Å Sieves	Hantzsch Ester	60	24	-
6	HCO_2NH_4 and NEt_3	4Å Sieves	$Na(BH_3CN)$	75	24	-
7	NH_3 in EtOH	$Ti(OEt)_4$	$NaBH_4$	75	48	74

It was found that addition of ammonia in ethanol with titanium ethoxide[15] for 14 hours, followed by addition of sodium borohydride and stirring at room temperature for a further 8 hours, and finally the addition of acetone (to remove any remaining active hydride) and 30 equivalents of acetic acid followed by heating the reaction mixture at reflux for 48 hours gave a clean reaction as monitored by TLC to quinolizidine 7, giving a 74% yield after purification by column chromatography over Brockmann Grade (III) neutral alumina. The tandem reductive amination/double intramolecular Michael addition generates 6 new bonds, 2 stereogenic centres and two rings, giving a single diastereomer.

In conclusion, we have increased the yield of our original hippodamine synthesis and reduced the number of steps required using a two-directional cross-metathesis of dialkene 4 with ethyl acrylate. We have also reported a new tandem reductive amination/double intramolecular Michael addition, which forms directly the quinolizidine core of hippodamine in a single step from a symmetrical keto-diester linear precursor. This new tandem reaction also reduces the number of steps for the synthesis of hippodamine to seven, and also removes any protecting group chemistry from the synthetic sequence and reduces waste whilst equalling the yield of the previous approach.

Acknowledgements

The authors wish to thank Leverhulme Trust (MR), AstraZeneca (AFN, CASE award) and EPSRC (RAS, Advanced Research Fellowship) for funding and

EPSRC Mass spectrometry service, Swansea for carrying out some of the high resolution mass-spectra.

References

1. King AG, Meinwald J: Chem Rev. 1996, 96:1105–1122.
2. Al Abassi S, Birkett MA, Pettersson J, Pickett JA, Wadhams LJ, Woodcock CM: J Chem Ecol. 2001, 27:33–43.
3. Tursch B, Daloze D, Pasteels A, Cravador A, Braekman JC, Hootele C, Zimmermann D: Bull Soc Chim Belg. 1972, 81:649–650.
4. Tursch B, Daloze D, Braekman JC, Hootele C, Cravador A, Losman D, Karlsson R: Tetrahedron Lett. 1974, 409–412.
5. Mueller RH, Thompson ME: Tetrahedron Lett. 1980, 21:1093–1096.
6. Mueller RH, Thompson ME, DiPardo RM: J Org Chem. 1984, 49:2217–2230.
7. Ayer WA, Dawe R, Eisner RA, Furuichi K: Can J Chem. 1976, 54:473–481.
8. Adams DR, Carruthers W, Crowley PJ: J Chem Soc Commun. 1991, 1261–1263.
9. Rejzek M, Hughes DL, Stockman RA: Org Biomol Chem. 2005, 3:73–83.
10. Stockman RA, Rejzek M: Tetrahedron Lett. 2002, 43:6505–6506.
11. Karatholuvhu MS, Sinclair A, Newton AF, Alcaraz ML, Stockman RA, Fuchs PL: J Am Chem Soc. 2006, 128:12656–12657.
12. Garber SB, Kingsbury JS, Gray BL, Hoveyda AH: J Am Chem Soc. 2000, 122:8168–8179.
13. Fürstner A, Thiel OR, Kindler N, Bartowska B: J Org Chem. 2000, 65:7990–7995.
14. Arini LG, Szeto P, Hughes DL, Stockman RA: Tetrahedon Lett. 2004, 45:8371–8374.
15. Miriyala B, Bhattacharyya S, Williamson JS: Tetrahedron. 2004, 60:1463–1471.

Accelerations in the Lactonization of Trimethyl Lock Systems are Due to Proximity Orientation and not to Strain Effects

Rafik Karaman

ABSTRACT

DFT at B3LYP/6-31G(d,p) and HF at 6-31G and AM1 semiempirical calculations of thermodynamic and kinetic parameters for the trimethyl lock system (an important enzyme model) indicate that the remarkable enhancement in the lactonizations is largely the result of a proximity orientation as opposed to the currently advanced strain effect.

Introduction

The study of enzyme mechanisms is of vast importance and has become central to both bioorganic chemistry and computational chemistry. Studies in this field by Bruice, Benkovic, Jencks, and Bender over the past 40 years have contributed largely to understanding the mode and scope by which certain enzymes achieve their catalytic activities [1]. These and studies of many others have been carried out in order to understand enzymatic catalysis which is characterized by high substrate specificity, chemoselectivity, and stereospecificity along with large rate enhancements. Among these are (1) the proximity model of Bruice [2, 3] on the intramolecular cyclization of dicarboxylic semiesters, (2) the "orbital steering" theory suggested by Koshland [4, 5], (3) the "spatiotemporal hypothesis" devised by Menger [6–9] which describes the importance of the distance between the two reactive centers in determining whether a reaction is inter- or intramolecular, and (4) the gem-trimethyl lock (stereopopulation control) proposal of Cohen [10–12].

Reaction models for mimicing enzyme catalysis usually fail to reach a desirable rate enhancement due to a lack of a high degree of conformational restrictions comparable to that existing in the enzyme-substrate complex [1]. In 1970, Cohen studied the lactonization of a series of hydroxyhydrocinnamic acids and found accelerations above 1015 when compared to the intermolecular cyclization of the corresponding counterparts. He attributed this large enhancement to what he called "stereopopulation control" [10–12]. Cohen's interpretation of his data was criticized by various researchers who claimed that the high rate enhancement results from relief in strain energy during the lactonization of the hydroxyhydrocinnamic acid and not because of a stereopopulation control within the trimethyl lock system [13–15].

Recently, we have been engaged in exploring the driving force(s) behind the remarkable acceleration rates in some intramolecular reactions [16–19]. Using ab initio molecular orbital at different levels, molecular mechanics, and semiempirical molecular orbital methods, we studied the thermodynamic and kinetic behavior of the lactonization of some hydroxy acids, the cyclization reactions of Bruice's di-carboxylic semiesters, the proton transfer reaction in Menger's system, and the cyclization of some ω-bromoalkanecarboxylate anions. The results from these studies revealed the following main conclusions. (1) Rate enhancement in intramolecular reactions can be driven by proximity orientation which is due to strain effects or by proximity that is not related to strain effects of a starting material and/or a corresponding transition state. For example, our study on cyclization of Bruice's dicarboxylic semiesters reveals that the activation energy in these systems is dependent on the difference in the strain energies of the transition states and the reactants, and there is no relationship between the cyclization rate and the distance between the nucleophile and the electrophile, and the reactivity extent of the semiester system is linearly correlated with the strain energy difference between the transition state and the reactant. On the other hand, acid lactonization of hydroxy acids reveals that the enhancement in rates of the lactonization in this system stems from the close proximity of the electrophile to the nucleophile. Further, it shows that the rate of the lactonization reaction is solely dependent on the ratio between the angle of attack of the nucleophile and the distance between the two reacting centers. This is in accordance with Menger's "spatiotemporal hypothesis" that relates distance between the nucleophile and the electrophile to the rate of the reaction. (2) Significant rate enhancements in intramolecular reactions are due to both enthalpic and entropic effects and not only due to enthalpic effects as was proposed by Bruice. (3) The nature of the reaction (intermolecular or intramolecular) is largely dependent on the distance between the two reacting centers. For example, our ab initio calculations on Menger's system show that when the distance between the two reacting centers is 2.4 Å, the reaction is intramolecular, whereas, when the distance is 3 Å, the reaction prefers the intermolecular process. Further, our study shows that the proximity between the nucleophile and electrophile is largely dependent on the strain energy of the system. For a strained system, the distance between the two reacting centers is shorter than that in unstrained systems [16–19].

In this letter, we describe the DFT at B3LYP/6-31G (d,p) and the RHF at 6-31G levels, as well as AM1 semiempirical calculations results (thermodynamics and kinetics) for the acid-catalyzed and uncatalyzed lactonization reactions of the trimethyl lock system. Our goal was to establish the driving force behind the remarkable accelerations of the intramolecular reaction in the tri-methyl lock system 1c in Figure.

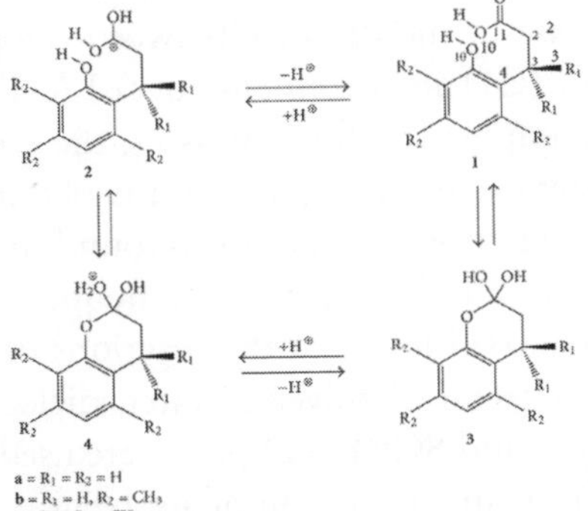

Figure 1. Hydroxyhydrocnnamc acids.

Methods

The DFT, HF, and AM1 calculations were carried out using the quantum chemical package Gaussian-98 [20]. The MM2 molecular mechanics strain energy calculations were performed using Allinger's MM2 program installed in Chem 3D Ultra 8.0 [21]. The starting geometries of all the molecules in this study were obtained using the Argus Lab program [22]. The calculations were carried out based on the restricted Hartree-Fock (RHF) method with full optimization of all geometrical variables (bond lengths, bond angles, and dihedral angles) [23]. To avoid results with local minima optimization, the keyword Freq Opt = (Z-matrix, MaxCycle = 300, CalcAll) GFINPUT IOP(6/7 = 3) was sed in the input files of the starting geometries. The geometry optimizations included estimations of second derivatives (Hessian matrix) for each of the 3n–6 parameters in each species (2n–3 for planar structures) [24]. DEP analytical gradients were used throughout the optimization. Geometries were optimized in internal coordinates and were terminated when Herbert's test was satisfied in the Broyden-Fletcher-Goldfarb-Shanno (BFGS) method.

An energy minimum (a stable compound or a reactive intermediate) has no negative vibrational force constant. A transition state is a saddle point which has only one negative vibrational force constant [25].The transition state structures were verified by their only one negative frequency. The verification was accomplished by viewing the frequency results via the Molden program [26]. The "reaction coordinate method" [27–29] was used to calculate the activation energy for the lactonization processes of hydroxyl acids 1a–1d. In this method, the value of one bond is limited for the appropriate degree of freedom while all other variables are optimized. The activation energy values for the cyclization reactions were calculated from the difference in the energies of the global optimum structures for the reactants 1a–1d and the derived transition states of the cyclization reactions. The transition state structures for the cyclization reactions of 1a–1d were obtained from the decrease in the distance between the phenolic oxygen (O10) and the carbonyl carbon (C1) in increments of 0.1 Å. Full optimization of the transition states was accomplished after removing the constrains imposed while executing the energy profile. The frequency results obtained from the optimization were viewed by Molden program and it was found that all the transition state structures, studied here, have only one negative frequency. The DFT at B3LYP/6-31G (d,p) and HF at 6-31G levels of the reactions of 1a–1d were calculated with and without the inclusion of solvent (water, dielectric constant = 78.39). The keywords SCF = Tight and SCRF = Dipole were used in the input files when calculating energies with the incorporation of a solvent.

Results and Discussion

In order to determine whether stereopopulation control, suggested by Cohen et al. [10–12] or conventional relief of strain energy, proposed by Winans and Wilcox [13] and supported by a theoretical study by Houk et al. [15] is the driving force for the enzyme-like accelerelation in 1c, we have calculated the AM1 thermodynamic properties for the lactonization of hydroxyhydrocinnamic acids 1a–1d (Figure 1) using Gaussian 98 version 3.0 [20] available at our Al-Quds computer center. These thermodynamic calculations were performed in order to prove that rates of reactions cannot be predicted from free energy changes (MM2 strain energy, thermodynamic parameter) as was previously reported by Winans and Wilcox [13].

The AM1 calculations of the enthalpic and entropic energies of the lactonization processes of hydroxy acids having a trimethyl lock (1c) and lacking the trimethyl lock (1a) (see Table 1) reveal that the difference in the free energy values between the lactonization reactions of the two systems (ΔΔG) is 9.7 kcal/mol (ΔΔG = ΔG tetrahedral intermediate-ΔG hydroxyl acid). Based on this value, the calculated rate enhancement of the lactonization of 1c compared to that of 1a should be 1.1 × 107, which is 104-fold less than the experimentally determined value (see Table 1) [10–12]. Further, the MM2 calculated difference in strain energies (ΔEs3,1=Es3–Es1, for the numbering see Figure 1) in the two systems 1a and 1c is about 7 kcal/mol, which is equal to 1.2 × 105 (106-fold less than the experimentally value). This result excludes the notion that the remarkable rate enhancement in the lactonization of 1c is solely accommodated by conventional relief of strain as was reported by the groups of Houk and Wilcox [13–15].

Table 1. AM1 and MM2 calculated thermodynamic properties of hydroxyhydrocinnamic acid derivatives. (All values were calculated by AM1 method except for the values of Es which were calculated by MM2 method.)

Structure	$\Delta\Delta \mathbf{H}_{f3,1}$	$\mathbf{T\Delta S}_{3,1}$	$\boldsymbol{\Delta Es}_{3,1}$	$\Delta\Delta\mathbf{G}_{3,1}$	log $\mathbf{k}^{b}_{exp}$
PAA [a]	−1.95	−4.82	14.27	2.87	−10.000
1a	−5.25	−3.96	10.54	−1.29	−5.226
1b	−4.82	−1.77	9.95	−3.05	−5.206
1c	−12.10	−1.08	3.49	−11.03	5.771
1d	−9.39	−1.77	7.35	−7.62	−1.581

[a] PAA Stands for the bimolecular reaction of phenol and acetic acid.
[b] Taken from reference [10–12]. The numbers 1 and 3 refer to structures 1 and 3 (see Figure 1).

To test whether the conformational restriction plays an important role in the rate acceleration of the lactonization processes of hydroxy acids 1a and 1c, calculations of the rotational barrier around the C2–C3 bond were executed. Note that the C1–O10 distance in hydroxyhydrocinnamic acids is dependent upon the C1/C2/C3/C4 dihedral angle. This C1–O10 distance should be crucial for enhancing the lactonization rate according to the stereopopulation control (conformational locking) suggested by Cohen [10–12]. When plotting the C1–O10 distance in 1a and 1c against the heats of formation of the resulting conformations (see Figure 2), an important result emerges. In 1c, the shortest C1–O10 distance values correspond to the most stable conformations, whereas for 1a the shortest distance values correspond to the highest energy conformations.

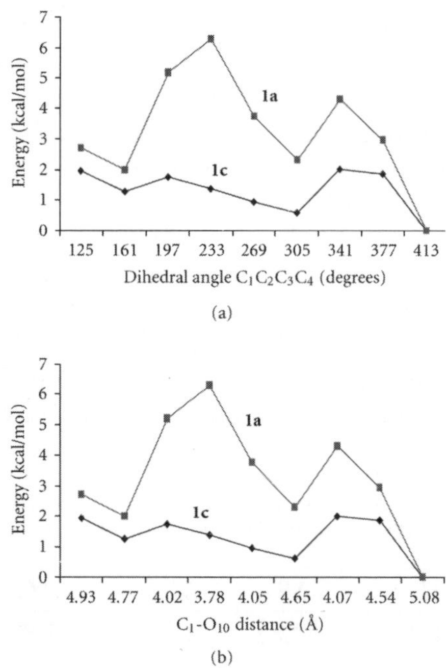

Figure 2. (2a) a plot of energy versus dihedral angle C1C2C3C4 in 2a and 2c. (2b) a plot of energy versus C1–O10 distance in 2a and 2c.

Further, calculations of the thermodynamic parameters (ΔG, ΔH and S) of each of the energetically most stable conformers in 1a, as well as those for the conformer with the shortest C1–O10 distance, reveal that for 1a to reside in the most productive position to react intramolecularly requires a ΔG = 10.73 kcal/mol (composed of ΔH = 4.91 kcal/mol and -TΔS = 5.82 kcal/mol). In contrast, 1c requires ΔG = 3.93 kcal/mol (ΔH = 1.85 kcal/mol and-TΔS = 2.08 kcal/mol) in order to fulfill the same task. The roughly 7 kcal/mol difference in ground-state free energy for the two systems is equivalent to about 1.4 × 106 in rate enhancement of 1c over 1a. Again, the calculations show clearly that the conformational locking is not the critical factor in the rate acceleration seen with 1c (comprising only a calculated 106 value versus an experimental 1011 value) [10–12].

In light of the results of the AM1 calculations that show neither the steric effects suggested by Winans and Wilcox [13] nor the conformational locking suggested by Cohen [10–12] are the main driving force for the enhancement in the lactonization of 1c, we have calculated, using the "reaction coordinate" method, the activation energy values (ΔΔG‡) for the formation and the collapse of the tetrahedral intermediate involved in the lactonization process. The DFT at B3LYP/6-31G (d,p) and HF/6-31G levels as well as the AM1 activation energy values were calculated with and without the inclusion of solvent (water, dielectric constant 78.39) and the results obtained indicate that the effect of water on the relative rate values is negligible (1.2–1.8 kcal/mol for the three different lactonization processes of 1a, 1c, and 1d). This is in accordance with previously reported studies of Houk et al. on lactonization of hydroxy acids, that indicate that the solvation effect more-or-less cancels out when comparing reactivities of species having the same structural features (even though the absolute rate constants cannot be evaluated) [13–15]. Thus, reaction rates were computed from the activation energy results for both the acid-catalyzed and uncatalyzed lactonization processes of hydroxy acids 1a–1d. The calculated rates are linearly related to the experimentally determined rate values, the correlations which are shown in (1) (for the acid-catalyzed process) and in (2) (for the un-catalyzed process). The AM1 calculations results of the activation energy values for the lactonization processes of hydroxy acids 1a–1d were correlated very well with the values obtained using the ab initio method at the DFT B3LYP/6-31G (d,p) and RHF/6-31G, and the correlation results are depicted in (3)-(4), respectively:

$$\log\left(\frac{k_{1x}}{k_{1a}}\right)_{calc} = 0.9911 \log k_{\exp} - 0.1238, \qquad R = 0.99, \tag{1}$$

where k_{1x} is the cyclization rate of hydroxy acid 1b-1d, and k_{1a} is the cyclization rate of hydroxy acid 1a;

$$\log\left(\frac{k_{2x}}{k_{2a}}\right)_{calc} = 0.9768 \log k_{\exp} - 0.6725, \qquad R = 0.99, \tag{2}$$

where k_{2x} is the cyclization rate of protonated hydroxy acid 2b-2d, and k_{2a} is the cyclization rate of protonated hydroxyl acid 2a;

$$\Delta\Delta G^{\ddagger}(HF) = 0.9216\ \Delta\Delta G^{\ddagger}\ (AM1) + 13.68, \qquad R = 0.99, \tag{3}$$

where$\Delta\Delta G^{\ddagger}$ (HF) and$\Delta\Delta G^{\ddagger}$ (AM1) are the calculated activation energy values by RHF/6-31G and AM1 methods, respectively;

$$\Delta\Delta G^{\ddagger}\ (B3LYP) = 1.1878\ \Delta\Delta G^{\ddagger}\ (AM1) + 0.9788, \qquad R = 0.99, \tag{4}$$

where $\Delta\Delta G^{\ddagger}$ (B3LYP) is the calculated activation energy values by B3LYP/6-31G (d,p).

The calculated rate values of the lactonization processes of 1a and 1c using (2) are 104 and 1014, respectively. This gives a predicted rate enhancement of about 1010 which is in good agreement with the experimentally determined value (1011). Our calculations also reveal that the rate-limiting step is formation rather than collapse of the tetrahedral intermediate [16–19], in opposition to the conclusions reported by Houk et al. [15].

It is accepted that strain-accelerated reactions occur for compounds that, by necessity, are rigid with high bond-rotation barriers that exceed those required for the reaction. Further, such compounds have distorted bond distances or/and bond angles when compared to strain-free compounds [1]. Our theoretical calculations indeed predict distortion of bond angles and bond distances in the phenyl ring of 1c which is supported by the X-ray crystal structure of the corresponding alcohol [30]. However, the same distortions with the same magnitude are observed with 1b and to some extent with the corresponding tetrahedral intermediates 3c and 3b. If the acceleration is due to strain relief, we should see comparable rates for 1c and 1b, but actually the lactonization rate of 1c is 1010 times faster than that of 1b. This suggests no significant strain relief upon the lactonization of 1c.

The second convincing point excluding strain as the main driving force for the rate acceleration is that in 1c the rotation barrier around C2-C3 is found to be 3 kcal/ mol smaller than that of 1a. This surprising conclusion arises from the fact that 1c has a stronger intramolecular carboxyl/hydroxyl hydrogen-bonding than does 1a. If Winans and Wilcox were correct [13] and the rate acceleration is indeed due to strain-relief, then the C2-C3 bond in 1a should be more affordable to rotation than that of 1c. In other words, if strain-accelerated reactions occur in compounds that are rigid with high bond rotation barriers, then the rate comparison between 1a and 1c cannot be due to strain effects.

For a better understanding the lactonization process, we conducted reaction-coordinate calculations for 1a and 1c (when the C1–O10 distance is in the range of 2.5–1.5 Å) by calculating the change in energy as a function of the O10/C1/C2 attack angle (α) and the C1–C10 distance between the two reacting centers (r). An excellent correlation was observed between the entalpic energy E and α/r the ratio between the attack angle and the distance. Further, it was found that the slope (S) follows the order S (1a)>S (1d)>S (1c). This result suggests that the energy needed to increase the value of angle α to reach the optimal value for formation of the transition state is less in the case of 1c when compared to 1a. This in turn indicates that the approach of the hydroxyl group to the carbonyl carbon in the case of 1c is much easier than in the case of 1a.

Conclusions

The combined results indicate that the driving force in the first and second stages of the approach (C1–O10 = 4–2.5 Å, and 2.5–1.5 Å), is due to a proximity effect. Hydroxy acids that are rigid in the organized state have lower activation energies (such as 1c) than those with less rigidity (such as 1a). It is worthy to note that Mengerhas advocated abandoning thermodynamic models involving entropy in favor of distance effects on rate, and in fact he has derived an equation relating rate and distance [31, 32].

Acknowledgements

The author thanks the Karamans Co. and the German-Palestinian-Israeli fund agency for support of our hardware computational facilities. Special thanks go to Dr. Omar Deeb and Sherin Alfalah and Donia Karaman for support in computational software and technical assistance. His sincere appreciations are given to Professor Fred Menger (Emory University, Ga, USA) for helpful discussions.

References

1. A. W. Czarink, “Intramolecularity: Proximity and Strain,” in Mechanistic Principles of Enzyme Activity, J. F. Liebman and A. Greenberg, Eds., VCH, New York, NY, USA, 1988.
2. T. C. Bruice and U. K. Pandit, “The effect of geminal substitution ring size and rotamer distribution on the intramolecular nucleophilic catalysis of the hydrolysis of monophenyl esters of dibasic acids and the solvolysis of the intermediate anhydrides,” Journal of the American Chemical Society, vol. 82, no. 22, pp. 5858–5865, 1960.
3. T. C. Bruice and U. K. Pandit, “Intramolecular models depicting the kinetic importance of “Fit” in enzymatic catalysis,” Proceedings of the National

Academy of Sciences of the United States of America, vol. 46, no. 4, pp. 402–404, 1960.

4. A. Dafforn and D. E. Koshland, Jr., "Theoretical aspects of orbital steering," Proceedings of the National Academy of Sciences of the United States of America, vol. 68, no. 10, pp. 2463–2467, 1971.
5. G. A. Dafforn and D. E. Koshland, Jr., "The sensitivity of intramolecular reactions to the orientation of the reacting atoms," Bioorganic Chemistry, vol. 1, no. 1-2, pp. 129–139, 1971.
6. F. M. Menger, "On the source of intramolecular and enzymatic reactivity," Accounts of Chemical Research, vol. 18, no. 5, pp. 128–134, 1985.
7. F. M. Menger, J. F. Chow, H. Kaiserman, and P. C. Vasquez, "Directionality of proton transfer in solution. Three systems of known angularity," Journal of the American Chemical Society, vol. 105, no. 15, pp. 4996–5002, 1983.
8. F. M. Menger, "Directionality of organic reactions in solution," Tetrahedron, vol. 39, no. 7, pp. 1013–1040, 1983.
9. F. M. Menger, J. Grossman, and D. C. Liotta, "Transition-state pliability in nitrogen-to-nitrogen proton transfer," Journal of Organic Chemistry, vol. 48, no. 6, pp. 905–907, 1983.
10. S. Milstien and L. A. Cohen, "Concurrent general-acid and general-base catalysis of esterification," Journal of the American Chemical Society, vol. 92, no. 14, pp. 4377–4382, 1970.
11. S. Milstien and L. A. Cohen, "Rate acceleration by stereopopulation control: models for enzyme action," Proceedings of the National Academy of Sciences of the United States of America, vol. 67, no. 3, pp. 1143–1147, 1970.
12. S. Milstien and L. A. Cohen, "Stereopopulation control. I. Rate enhancement in the lactonizations of o-hydroxyhydrocinnamic acids," Journal of the American Chemical Society, vol. 94, no. 26, pp. 9158–9165, 1972.
13. R. E. Winans and C. F. Wilcox, Jr., "Comparison of stereopopulation control with conventional steric effects in lactonization of hydrocoumarinic acids," Journal of the American Chemical Society, vol. 98, no. 14, pp. 4281–4285, 1976.
14. A. E. Dorigo and K. N. Houk, "The origin of proximity effects on reactivity: a modified MM2 model for the rates of acid-catalyzed lactonizations of hydroxy acids," Journal of the American Chemical Society, vol. 109, no. 12, pp. 3698–3708, 1987.

15. K. N. Houk, J. A. Tucker, and A. E. Dorigo, "Quantitative modeling of proximity effects on organic reactivity," Accounts of Chemical Research, vol. 23, no. 4, pp. 107–113, 1990.

16. R. Karaman, "Analysis of Menger's 'spatiotemporal hypothesis'," Tetrahedron Letters, vol. 49, no. 41, pp. 5998–6002, 2008.

17. R. Karaman, "Reevaluation of Bruice's proximity orientation," Tetrahedron Letters, vol. 50, no. 4, pp. 452–456, 2009.

18. R. Karaman, "A new mathematical equation relating activation energy to bond angle and distance: a key for understanding the role of acceleration in lactonization of the trimethyl lock system," Bioorganic Chemistry, vol. 37, no. 1, pp. 11–25, 2009.

19. R. Karaman, submitted to Bioorganic Chemistry.

20. http://www.gaussian.com.

21. U. Burker and N. L. Allinger, Molecular Mechanics, American Chemical Society, Washington, DC, USA, 1982.

22. C. J. Casewit, K. S. Colwell, and A. K. Rappé, "Application of a universal force field to main group compounds," Journal of the American Chemical Society, vol. 114, no. 25, pp. 10046–10053, 1992.

23. M. J. S. Dewar, E. G. Zoebisch, E. F. Healy, and J. J. P. Stewart, "AM1: a new general purpose quantum mechanical molecular model," Journal of the American Chemical Society, vol. 107, no. 13, pp. 3902–3909, 1985.

24. M. J. S. Dewar, G. P. Ford, M. L. McKee, H. S. Rzepa, W. Thiel, and Y. Yamaguchi, "Semiempirical calculations of molecular vibrational frequencies: the MNDO method," Journal of Molecular Structure, vol. 43, no. 1, pp. 135–138, 1978.

25. J. N. Murrell and K. J. Laidler, "Symmetries of activated complexes," Transactions of the Faraday Society, vol. 64, pp. 371–377, 1968.

26. http://www.cmbi.kun.nl/~schaft/molden/molden.html.

27. A. Goldblum and G. H. Loew, "Quantum chemical studies of model cytochrome P450 oxidations of amines. 1. MNDO pathways for alkylamine reactions with singlet and triplet oxygen," Journal of the American Chemical Society, vol. 107, no. 14, pp. 4265–4272, 1985.

28. K. Muller, "Reaction paths on multidimensional energy hypersurfaces," Angewandte Chemie International Edition in English, vol. 19, no. 1, pp. 1–13, 1980.

29. M. J. S. Dewar and S. Kirschner, "MINDO [modified intermediate neglect of differential overlap]/2 study of aromatic ("allowed") electrocyclic reactions of cyclopropyl and cyclobutene," Journal of the American Chemical Society, vol. 93, no. 17, pp. 4290–4291, 1971.

30. J. M. Karle and I. L. Karle, "Correlation of reaction rate acceleration with rotational restriction. Crystal-structure analysis of compounds with a trialkyl lock," Journal of the American Chemical Society, vol. 94, no. 26, pp. 9182–9189, 1972.

31. F. M. Menger, A. L. Galloway, and D. G. Musaev, "Relationship between rate and distance," Chemical Communications, vol. 9, no. 18, pp. 2370–2371, 2003.

32. F. M. Menger, "An alternative view of enzyme catalysis," Pure and Applied Chemistry, vol. 77, no. 11, pp. 1873–1886, 2005.

Unexpected Degradation of the Bisphosphonate P-C-P Bridge Under Mild Conditions

P. A. Turhanen and J. J. Vepsäläinen

ABSTRACT

Unexpected degradation of the P-C-P bridge from novel bisphosphonate derivative 1a and known etidronate trimethyl ester (1b) has been observed under mild reaction conditions. A proposed reaction mechanism for the unexpected degradation of 1a and 1b is also reported.

Background

Bisphosphonates (BPs) are analogs of naturally occurring pyrophosphate, where the chemically and enzymatically labile P-O-P bridge has been replaced with a P-C-P bridge, making these compounds relatively resistant to chemical hydrolysis and completely resistant to enzymatic hydrolysis (Figure 1).[1-5] These BP compounds bind strongly to calcium phosphate and inhibit its formation, aggregation and dissolution.[6] The affinity for the bone mineral represents the basis for their use in the treatment of many diseases associated with increased bone resorption, such as metastatic bone disease, Paget's disease and osteoporosis.[1-6] As described above, the BPs have been used for decades in the therapy of bone diseases but recently these compounds have been found to be active in many other fields, such as in the treatment of parasitic diseases [7-11] and atherosclerosis.[12] Furthermore, the BPs have been shown to be effective against calcifying nanoparticles (CNPs, known also as nanobacteria) which may be responsible for several human diseases where calcium phosphate deposition is a hallmark, e.g. cardiovascular diseases, kidney stones, urological diseases, e.g. prostatitis, many cancers and various forms of autoimmune diseases.[13,14] Therefore, it is very important to understand the chemistry of BPs in detail.

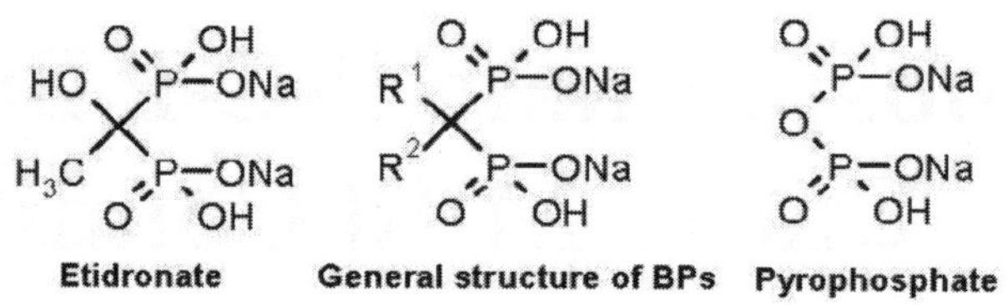

Figure 1. Structures of etidronate, pyrophosphate and general structure of bisphosphonates.

Etidronate, (1-hydroxyethylidene)-1,1-bisphosphonic acid (HEBPA) disodium salt, is one of the earliest synthesized and is the most extensively investigated BP compound, still being in clinical use today (Figure 1).[1-6,15] Our group has designed, synthesized and studied in vitro several different etidronate and alendronate derivatives to act as biodegradable prodrugs of these drugs.[16-23] During our ongoing study to prepare new, possibly bioreversible BP derivatives, we observed unexpected degradation of the P-C-P bridge under mild reaction conditions in two of the prepared etidronate derivatives. Earlier, Szymczak et al. [24] have described the formation of H-phosphonate (also known as phosphite) and phosphate components from a phosphonate-phosphate compound (same kind of structure as 8 in Scheme 2) either in CH3CN/Et3N/H2O (v/v) or phosphate buffer, pH 7.4 at 37°C. Szajnman et al. [25] has reported loss of two molecules of phosphite in tetraethyl oxirane-2,2-diylbis(phosphonate); however the kind of degradation which we will discuss in this article has not been previously reported.

Scheme 1. Preparation of BP derivative 1a (R2 = Et, R3 = C(O)OEt) and its degradation to acetate 6 and phosphites 2–4. Reagents and conditions: i) excess ClC(O)OEt, 6 equiv Na_2CO_3, reflux, overnight, 55%; ii) 4 equiv NaOH (40% NaOH in H_2O), MeOH, 30 min, rt.

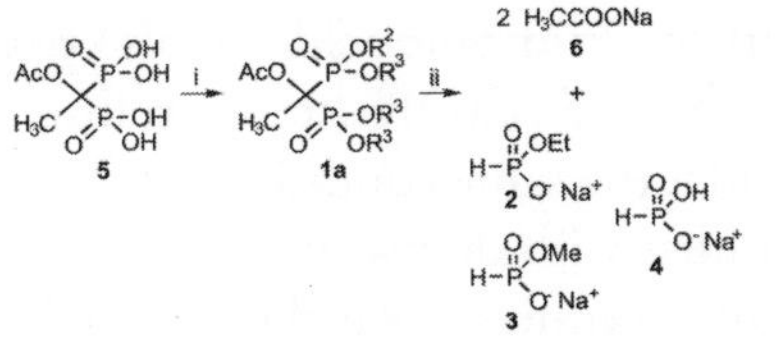

Scheme 2. Degradation of trimethyl ester of etidronate (1b) and stability of the tetramethyl ester (1c, R=R'=Me) and P,P'-dimethyl ester (1d, R=Me, R'=Na+) of etidronate. Reagents and conditions: i) 1 drop of 6 M NaOH, H_2O, 1 h, rt, (measured pH was ≥11); ii) 5 equiv triethyl amine, H_2O, 1 h, 60°C; iii) when R=R'=Me (1c), 1 eq, triethylamine, H_2O, 10 min. ca. 98% conversion. iv) when R=Me, R'=Na+ (1d), 5 equiv NaOH, H_2O, overnight, reflux.

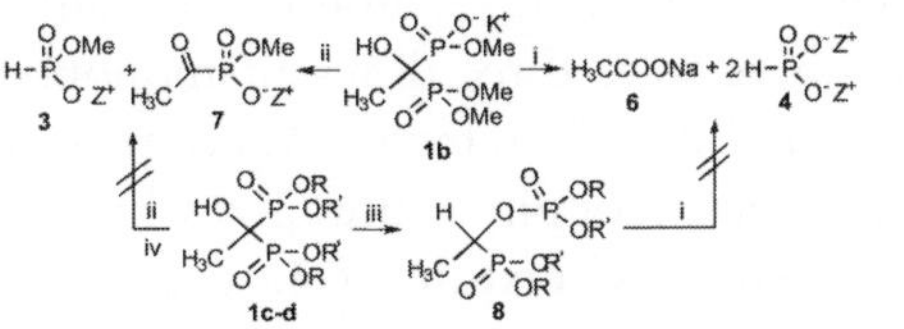

Results and Discussion

As mentioned in the introduction, the P-C-P bridge of BPs has been reported to be relatively stable against chemical hydrolysis, however here we report the unexpectedly easy degradation of two etidronate derivatives into acetate and phosphite moieties. In our ongoing study to prepare novel biodegradable BP derivatives, a new carbonate derivative of etidronate was synthesised. The synthesis was started from the known acetylated etidronic acid [21] (5, see Scheme 1) by treating it with ethyl chloroformate and sodium carbonate. The NMR spectroscopy results were surprising since they pointed to the formation of a novel etidronate derivative 1a (see Scheme 1). In the 31P NMR spectrum, there were four doublets (1:1:1:1) due to the presence of two diastereomers. The ^1H NMR spectrum contained two complicated splitting patterns at approx. 4.46 and 4.28 ppm, their integral ratio was 1 to 3, respectively, indicating two different kinds of $-OCH_2$ groups in ratio 1:3. After inspection of ^{13}C NMR spectra and the ESI-MS results, we concluded that the prepared molecule had the unanticipated structure of 1a and not the expected structure where $R^2=R^3=C(O)OEt$ (see Scheme 1). To confirm the selective formation of 1a, the synthesis was repeated several times, but the result was always the same (formation of 1a was observed in all experiments), though in some experiments a transesterification of the acetyl group to C(O)OEt group was observed in yields of 0–13% as confirmed by the ^{1}H and ^{31}P NMR spectra. We were unable to provide any direct explanation for the variation in the transesterification proportion. Etidronic acid was also tested as a starting material to prepare a derivative such as 1a [C(O)OEt group instead of Ac group], but the reaction did not occur under the same conditions as those used in the preparation of 1a. Our subsequent studies with derivative 1a led us to another very surprising result, which occurred when 4 equiv of NaOH (40% NaOH in H_2O) were added to the solution of 1a in MeOH and stirred for 30 minutes at room temperature. After evaporation of the reaction mixture to dryness, the residue contained almost exclusively (>95% degradation was observed) sodium acetate 6 and phosphites 2–4 (compound 4 can be also called phosphorous acid monosodium salt) as can be seen in Scheme 1. Compounds 2–4 were readily characterized by their P-H chemical shifts and characteristic $^{1}J_{HP}$ coupling constants (ca. 600 Hz). In the 31P NMR spectrum, there were three different monophosphorus components confirmed to be compounds 2–4. Two moles of acetate 6 were detected compared to one mole of the total amount of phosphites 2–4 which was the expected result. Interestingly, the decomposition mixture of 1a contained not only monoethyl phosphite 2 and phosphite 4 but also monomethyl derivative 3 (according the ^{31}P NMR spectrum, the ratio was approx.: 1:0.86:1). The formation of this monomethyl phosphite 3 under the conditions used (see Scheme 1, procedure ii) can be explained based on: 1) partial transesterification of bisphosphonate 1a before degradation of P-C-P bridge, 2) partial esterification of phosphonate group after the carbonate groups (R3) decomposition from compound 1a (this is proposed to occur rapidly after the addition of 40% NaOH) and before the degradation of P-C-P bridge, 3) partial transesterification of 2, and 4) esterification of 4 (see Scheme 1).

These unexpected degradation results we observed for the HEBPA derivative 1a led us to examine what would happen to more simple derivatives of HEBPA, such as trimethyl (1b), tetramethyl (1c) and P,P'-dimethyl (1d) esters of etidronate under the same kinds of conditions (see Scheme 2). Compounds 1b-d were prepared as reported elsewhere.[20,26-28] Again very surprising results were obtained. Trimethyl ester of etidronate (1b) was degraded to the acetate 6 and phosphorous acid salt 4, under even milder conditions than the degradation of 1a (50 mg of 1b in 1 ml H2O and 1 drop of 6 M NaOH was stirred for 1 hour at rt; measured pH was ≥11; see Scheme 2). Tetramethyl (1c) or P,P'-dimethyl ester were not degraded under the same conditions, only the formation of phosphonate-phosphate derivative 8 from 1c was observed as expected in the light of the earlier results concern the rearrangement process.[17,20,21,29-32] This rearrangement of 1c to 8 was observed to happen rapidly and almost completely (98% conversion) when 1 equiv of triethyl amine was present in water (see Scheme 2).

Compound 1b was selectively degraded to the phosphite 3 and acetyl phosphonate 7 when 5 equiv of triethyl amine was used in H2O (see Scheme 2). Dialkyl acetylphosphonates and dialkyl phosphites are common starting materials for the synthesis of tetraalkyl esters of HEBPA, [26,27] but this is the first time when the "reverse" synthesis has been reported. P,P'-dimethyl ester 1d did not degrade to compounds 3 and 7 or 6 and 4 even when refluxed overnight with 5 equiv NaOH in H2O.

The decomposition mechanism for 1b can be explained in two ways; either via a decomposition mechanism resembling the reversible route of the formation of tetraesters (see Scheme 3 route a), since e.g., 1c, are prepared from phosphites, H-P(O)(OMe)2, and phosphonates, MeCOP(O)(OMe)2, or route b resembling the rearrangement process.[19] The driving force in both reactions is the formation of three charged molecules from one P-C-P compound since this is a highly entropically favoured process. Decomposition of the first P-C bond starts with deprotonation of the hydroxyl group followed by elimination of the methyl phosphite and the formation of ketone (route a) or by nucleophilic attack of the oxygen of the ionized phosphate on the bridging carbon to release dimethyl phosphite and the oxirane ring containing derivative (route b). In route a, water or hydroxide ion attacks the carbonyl carbon and P-C bond cleavage occurs giving rise to acetic acid and dimethyl phosphite which can undergo a further reaction with water or hydroxide to give methyl phosphite. In route b, the attack of water on the carbon of oxirane ring yields hydrate followed by elimination of methyl phosphite and acetic acid. We believe that route a is more probable, since during the reaction with a weaker base, such as triethylamine, only the first P-C bond is cleaved and products 3 and 7 are observed. On the other hand, decomposition of 1a is more likely to follow route b.

Scheme 3. Proposed reaction mechanism for 1b decomposition.

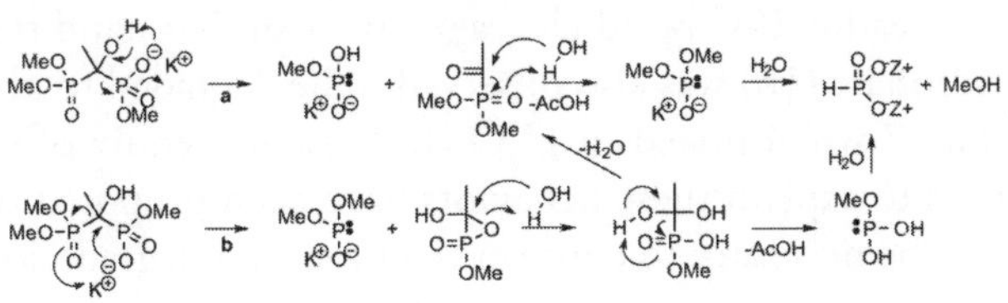

The proposed decomposition mechanism for 1a (see Scheme 4) is more complicated. The reaction starts with the hydrolysis of one carbonate ester leading to a monoanion comparable to 1b. After this step, the decomposition can continue following routes that are similar to either route a or route b in Scheme 3. The other possibility, route b (in Scheme 4), is a nucleophilic attack of oxygen to the bridging carbon and the formation of an oxirane ring containing derivative, since the adjacent acetate group is a rather good leaving group. Subsequently, P-C-bond decomposition will follow the same mechanism as reported in Scheme 3.

Scheme 4. Proposed reaction mechanism for 1a decomposition.

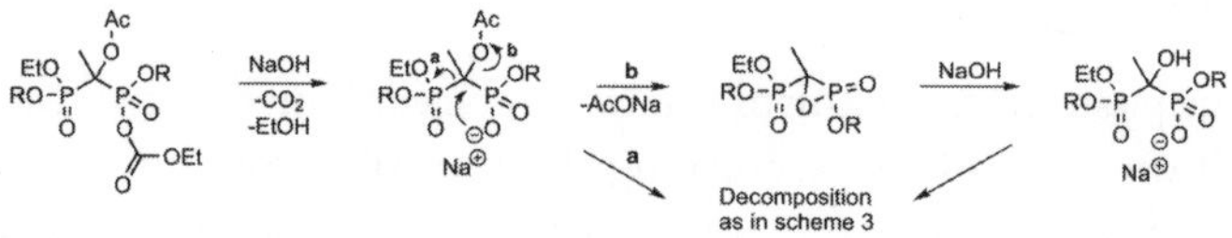

The initial reaction in Scheme 3 also explains the formation of rearranged product 8 from tetraester 1c (this rearrangement is proposed to happen via oxirane ring), [19] since the charged oxygen is a good nucleophile compared to OH-group and far better than oxygen bound to phosphorus with a douple bond (P=O).

All of the compounds were easily identified by their 1H, 13C and 31P NMR spectra. In 31P NMR signals for the phosphites 2, 3 and 4 were 6.26 ppm, 8.47 ppm and 5.81 ppm, respectively and 0.10 ppm for acetylphosphonate 7. These values were comparable to those reported earlier.[27]

Conclusion

In conclusion, a novel carbonate derivative of etidronate (1a) was prepared by the reaction of acetylated etidronic acid with ethyl chloroformate and sodium carbonate. Compound 1a was found to undergo remarkably facile cleavage of the P-C bond under mild basic conditions. The trimethyl ester of etidronate (1b) was also found to undergo readily P-C bond cleavage under similar conditions. The trimethyl ester of etidronate (1b) was also observed to be degraded to phosphite 3 and acetylphosphonate 7 when mixed in H_2O containing 5 equiv of triethyl amine. Some mechanisms to explain these behaviors have been proposed though further investigations will be necessary to confirm the proposed degradation pathways.

Acknowledgements

Authors would like to thank Mrs. Maritta Salminkoski for her expert technical assistance for performing most of the experiments reported here, and Mrs. Katja Höppi for the ESI-MS analysis.

References

1. leisch H: Bisphosphonates in Bone Disease: From the Laboratory to the Patient. The Parthenon Publishing Group Inc.: New York; 1995.
2. Papapoulos SE, Landman JO, Bijvoet OLM, Löwik CWGM, Valkema R, Pauwels EKJ, Vermeij P: Bone. 1992, 13:S41–S49.
3. Yates AJ, Rodan GA: DDT. 1998, 3:69–78.
4. Socrates E, Papapoulos MD: Am J Med. 1993, 95:48S–52S.
5. Giannini S, D'Angelo A, Sartori L, Passeri G, Garbonare LD, Crebaldi C: Obst Gynecol. 1996, 88:431–436.
6. Fleich H: Drugs. 1991, 42:919–944.
7. Szajnman SH, Montalvetti A, Wang Y, Docampo R, Rodriguez JB: Bioorg Med Chem Lett. 2003, 13:3231–3235.
8. Szajnman SH, Bailey BN, Docampo R, Rodriguez JB: Bioorg Med Chem Lett. 2001, 11:789–792.
9. Martin MB, Sanders JM, Kendrick H, Luca-Fradley K, Lewis JC, Grimley JS, Van Brussel EM, Olsen JR, Meints GA, Burzynska A, Kafarski P, Croft SL, Oldfield E: J Med Chem. 2002, 45:2904–2914.
10. Garzoni LR, Caldera A, Nazareth L, Meirelles M, Castro SL, Docampo R, Meints GA, Oldfield E, Urbina JA: Int J Antimicrob Agents. 2004, 23:273–285.
11. Garzoni LR, Waghabi MC, Baptista MM, Castro SL, Nazareth L, Meirelles M, Britto CC, Docampo R, Oldfield E, Urbina JA: Int J Antimicrob Agents. 2004, 23:286–290.
12. Ylitalo R: Gen Pharmacol. 2002, 35:287–296.
13. Kajander EO: Lett Appl Microbiol. 2006, 42:549–552.
14. Aho K, Soininen T, Turhanen PA, Kajander EO, Vepsäläinen JJ, unpublished results.
15. Major PP, Lipton A, Berenson J, Hortobagyi G: Cancer. 2000, 88:6–14.
16. Turhanen PA, Niemi R, Peräkylä M, Järvinen T, Vepsäläinen JJ: Org Biomol Chem. 2003, 1:3223–3226.
17. Turhanen PA, Vepsäläinen JJ: Synthesis. 2005, 13:2119–2121.
18. Turhanen PA, Vepsäläinen JJ: Synthesis. 2005, 18:3063–3066.
19. Niemi R, Turhanen P, Vepsäläinen J, Taipale H, Järvinen T: Eur J Pharm Sci. 2000, 11:173–180.

20. Turhanen PA, Ahlgren MJ, Järvinen T, Vepsäläinen JJ: Synthesis. 2001, 4: 633–637.

21. Turhanen PA, Vepsäläinen JJ: Synthesis. 2004, 7:992–994.

22. Turhanen PA, Vepsäläinen JJ: Beilstein J Org Chem. 2006, 2:2.

23. Vepsäläinen JJ: Curr Med Chem. 2002, 9:1201–1208.

24. Szymczak M, Szymańska A, Stawiński J, Boryski J, Kraszewski A: Org Lett. 2003, 5:3571–3573.

25. Szajnman SH, García Liñares G, Moro P, Rodriguez JB: Eur J Org Chem. 2005, 3687–3696.

26. Nicholson DA, Vaughn H: J Org Chem. 1971, 36:3843–3845.

27. Turhanen PA, Ahlgren MJ, Järvinen T, Vepsäläinen JJ: Phosphorus Sulfur Silicon. 2001, 170:115–133.

28. Van Gelder JM, Breuer E, Ornoy A, Schlossman A, Patlas N, Golomb G: Bone. 1995, 16:511–520.

29. Vachal P, Hale JJ, Lu Z, Streckfuss EC, Mills SG, MacCoss M, Yin DH, Algayer K, Manser K, Kesisoglou F, Ghosh S, Alani LL: J Med Chem. 2006, 49:3060–3063.

30. Fitch SJ, Moedritzer K: J Am Chem Soc. 1962, 84:1876–1879.

31. Ruel R, Bouvier J-P, Young RN: J Org Chem. 1995, 60:5209–5213.

32. Tromelin A, El Manouni D, Burgada R: Phosphorus and Sulfur. 1986, 27:301–312.

Synthesis of Benzo[b] fluorenone Nuclei of Stealthins

Sujit Kumar Ghorai, Saroj Ranjan De, Raju Karmakar, Nirmal Kumar Hazra and Dipakranjan Mal

ABSTRACT

Two routes, one based on a Michael-initiated aldol condensation and the other on an intramoleculer carbonyl-ene reaction, have been found to be feasible for an entry to benzo[b]fluorenones. Reaction of 4,9-dimethoxybenz[f]indenone with nitromethane in the presence of DBU gave the corresponding Michael adduct, which afforded 2-methyl-5,10-dimethoxybenzo[b]fluorenone on reaction with methacrolein under a variety of basic conditions. Similarly, 2-methallyl-4,9-dimethoxybenz[f]indenone reacted with nitromethane to give the corresponding Michael adduct, Nef reaction of which furnished 3-formyl-2-methyl-4,9-dimethoxybenz[f]indanone. This underwent ene-cyclization under the influence of SnCl4. 5H2O, and yielded 2-methyl-5,10-dimethoxybenzo[b]fluorenone.

Introduction

Stealthins A (1a) and B (1b), isolated from Streptomyces viridochromogenes as potent radical scavengers, are the first known members of natural benzo[b]fluorenones [1]. Interest in this group of compounds grew considerably due to the identification of structurally allied natural products, stealthin C (1c) [2], kinafluorenone (2) [3], prekinamycin (3) [4], and kinamycin antibiotics (e.g., kinamycin D, 4) [5] (Figure 1). Their synthesis became an active area of research since 1996 [6–19]. In line with the Ishikawa approach [19], we intended to explore the chemistry of benz[f]indenones (e.g., 5b) to establish new synthetic routes to functionalized benzo[b]fluorenones [20]. Herein, we report regiospecific construction of the D-ring of benzo[b]fluorenones (e.g., 6) from the corresponding benz[f] indenones.

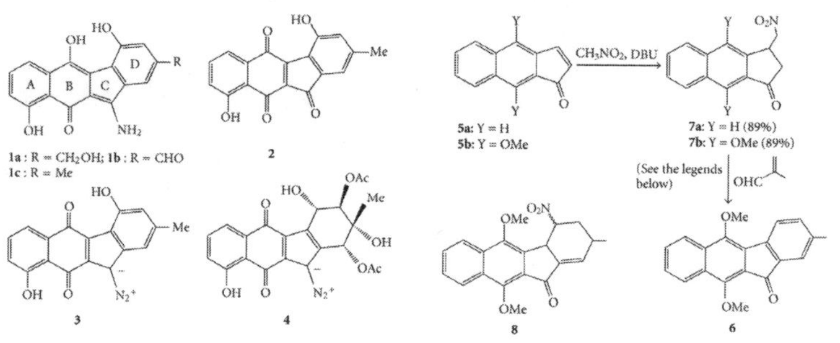

Figure 1. Naturally occurring benzo[b]fluorenones.

Results and Discussion

Initial studies were focused on the utilization of the readily accessible benz[f] indenones 5a and 5b [21]. DBU-promoted Michael addition of nitromethane to the indenones furnished indenones 7a and 7b, respectively. The intended annulation of 7b with methacrolein was then studied with different base-solvent systems (Scheme 1). But, none of the attempts gave desired product 8. Instead, most of the methods produced benzo[b]fluorenone 6 in low yields. The best yield was 25%, which was obtained with DBU in benzene. The presence of singlet at δ 2.23 for Ar-CH_3 in H1 NMR spectrum was indicative of the structure 6. It is probable that the compound 6 was formed from tetracycle 8 through elimination of HNO_2. Considering the fact that even a weak base such as n-Bu3P caused the elimination of NO_2 group, we examined the route (Scheme 2), involving an acid-catalyzed cyclization. The d4 Synthon equivalent 9 was prepared in two steps from methacrolein by adaptation of Miyakoshi protocol [22] (Scheme 2). Conjugate addition of the reagent 9 to benz[f]indenone 5b in the presence of DBU provided 10 in good yield. ^{1}H NMR spectrum of the product indicated the formation of a 1:1 mixture of diastereomers. When treated with 1 N HCl, the mixture afforded the expected product 8 in trace amount, the major product being 6 (25%). Repeated attempts to optimize the transformation of 10 into 8 were of no avail.

Scheme 1. Michael-aldol sequence.

(i) DBU, 0°C-rt, 24 h, **6**, 25%;
(ii) Et_3N, $CHCl_3$, 0°C-rt, 24 h, **6**, 18%;
(iii) i-Pr_2NH, $CHCl_3$, reflux, 6 h, **6**, 15%;
(iv) n-Bu_3P, C_6H_6, 24 h, 0°C-rt, **6**, 18%;
(v) t-BuOLi, THF, −65°C-rt, intractable mixture;
(vi) t-BuOK, t-BuOH, 0°C-rt, intractable mixture

Scheme 2. Michael-aldol sequence.

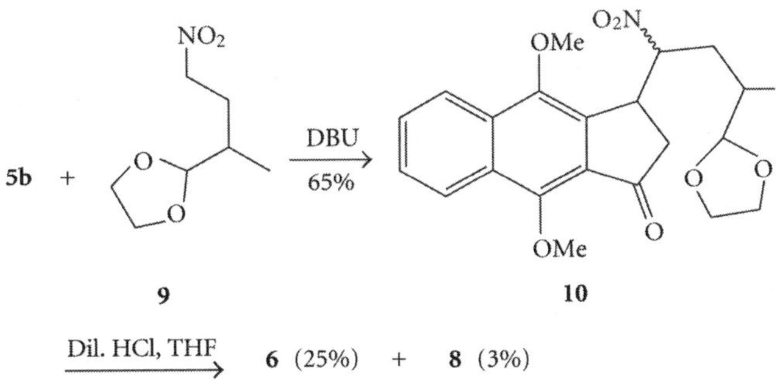

As an alternative avenue, the strategy (Scheme 3) based upon the intramolecular carbonyl-ene reaction [23] of 11 was undertaken. Preparation of the key precursor 11 is depicted in Scheme 3. LDA-promoted allylation of 12 [21] with allyl bromide 13a furnished 14a. Characteristic multiplets at δ 5.6 and two doublets at δ 5.1 and δ 4.87 in 1H NMR spectrum were in complete agreement with the structure 14a. The cis-relationship between the angular allyl group and the methano bridge was inferred by comparing the NMR signals of C-4a H of an angularly methylated analog [24]. Flash vacuum pyrolysis (FVP) of the adduct 14a at 475°C/0.01 mm gave enone 5c in sufficiently pure form for the next use. It was then subjected to conjugate addition with nitromethane in the presence of DBU to produce allylated nitro adduct 15a as a single isomer in 92% yield (Scheme 3). Similarly, precursor 15b was obtained in three steps from 12. Methallylation of 12 with methallyl iodide 13b in the presence of LDA gave 14b (91%), FVP of which furnished 5d. Addition of nitromethane to 5d in the presence of DBU-furnished intermediate 15b in 80% yield. Nef reaction [25] of 15b with NaOMe and TiCl3-buffer provided aldehyde 11 in moderate yield (50%). The singlet at δ 9.94 in 1H NMR spectrum and the band at 1718 cm−1 in IR spectrum confirmed the presence of CHO functionality in 11. When a solution of the aldehyde in dichloromethane was treated with SnCl4 5H2O [26], D-ring aromatized compound 6 was formed in 45% yield.

Scheme 3. Intramolecular carbonyl-ene approach.

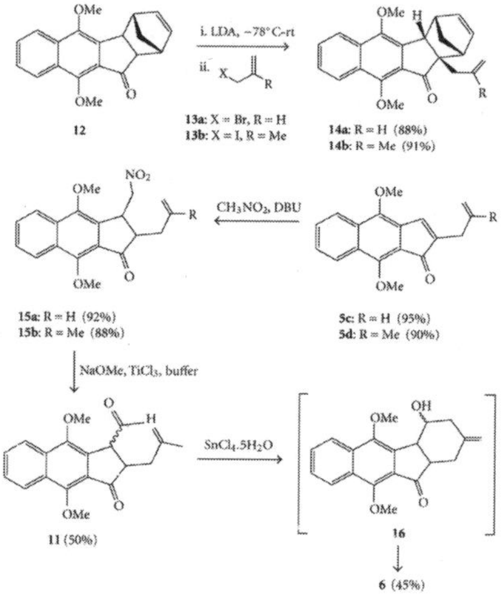

Conclusion

We have validated two synthetic routes to benzo[b]fluoren-11-ones from benz[f] indenones 5. The intramolecular carbonyl-ene reaction of the intermediate 11 (Scheme 3) proved to be better pathway than the tandem Michael-aldol route (Scheme 2) to benzo[b]fluoren-11-one 6.

Experimental

The general experimental is described in [27].

Benz(f)inden-1-One 5c

This compound was prepared from 14a. Mp: 80–85°C; yellow; yield 93%; IR (cm^{-1}): 1684; ^{1}H NMR (200 MHz), 8.20 (d, 1H, J = 8.1), 7.97 (d, 1H, J = 8.1), 7.57–7.41 (m, 3H), 6.01–5.88 (m, 1H), 5.23–5.12 (m, 2H), 4.28 (s, 3H), 4.02

(s, 3H), 3.11–3.07 (m, 2H); C13 NMR (50 MHz): 193.99, 152.2, 144.6, 141.5, 139.8, 134.5, 132.9, 131.1, 129.4, 127.1, 126.2, 125.7, 123.0, 117.1, 115.5, 62.8, 62.7, 29.5; MS (m/z): 280 (M+, 100%), 265, 250, 223, 178, 165.

Benz(f)inden-1-One 5d

This compound was prepared as a yellow oil in 90% yield from 14b. IR (cm^{-1}): 1689; H1 NMR (300 MHz): 8.10 (d, 1H, J = 8.1), 7.97 (d, 1H, J = 8.1), 7.60–7.40 (m, 3H), 4.85 (d, 2H, J = 10.8), 4.28 (s, 3H), 4.02 (s, 3H), 3.04 (s, 2H), 1.79 (s, 3H); C13 NMR (75 MHz): 192.9, 152.0, 144.5, 142.6, 140.9, 140.4, 132.7, 131.0, 129.3, 127.0, 126.0, 125.6, 122.9, 115.3, 112.5, 62.9, 62.7, 33.3, 22.5; MS (m/z): 294 (M+, 100%), 279, 263, 236, 165, 152, 139.

Benzo(b)fluorenone 6: Method A

To a stirred solution of the nitro compound 7a (100 mg, 0.33 mmol), and methacrolein (60 mg, 0.86 mmol) in benzene (5 mL) at 0°C was added DBU (10 mg, 0.066 mmol). Stirring was continued for 24 hours at rt. The reaction mixture was diluted with diethyl ether (50 mL), washed with saturated sodium bicarbonate solution (10 mL) and then with brine (10 mL). The organic phase was dried (Na_2SO_4) and concentrated. The resulting residue was purified by preparative TLC to give a yellow crystalline solid of 6 (26 mg, 25%).

Method B

To a stirred solution of aldehyde 11 (50 mg, 0.154 mmol) in dichloromethane (6 mL) was added SnCl4 · 5H2O (5 mg) under nitrogen atmosphere. The stirring was continued for 30 hours. After usual work up of the reaction mixture, the residue was purified by preparative TLC to provide 6 (21 mg, 45%). Mp: 150-151°C; IR (cm^{-1}): 1695; ^{1}H NMR (200 MHz): 8.29 (d, 1H, J = 7.4), 8.04 (d, 1H, J = 7.4), 7.88 (d, 1H, J = 7.7), 7.65–7.36 (m, 4H), 4.28 (s, 3H), 4.00 (s, 3H), 2.42 (s, 3H); ^{13}C NMR (50 MHz): 190.4, 153.7, 146.4, 140.0, 138.8, 136.4, 135.4, 133.6, 130.8, 129.4, 127.8, 126.7, 125.5, 124.4, 124.1, 122.4, 119.9, 63.1, 61.1, 21.3; MS (m/z): 304 (M+, 100%), 289, 218, 189, 149, 57.

Compound 7a

This was prepared from benzindenone 5a and nitromethane in 89% yield according to the procedure described earlier [25]. Mp: 124°C; IR (cm^{-1}): 1711; ^{1}H

NMR (200 MHz): 8.37 (s, 1H, ArH) 8.03 (d, 1H, J = 8.1), 7.89–7.87 (m, 2H), 7.68–7.54 (m, 2H), 4.91–4.82 (ABq, 1H, J = 12.8, 5.8), 4.64–4.53 (ABq, 1H, J = 12.8, 5.8), 4.42–4.34 (m, 1H), 3.21–3.07 (ABq, 1H, J = 19.2, 8.2), 2.74–2.63 (ABq, 1H, J = 19.2, 3.9).

Compound 7b

This was prepared from 5b, following the procedure described for compound 7a. Mp: 179-180°C; white solid; yield: 89%; IR (cm^{-1}): 1715; H1 NMR (200 MHz): 8.40 (d, 1H, J = 8.2), 8.10 (d, 1H, J = 8.5), 7.74–7.55 (m, 2H), 5.34–5.29 (m, 1H), 4.43–4.38 (m, 2H), 4.18 (s, 3H), 4.03 (s, 3H) 3.17–3.04 (m, 1H), 2.76–2.65 (m, 1H); ^{13}C NMR (50 MHz): 200.4, 152.9, 148.6, 134.1, 132.7, 129.7, 126.9, 125.3, 122.7, 121.8, 77.8, 77.2, 63.4, 61.8, 42.3, 34.4; MS (m/z): 301 (M+), 254, 239, 197, 141, 115.

Compound 8

To a mixture of enone 5b (0.178 g, 0.740 mmol) and 1,1-ethanediyldioxy-2-methyl-4-nitrobutane 9 (0.389 g, 2.22 mmol) in CH_2Cl_2 (4 mL) was added DBU (12 mg, 0.078 mmol) and the mixture was stirred at rt for 6 hours. It was then concentrated and purified by column chromatography to afford 10 as an oil (0.2 g, 65%). ^1H NMR spectrum revealed the presence of two isomers as indicated by three signals δ 4.15, 4.08, and 4.05, corresponding to the methoxy groups. The peak at δ 4.15 was not resolved. The methine hydrogens of $CHNO_2$ appeared at δ 5.19. To a stirred solution of above nitro acetal 10 (100 mg, 0.24 mmol) in THF (8 mL) was added 10% HCl (1 mL) solution. Stirring was continued for 20 hours. After usual work up of the reaction mixture, the residue was chromatographed to afford 6 (18 mg, 25%) and 8 (2.5 mg, 3%).

Compound 9

Yield: 50%; colorless oil; IR (cm^{-1}): 1541; H1 NMR (200 MHz): 4.66 (d, 1H, J = 4), 4.46 (t, 2H, J = 6), 3.98–3.79 (m, 4H), 2.30–1.77 (m, 3H), 1.00 (d, 3H, J = 6.7); C13 NMR (50 MHz): 106.8, 74.1, 65.1, 65.0, 34.3, 29.0, 14.7.

Compound 11

This was prepared from compound 15b by Nef reaction [25]. Yield: 50%; purity > 80%; ^1H NMR (500 MHz): 9.94 (s, 1H), 8.40 (d, 1H, J = 8.4), 8.10 (d, 1H,

J = 8.4), 7.69 (t, 1H, J = 8.2), 7.52 (t, 1H, J = 8.2), 4.87 (s, 1H), 4.78 (s, 2H), 4.21 (s, 3H), 4.16 (brs, 1H), 4.02 (s, 3H) 3.31–3.27 (m, 1H); 2.78–2.75 (ABq, 1H, J = 14.0, 4.3) 1.75 (s, 3H).

Compound 12

^{1}H NMR (200 MHz): 8.37 (d, 1H, J = 8.0, 1H), 8.05 (d, 1H, J = 8.0, 1H), 7.68–7.51 (m, 2H), 6.85 (brs, 1 H), 4.80–4.60 (m, 1H), 4.32 (s, 3H), 3.70 (s, 3H), 2.85 (brs, 1H), 2.22–2.15 (m, 1H), 1.5-1.4 (m, 2H), 1.25–1.21 (m, 3H).

Compound 14a

This was prepared as thick brownish oil in 88% yield from pentacycle 12, following an earlier method [24]. IR (cm^{-1}): 1705; H1 NMR (300 MHz): 8.33 (d, 1H, J = 8.7), 8.06 (d, 1H, J = 8.4), 7.61 (m, 1H), 7.49 (m, 1H), 6.05–6.02 (m, 1H), 5.65–5.50 (m, 2H), 5.07 (d, 1H, J = 16.8), 4.85 (d, 1H, J = 10.2), 4.08 (s, 3H), 4.03 (s, 3H), 3.78 (d, 1H, J = 4.2), 3.45 (brs, 1H), 2.89–2.91 (m, 2H), 2.45–2.38 (m, 1H), 2.0 (ABd, 1H, J = 8.7), 1.80 (ABd, 1H, J = 8.7); ^{13}C NMR (75 MHz): 207.1, 151.2, 147.6, 138.0, 135.2, 135.1, 134.3, 132.4, 128.9, 126.6, 125.8, 125.0, 121.8, 117.5, 63.0, 62.4, 62.1, 50.8, 50.9, 47.0, 46.6, 41.6.

Compound 14b

This was prepared from pentacycle 12, following the procedure adopted for compound 14a. Yield: 89%; thick oil; IR (cm^{-1}): 1705; ^{1}H NMR (400 MHz): 8.32 (d, 1H, J = 8.1), 8.07 (d, 1H, J = 8.1), 7.62 (br t, 1H), 7.49 (br t, 1H), 6.00 (dd, 1H, J = 2.8, 5.6), 5.49 (dd, 1H, J = 2.8, 5.6), 4.70 (brs, 1H), 4.66 (brs, 1H), 4.08 (s, 3H), 4.03 (s, 3H), 3.93 (d, 1H, J = 4), 3.44 (brs, 1H), 3.05 (ABd, 1H, J = 13.8), 2.94 (brs, 1H), 2.41 (ABd, 1H, J = 13.8), 1.98 (ABd, 1H, J = 8.8), 1.79 (ABd, 1H, J = 8.8), 1.46 (s, 3H); ^{13}C NMR (50 MHz): 207.4, 151.0, 147.6, 143.1, 138.3, 135.5, 135.3, 134.6, 132.3, 128.8, 126.7, 125.8, 125.0, 121.8, 114.3, 62.8, 61.8, 61.6, 52.5, 50.8, 46.8, 46.1, 45.7, 23.8.

Compound 15a

This was prepared from 5c, following the procedure described for compound 7a. Mp: 98-99°C; white solid; yield: 92%; IR (cm^{-1}): 1712; ^{1}H NMR (300 MHz): 8.40 (d, 1H, J = 8.7), 8.10 (d, 1H, J = 8.4), 7.72–7.66 (m, 1H), 7.58 (m, 1H), 5.75–5.61 (m, 1H), 5.23–5.04 (m, 3H), 4.54–4.47 (m, 1H), 4.20 (s, 3H), 4.10

4.05 (m, 1H), 4.01 (s, 3H), 2.83–2.78 (m, 1H); 2.65–2.51 (m, 2H); ^{13}C NMR (75 MHz): 202.7, 153.1, 148.7, 133.8, 133.3, 133.0, 129.8, 127.0, 125.4, 122.3, 121.9, 118.8, 77.7, 77.3, 63.4, 61.8, 52.2, 39.4, 36.1; MS (m/z): 341 (M+), 307, 290, 280 (100%), 265, 165.

Compound 15b

This was prepared from 5d, following the procedure adopted for compound 7a. Mp: 122-123°C; white solid; yield: 90%; IR (cm^{-1}): 1707; H1 NMR (300 MHz): 8.41 (d, 1H, J = 8.4), 8.10 (d, 1H, J = 8.4), 7.72–7.60 (m, 1H), 7.61–7.55 (m, 1H), 5.07 (dd, 1H, J = 12.8, 4.2), 4.88 (s, 1H), 4.76 (s, 1H), 4.59 (dd, 1H, J = 12.8, 8.7), 4.20 (s, 3H), 4.15–3.90 (m, 1H), 4.00 (s, 3H), 2.95–2.88 (m, 1H); 2.63 (dd, 1H, J = 13.8, 5.1), 2.34 (dd, 1H, J =13.8, 9.0), 1.73 (s, 3H); ^{13}C NMR (75 MHz): 202.8, 153.3, 148.8, 142.5, 133.3, 132.9, 129.9, 129.8, 126.9, 125.3, 122.0, 114.2, 77.8, 63.3, 61.9, 50.8, 40.5, 39.9, 21.9. MS (m/z): 355 (M+), 319, 304, 294 (100%), 279, 265, 253, 236, 223, 165, 152, 139; anal. calcd for $C_{20}H_{21}NO_5$: C, 67.59; H, 5.96; N, 3.94, found C, 67.51; H, 5.93; N, 3.93.

Acknowledgements

This work was supported by the Council of Scientific and Industrial Research (CSIR) and the Department of Science and Technology, New Delhi. The second and the third authors are grateful to the CSIR, for their research fellowships.

References

1. K. Shin-ya, K. Furihata, Y. Teshima, Y. Hayakawa, and H. Seto, "Structures of stealthins A and B, new free radical scavengers of microbial origin," Tetrahedron Letters, vol. 33, no. 46, pp. 7025–7028, 1992.
2. S. J. Gould, N. Tamayo, C. R. Melville, and M. C. Cone, "Revised structures for the kinamycin antibiotics: 5-diazobenzo[b]fluorenes rather than benzo[b] carbazole cyanamidines," Journal of the American Chemical Society, vol. 116, no. 5, pp. 2207–2208, 1994.
3. M. C. Cone, C. R. Melville, M. P. Gore, and S. J. Gould, "Kinafluorenone, a benzo[b]fluorenone isolated from the kinamycin producer Streptomyces murayamaensis," The Journal of Organic Chemistry, vol. 58, no. 5, pp. 1058–1061, 1993.

4. S. J. Gould, J. Chen, M. C. Cone, M. P. Gore, C. R. Melville, and N. Tamayo, "Identification of prekinamycin in extracts of Streptomyces murayamaensis," The Journal of Organic Chemistry, vol. 61, no. 17, pp. 5720–5721, 1996.
5. J. Marco-Contelles and M. T. Molina, "Naturally occurring diazo compounds: the kinamycins," Current Organic Chemistry, vol. 7, no. 14, pp. 1433–1442, 2003.
6. D. Mal and N. K. Hazra, "The first approach to kinamycin antibiotics: synthesis of kinafluorenone scaffold," Tetrahedron Letters, vol. 37, no. 15, pp. 2641–2642, 1996.
7. F. M. Hauser and M. Zhou, "Total synthesis of the structure proposed for prekinamycin," The Journal of Organic Chemistry, vol. 61, no. 17, p. 5722, 1996.
8. S.-I. Mohri, M. Stefinovic, and V. Snieckus, "Combined directed Ortho-, remote-metalation and cross-coupling strategies. Concise syntheses of the kinamycin biosynthetic grid antibiotics phenanthroviridin aglycon and kinobscurinone," The Journal of Organic Chemistry, vol. 62, no. 21, pp. 7072–7073, 1997.
9. G. Qabaja and G. B. Jones, "Annnulattion strategies for benzo[b]fluorene synthesis: efficient routes to the kinafluorenone and WS-5995 antibiotics," The Journal of Organic Chemistry, vol. 65, no. 21, pp. 7187–7194, 2000.
10. E. González-Cantalapiedra, Ó. de Frutos, C. Atienza, C. Mateo, and A. M. Echavarren, "Synthesis of the benzo[b]fluorene core of the kinamycins by arylalkyne-allene and arylalkyne-alkyne cycloadditions," European Journal of Organic Chemistry, vol. 2006, no. 6, pp. 1430–1443, 2006.
11. V. B. Birman, Z. Zhao, and L. Guo, "Benzo[b]fluorenes via indanone dianion annulation. A short synthesis of prekinamycin," Organic Letters, vol. 9, no. 7, pp. 1223–1225, 2007.
12. A. Martínez, J. C. Barcia, A. M. Estévez, et al., "A novel approach to the synthesis of benzo[b]fluoren-11-ones," Tetrahedron Letters, vol. 48, no. 12, pp. 2147–2149, 2007.
13. X. Lei and J. A. Porco, Jr., "Total synthesis of the diazobenzofluorene antibiotic (–)-kinamycin C 1," Journal of the American Chemical Society, vol. 128, no. 46, pp. 14790–14791, 2006.
14. K. C. Nicolaou, H. Li, A. L. Nold, D. Pappo, and A. Lenzen, "Total synthesis of kinamycins C, F, and J," Journal of the American Chemical Society, vol. 129, no. 34, pp. 10356–10357, 2007.

15. I. Hussain, M. A. Yawer, M. Lau, et al., "Regioselective synthesis of fluorinated phenols, biaryls, 6H-benzo[c]chromen-6-ones and fluorenones based on formal [3+3] cyclizations of 1,3-bis(silyl enol ethers)," European Journal of Organic Chemistry, vol. 2008, no. 3, pp. 503–518, 2008.

16. S. Reim, M. Lau, and P. Langer, "Synthesis of fluorenones based on a '[3+3] cyclization/Suzuki cross-coupling/Friedel-Crafts acylation' strategy," Tetrahedron Letters, vol. 47, no. 38, pp. 6903–6905, 2006.

17. W. Williams, X. Sun, and D. Jebaratnam, "Synthetic studies on the kinamycin family of antibiotics: synthesis of 2-(Diazobenzyl)-p-naphthoquinone, 1,7-Dideoxy-3-demethylprekinamycin, and 1-deoxy-3-demethylprekinamycin," The Journal of Organic Chemistry, vol. 62, no. 13, pp. 4364–4369, 1997.

18. N. Chen, M. B. Carrière, R. S. Laufer, N. J. Taylor, and G. I. Dmitrienko, "A biogenetically-lnspired synthesis of a ring-D model of kinamycin F: insights into the conformation of ring D," Organic Letters, vol. 10, no. 3, pp. 381–384, 2008.

19. N. Etomi, T. Kumamoto, W. Nakanishi, and T. Ishikawa, "Diels-Alder reactions using 4,7-dioxygenated indanones as dienophiles for regioselective construction of oxygenated 2,3-dihydrobenz[f]indenone skeleton," Beilstein Journal of Organic Chemistry, vol. 4, no. 15, pp. 1–8, 2008.

20. D. Mal, S. K. Ghorai, and N. K. Hazra, "A convenient synthesis of 4,8,9-trimethoxybenz[f]indenone, a potential BCD ring intermediate for stealthins and kinamycins," Indian Journal of Chemistry, Section B, vol. 40, no. 10, pp. 994–996, 2001.

21. D. Mal, N. K. Hazra, K. V. S. N. Murty, and G. Majumdar, "Benz[f]indenones: a novel synthesis by an anionic [4+2] cycloaddition/retro Diels-Alder pathway," Synlett, vol. 1995, no. 12, pp. 1239–1240, 1995.

22. T. Miyakoshi, "A convenient synthesis of 4-oxoalkanals," Synthesis, vol. 1986, no. 9, pp. 766–768, 1986.

23. M. L. Clarke and M. B. France, "The carbonyl ene reaction," Tetrahedron, vol. 64, no. 38, pp. 9003–9031, 2008.

24. A. Usman, I. A. Razak, S. Chantrapromma, et al., "1,4-methano-11a-methyl-4,4a,11,11a-tetrahydro-1H-benzo[b]fluoren-11-one," Acta Crystallographica, Section C, vol. 57, part 9, pp. 1118–1119, 2001.

25. S. K. Ghorai, N. K. Hazra, and D. Mal, "Facile synthesis of 4-functionalized cyclopentenones," Synthetic Communications, vol. 37, no. 12, pp. 1949–1956, 2007.

26. F. M. Hauser and D. Mal, "A novel route for stereospecific construction of the A ring of anthracyclinones: total synthesis of (±)-γ-citromycinone," Journal of the American Chemical Society, vol. 106, no. 6, pp. 1862–1863, 1984.

27. A. Patra, S. K. Ghorai, S. R. De, and D. Mal, "Regiospecific synthesis of benzo[b]fluorenones via ring contraction by benzil-benzilic acid rearrangement of benz[a]anthracene-5,6-diones," Synthesis, vol. 2006, no. 15, pp. 2556–2562, 2006.

Shape-Persistent Macrocycles with Intraannular Alkyl Groups: Some Structural Limits of Discotic Liquid Crystals with an Inverted Structure

S. Höger, J. Weber, A. Leppert and V. Enkelmann

ABSTRACT

The synthesis and thermal properties of new shape-persistent macrocycles of different sizes decorated with intraannular alkyl chains are described. The alkyl chain length is in all cases sufficient to cross the rings and to fill their interior completely. The investigation of the thermal behavior has shown that the smaller cycles do not exhibit thermotropic mesophases. Single crystal x-ray analysis indicates that the anisotropy in these compounds is too small to describe them as plates rather than spheres. For the larger macrocycles it is shown that longer adaptable substituents decrease the phase transition temperatures compared to previously described structures.

Background

During the past several years, the interest in the design and study of shape-persistent macrocycles with an interior in the nanometer regime has considerably increased. From the structural point of view, most compounds are based on the phenylene, phenylene acetylene or phenylene butadiynylene backbone, or they contain a mixture of these structural elements. However, apart from meeting the synthetic challenge, the supramolecular chemistry of rigid rings is currently investigated with considerable effort. For example, shape-persistent macrocycles can act as host structures for appropriate guest molecules; they can form 1D aggregates (in solution or gas phase) or regular 2D lattices after deposition at defined surfaces. [1-8] Furthermore, shape-persistent macrocycles are interesting mesogenes for discotic liquid crystalline materials. [9-12]

The common design principle of conventional discotic liquid crystals (LCs) is a more or less rigid core (disk-like or macrocyclic) with peripheral flexible side groups that point outward. [13-16] They can be used for a variety of different optic and electronic applications, for example as materials for photovoltaics (in the columnar phase) [17-19] or as compensation layers in display technology (in the nematic phase). [20-22]

Recently, we could show that shape-persistent macrocycles with fixed intraannular side chains (e.g. 1) can also exhibit liquid crystalline behavior. [23,24] Compound 1 is a shape-persistent macrocycle which is based on the phenylene-ethynylene-butadiynylene backbone. As it generally holds for rigid compounds, no matter if they are rod-like or cyclic, flexible side groups need to be added to keep these compounds tractable, i.e. soluble or/and meltable. For this purpose, 1 contains four long octadecyloxy side groups pointing inside the ring and eight propyloxy groups at the adaptable positions of the compound. The term adaptable means that the orientation of substituents at that position can be influenced by an external parameter. [25,26]

Compound 1 melts at 134°C to form a nematic mesophase that becomes isotropic at 159°C. However, compared to all previously reported discotic LCs, this compound is composed of a rigid periphery and the flexible side chains point inward. Hence, 1 can be described as a discotic LC with an inverted structure (figure 1). [23,24]

1

Figure 1. a) Design principle for common discotic liquid crystals; b) Design principle for discotic liquid crystals with an inverted structure; c) Structure of the shape-persistent macrocycle 1 containing extraannular methyl groups, intraannular octadecyloxy groups and adaptable propyloxy groups.

By comparing the structure of 1, and other shape-persistent macrocycles, with their thermal behavior we could identify some preliminary guidelines for the observation of liquid crystallinity in those compounds. Among these is the necessity of the rings to fill their interior more or less with their own alkyl chains and the absence of bulky peripheral side groups, both in order to prevent an interlocking of the rings. However, the number of compounds that follow this new design principle is still rather limited. This was a motivation for us to synthesize additional macrocycles and to explore their thermal behavior.

Here we describe the synthesis and the thermal behavior of several new shape-persistent macrocycles with fixed intraannular alkyl groups. Their investigation is a further step towards obtaining a clearer structure-property relationship in these materials. Although the new compounds described in this paper differ in size and backbone structure, the presence of intraannular alkyl groups that fill the ring interior completely is common for all macrocycles, thus making them potential candidates for the observation of thermotropic liquid crystallinity according to the new design principle described above.

Results and Discussion

Synthesis

Scheme 1 shows the synthesis of the macrocycles that have a reduced interior size compared to macrocycle 1. 3,5-Diiodo-4-methylbenzene (2) was treated with trimethylsilyl(TMS)acetylene under standard Hagihara-Sonogashira coupling conditions and subsequently deprotected with K_2CO_3 in MeOH/THF. As expected, the Pd-catalyzed coupling reaction runs under milder conditions and with higher yields compared to the preparation of 3 from the corresponding dibromo compound. [27] Reaction of 4 with an excess of 5, [28] coupling of 6 with

TMS-acetylene and deprotection, again with K_2CO_3 in MeOH/THF, gave the bisacetylenic half-ring 7. The oxidative dimerization of 7 was performed by slow addition of a solution of 7 in pyridine to a suspension of CuCl and $CuCl_2$ in the same solvent. Column chromatographic purification and repeated recrystallization from ethyl acetate gave the pure macrocycles not contaminated with higher oligomers, as determined by analytical gel permeation chromatography (GPC). The repeated purification process for all macrocycles was not optimized and is responsible for the rather low yield in the cyclization step.

Scheme 1. Synthesis of macrocycles with intraannular alkyl chains. i) TMS-acetylene, $PdCl_2(PPh_3)_2$, CuI, NEt3/THF (97%); ii) K_2CO_3, MeOH/THF (87%); iii) $PdCl_2(PPh_3)_2$, CuI, piperidine/THF (37–38%); iv) TMS-acetylene, $PdCl_2(PPh_3)_2$, CuI, piperidine/THF (76–82%); v) K_2CO_3, MeOH/THF (89–97%); vi) CuCl, $CuCl_2$, pyridine (12–16%).

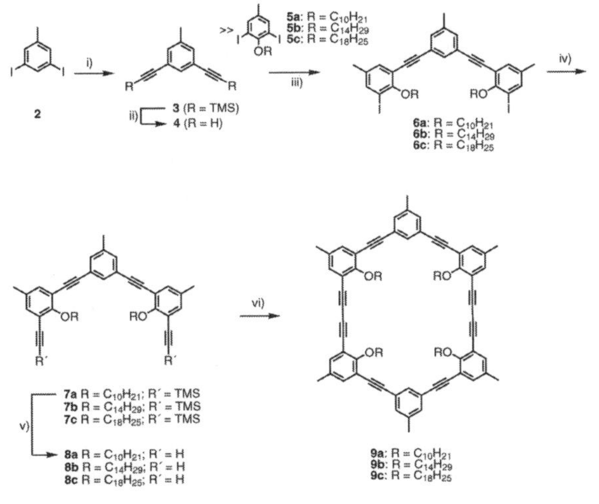

Our recent progress in the synthesis of functionalized polycyclic aromatic hydrocarbons (PAHs) opened the question about the influence of the presence of PAH substituents on the phase behavior of these macrocycles. [29,30] A PAH has previously been incorporated into a shape-persistent macrocycle as a part of the rigid backbone. [24] The investigation of the thermal behavior of that compound led to the indication that PAHs could stabilize the thermotropic mesophases. However, reports about shape-persistent macrocycles with extraannular PAH substituents are absent. If these compounds exhibit liquid crystallinity, the question about the mesogenic element (ring or PAH or both) arises. Additionally, biaxial nematic phases might be observable. [31,32] The synthesis of the corresponding PAH building block and the subsequent macrocycle synthesis is displayed in scheme 2.

Scheme 2. Synthesis of macrocycles with intraannular alkyl chains and extraannular PAH substituents. i) Pd(OAc)2, ligand, DBU, DMA (49%); BBr3, CH2Cl2 (99%); iii) I2/KI, ethylene diamine (quant.); iv) dimethyl sulfate, KOH, THF/water (50%); v) PdCl2(PPh3)2, CuI, piperidine/THF (71%); vi) Bu4NF, THF (97%); vii) CuCl, CuCl2, pyridine (28%).

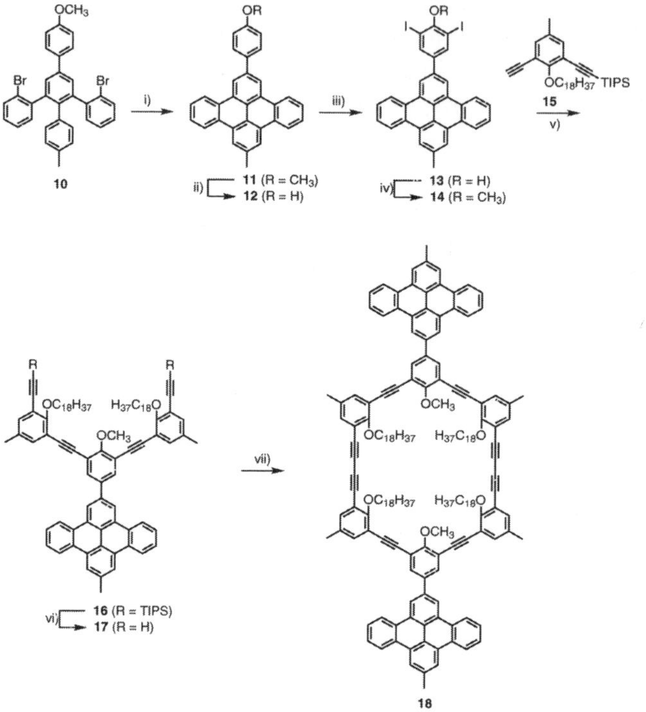

Treating the aromatic dibromo compound 10 with Pd(OAc)2 (20 mol%) and Cy2P(2,2'MeO)biph (30 mol%) in DMA gave the corresponding dibenzonaphthacene in 40–45% yield. Although a detailed investigation of the Pd-catalyzed dehydrohalogenation is yet not completed, the yield is reproducible higher than with Pd(PPh3)2Cl2, as we reported before. [29]11 was demethylated with BBr3, the resulting free phenol 12 iodinated with iodine and sodium iodide, and diiodide 13 realkylated with dimethylsulfate (model reactions with 4-methyl anisole showed that a direct diiodination could not be obtained under the conditions we used; for example, with N-iodo succinimide and FeCl3 we could obtain cleanly 2-iodo-4-methyl anisole). Pd-catalyzed Hagihara-Sonogashira coupling with 15, deprotection of the triisopropylsilyl (TIPS) groups and oxidative cyclodimerization under pseudo high-dilution conditions gave the macrocycle 18.

Based on the comparison of the thermal behavior of 9 and 18 with 1 we intended to prepare also isomers of the latter containing longer adaptable side groups, with and without extraannular PAH substituents (Scheme 3).

Pd-catalyzed coupling of the diiodo compound 14 with the mono protected bisacetylene 20, deprotection of the acetylenes with TBAF and subsequent coupling with an excess of the diiodo compound 5c gave the diiodid 23. Hagihara-Sonogashira coupling of 23 with an excess of TMS acetylene or TIPS acetylene, respectively, base (fluoride)-catalyzed removing of the silyl groups and cyclodimerization of the bisacetylenes 25, again under pseudo high-dilution conditions, gave the macrocycles 26a and 26b, respectively.

Scheme 3. Synthesis of macrocycles with adaptable hexyloxy groups. i) $PdCl_2(PPh_3)_2$, CuI, piperidine/THF (76–78%); ii) Bu4NF, THF (60–99%); iii) PdCl2(PPh3)2, CuI, piperidine/THF (46–48%); iv) TMS (TIPS)-acetylene, PdCl2(PPh3)2, CuI, piperidine/THF (64–98%); v) K_2CO_3, MeOH/THF or Bu4NF, THF (95–99%); vi) CuCl, $CuCl_2$, pyridine (7–23%).

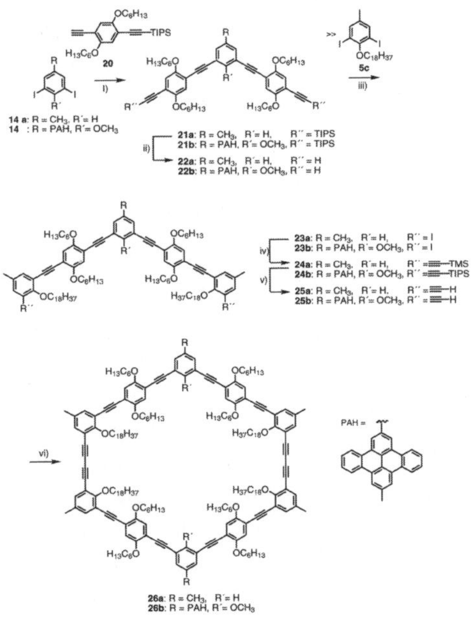

Thermal Behavior and X-ray Structure

The thermal behavior of the macrocycles 9 and 18 was of special interest in order to evaluate if macrocycles with a reduced interior size (compared to 1) also exhibit liquid crystallinity. All macrocycles 9 melt without decomposition (9a: 162°C; 9b: 132°C; 9c: 88°C). As expected, with increasing alkyl chain length the melting points of the compound decreases and in the latter case it is even below 100°C. However, the investigation of the three macrocycles 9 under the polarizing microscope showed that none of the compounds is thermotropic liquid crystalline. Under the assumption that the molecular structure in the solid state could help in understanding the absence of a mesophase, attempts were undertaken to perform a single crystal x-ray investigation on at least one of the compounds. By slow evaporation of a CH_2Cl_2 solution of 9a single crystals suitable for x-ray analysis could be obtained. 9a crystallizes as solvate with one molecule CH_2Cl_2 per macrocycle (not shown in figure 2).

The asymmetric unit of 9a contains two crystallographically independent (half) rings which are located on two centers of symmetry in the triclinic unit cell. These are not parallel to each other but inclined by an angle of about 25°. It should also be mentioned that the two macrocycles differ in the conformation of the alkyl chains. In one of them the side chains adopt an all-trans conformation whereas in the other two alkyl chains form a kink by a gauche conformation of the third bond. This is probably to fill the space between the rings effectively by enabling adjacent chains to be packed parall to each other. The top view shows that the internal void of both rings is already filled with the first few carbon atoms of the intraannular alkyl chains (Figure 2, left). The remaining portions of the alkyl chains are located above and below the rigid phenyl acetylene backbone of the rings which is in both cases relatively flat. Although the crystal packing shows that the alkyl chains are sandwiched between the macrocycles, there exist still some close contacts between the π-systems of adjacent macrocycles (e.g. C65–C72: 3.419 Å). An interlocking of the molecules through entanglement between rigid and flexible molecule parts cannot be observed. In addition, the extraannular methyl groups at the ring corners are also not bulky enough to extend into the interior of neighbored rings. Therefore, a physical crosslinking of the molecules, as it was supposed to be responsible for the absence of a mesophase in other shape-persistent rings, can be excluded. However, different to the previously investigated compounds the lateral extension of the macrocycles 9 is rather small. The distance between the two diacetylene bridges in 9 is around 1.2 nm compared to 2.3 nm in 1. Since the intraannular alykl chains are located above and below the macrocyclic framework, the anisotropy of the compound is remarkably lower as in 1, leading to an aspect ratio that does not support the mesophase formation. Interestingly, our expectation to observe LC behavior when extraannular PAH substituents were attached to this macrocycle was not fulfilled. Also for 18 no mesophase could be observed. Two reasons might account for that. First, the stronger interaction between the exocyclic parts of the rings leads to a significantly higher melting point (206°C), above the stability range of a possible mesophase. And second, the ability of the PAH substituents to rotate freely might lead to an even more unfavorable aspect ratio of the molecule. In any case, although only a few compounds with intraannular alkyl groups have been investigated, their thermal behavior indicates that the tendency of smaller rings to exhibit mesomorphic behavior is less pronounced compared to the larger rings (as e.g. 1). Investigations on derivatives of 9 with additional extraannular flexible side groups are ongoing and will be reported elsewhere.

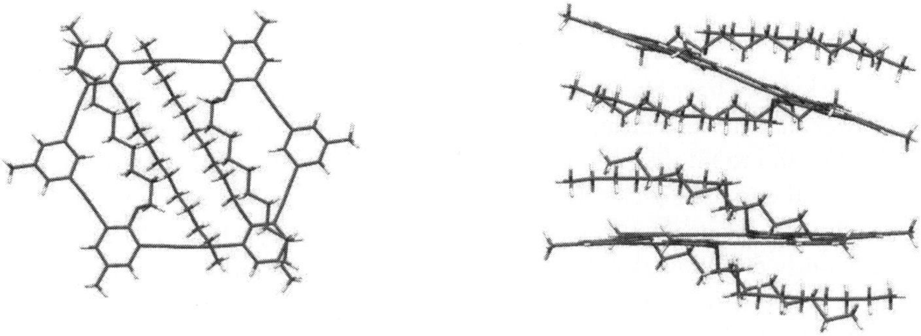

Figure 2. Single crystal X-ray structure of 9a (solvent not shown): a) Top view; b) Side view of two molecules of 9a in the crystal.

Apart from the size of the macrocycles and the size of the extra- and intraannular substituents, the thermal behavior of the compounds can be influenced by the adaptable substituents. Previous investigations have shown that the presence or absence of these substituents have a dramatic effect on the mesophase formation. [24] For example, when the eight adaptable propyloxy groups of 1 are removed, the material has a higher melting point (185°C) and does not exhibit a mesophase. Contrary, derivatives of 1 with longer adaptable side groups are yet not investigated. Two compounds of that structure are described in this work. Macrocycle 26a melts isotropically at 98°C (first heating, ΔH = 85.2 kJ/mol). However, cooling of the isotropic melt leads to the formation of a (monotropic) nematic phase at 90°C (ΔH = 2.2 kJ/mol) that recrystallizes at 34°C. Macrocycle 26b has a considerable higher melting point of 187°C (first heating, ΔH = 86.3 kJ/mol). Also in this case the melt is isotropic but becomes birefringent at 160°C, as observed by polarizing microscopy. This latter transition could not be observed by DSC. A recrystallization of the material during the cooling process could also not be observed, either by polarizing microscopy or by DSC measurements.

These data show that the adapable substituents are another parameter to fine tune the thermal behavior of discotic liquid crystals with an inverted structure. This parameter is absent in conventional discotic liquid crystals. As expected, longer adaptable alkyl substituents (compared to 1) decrease the melting point of the macrocycles but at the same time also narrow the temperature range of the mesophase so that for 26 only a monotropic mesophases could be observed. It is worth noting that also the attachment of PAH substituents does not in this particular case lead to an observation of an enantiotropic mesophase. This behavior is opposite to a previously reported example where a PAH unit incorporated into the macrocyclic backbone strongly stabilizes the mesophase.

Conclusion

In conclusion, we have prepared several new shape-persistent macrocycles of different sizes that contain intraannular alkyl chains of sufficient length to cross the whole rings and to fill their interior. The investigation of the thermal behavior has shown that the smaller cycles do not exhibit thermotropic mesophases. Although single crystal x-ray analysis has proven that these compounds fulfill in principle our previously described design principle for discotics with an inverted structure, the aspect ratio of the macrocycles with their alkyl surrounding is too small to describe them as plates rather than spheres.

For the larger macrocycles we could show that longer adaptable substituents decrease the phase transition temperatures. This parameter, absent in conventional discotics, is another tool to fine tune the thermal behavior of these materials.

Selected Crystallographic Data

Data collection with Mo Kα radiation on a Nonius KCCD diffratometer. Selected crystal data: 9a CH_2Cl_2 (T = 120 K): triclinic, P-1, a = 15.3188(6), b = 16.3517(6), c = 20.0689(7) Å, α = 88.9288(13), β = 69.1591(15), γ = 70.4677(14)°, V = 4399.0(3) Å^3, Z = 2, D_x = 1.089 g cm^{-3}, 54439 reflections measured, 15014 unique reflections (R_{int} = 0.072), 6315 reflections observed (I > 3σ(I)); the structure was solved by direct methods (Shelxs) and refined by full matrix least-squares analyses on F with anisotropic temperature factors for C, O, Cl. The H atoms were included with fixed isotropic temperature factors in the riding mode. R = 0.0390, R_w = 0.0464.

Acknowledgements

JW and AL contributed equally to this work. Financial Support by the Deutsche Forschungsgemeinschaft is gratefully acknowledged.

References

1. Moore JS: Acc Chem Res. 1997, 30:402–413.
2. Höger S: J Polym Sci Part A: Polym Chem. 1999, 37:2685–2698.

3. Haley MM, Pak JJ, Brand SC: Top Curr Chem. 1999, 201:81–130.
4. Grave C, Schlüter AD: Eur J Org Chem. 2002, 3075–3098.
5. Zhao D, Moore JS: Chem Commun. 2003, 807–818.
6. Höger S: Chem Eur J. 2004, 10:1320–1329.
7. Diederich F, Stang PJ, Tykwinsky R: Acetylene Chemistry. Weinheim: Wiley; 2005.
8. Zhang W, Moore JS: Angew Chem. 2006, 118:4524–4548. Angew. Chem. Int. Ed. 2006, 45: 4416–4439
9. Zhang J, Moore JS: J Am Chem Soc. 1994, 116:2655–2656.
10. Mindyuk O, Stetzer MR, Heiney PA, Nelson JC, Moore JS: Adv Mater. 1998, 10:1363–1366.
11. Seo SH, Seyler H, Peters JO, Jones TV, Kim TH, Chang JY, Tew GN: J Am Chem Soc. 2006, 128:9264–9265.
12. Pisula W, Kastler M, Yang C, Enkelmann V, Müllen K: Chem Asian J. 2007, 2:51–56.
13. Chandrasekhar S, Sadashiva BK, Suresh KA: Pramana. 1977, 9:471–80.
14. Demus D, Goodby J, Gray GW, Spiess H-W, Vill V: Handbook of Liquid Crystals. Volume 2B. Weinheim: Wiley; 1998.
15. Bushby RJ, Lozman OR: Curr Opin Colloid & Inter Sci. 2002, 7:343–354.
16. Collard DM, Lillya CP: J Am Chem Soc. 1991, 113:8577–8583.
17. Adam D, Schuhmacher P, Simmerer J, Haeussling L, Siemensmeyer K, Etzbach KH, Ringsdorf H, Haarer D: Nature. 1994, 371:141–143.
18. Boden N, Bushby RJ, Clements J, Movaghar B: J Mater Chem. 1999, 9:2081–2086.
19. Schmidt-Mende L, Fechtenkötter A, Müllen K, Moons E, Friend RH, MacKenzie JD: Science. 2001, 293:1119–1122.
20. Mori H, Itoh Y, Nishiura Y, Nakamura T, Shinagawa Y: Jpn J Appl Phys A. 1997, 36:143–147.
21. Okazaki M, Kawata K, Nishikawa H, Negoro M: Polym Adv Technol. 2000, 11:398–403.
22. Kumar S, Varshney SK: Angew Chem. 2000, 112:3270–3272. Angew. Chem. Int. Ed. 2000, 39: 3140–3142.
23. Höger S, Enkelmann V, Bonrad K, Tschierske C: Angew Chem. 2000, 112:3256–3258. Angew. Chem. Int. Ed. 2000, 39: 2268–2270.

24. Höger S, Cheng XH, Ramminger A-D, Enkelmann V, Rapp A, Mondeshki M, Schnell I: Angew Chem. 2005, 117:2862–2866. Angew. Chem. Int. Ed. 2005, 44: 2801–2805.
25. Morrison DL, Höger S: Chem Commun. 1996, 2313–2314.
26. Höger S, Morrison DL, Enkelmann V: J Am Chem Soc. 2002, 124:6734–6736.
27. MacDonal M-A, Puddephatt RJ, Yap GPA: Organometallics. 2000, 19:2194–2199.
28. Kang BS, Kim DH, Lim SM, Kim J, Seo M-L, Bark K-M, Shin SC: Macromolecules. 1997, 30:7196–7201.
29. Cheng XH, Höger S, Fenske D: Org Lett. 2003, 5:2587–2589.
30. Cheng X, Heyen AV, Mamdouh W, Ujii H, De Schryver F, Höger S, De Feyter S: Langmuir. 2007, 23:1281–1286.
31. Severing K, Saalwächter K: Phys Rev Lett. 2004, 92:125501/1–125501/4.
32. Prasad V, Kang S-W, Suresh KA, Joshi L, Wang Q, Kumar SJ: Am Chem Soc. 2005, 127:17224–17227.

Part 1. Reduction of S-Alkyl-Thionocarbonates and Related Compounds in the Presence of Trialkylboranes/Air

Jean Boivin and Van Tai Nguyen

ABSTRACT

A new, mild, and environment friendly process for the reduction of S-alkyl-thionocarbonates, iodides and related compounds to the corresponding hydrocarbons at room temperature with good to excellent yields is described. This method uses a trialkylborane in excess (Et_3B or Bu_3B) and air.

Background

The reduction of functional groups such as S-alkyl-dithiocarbonates (S-xanthates), iodides, O-alkyl-dithiocarbonates (O-xanthates) and related compounds to the corresponding alkanes is very important in organic synthesis, especially in natural products chemistry.[1,2] Deoxygenation (Barton-McCombie reaction) has been largely used to handle sensitive compounds such as sugars.[3] On the other hand, iodides and S-alkylxanthates are useful compounds that easily produce carbon radicals involved in radical chain reactions (cyclisation, intermolecular addition onto an olefin, etc) in which trapping of the final radical results in the transfer of the iodine atom or of the xanthate moiety.[4,5] To date, the most widely used reductive method to remove these functional groups that become superfluous at the end of the reaction process, is based on the Bu3SnH/AIBN combination that operates at 80°C or above. The main virtue of this method relies on its versatility and its efficiency. However, this procedure suffers from crippling drawbacks in terms of toxicity, cost, disposal, and tedious purification to remove tin residues.[6-10] However, we recently faced limits to the use of some of these methods for reductive removal of S-alkyl O-ethyl dithiocarbonates and we therefore proposed diethyl phosphite/DLP and H3PO2/Et3N/AIBN as attractive reagents for this purpose that circumvent these impediments.[11]

The core of the results reported in this article and the following parts,[12,13] has been presented as communications at the Xth Symposium of the « Institut de Chimie des Substances Naturelles », 1–3 June 2005, Gif-sur-Yvette, France and at the "1st German-French Congress in Organic Chemistry", Goslar, September 7th – 11th, 2005. The recent observation made by Wood and his colleagues that O-alkylxanthates may be deoxygenated by the combination Et3B/air/H2O under similar conditions (published on the Web on 08/18/2005),[14] the subsequent work by Renaud and coworkers who studied the reduction of B-alkylcatecholboranes under very similar conditions,[15] as well as the very recent kinetic study published by Newcomb,[16] prompt us to report in detail our findings in this area as a series of three articles. In this first article we report that trialkylboranes are useful reagents to achieve reduction of several radicophilic groups. A careful investigation of the process led us to develop a new method for the reduction of S-alkylxanthates, iodides and similar compounds to the corresponding hydrocarbons that requires a trialkylborane in excess and air as initiator, at 20°C, or even at lower temperature.[17,18]

Results and Discussion

During the last two decades, the number of applications of Et_3B as radical initiator has increased tremendously. As Et_3B is known to produce radicals at 20°C, or even much lower temperature (-78°C), we thought that trialkylboranes in the presence of dioxygen would be good candidates to trigger the radical addition of α-acyl S-xanthates onto olefins. When testing this hypothesis, we observed that compound

1a (Figure 1) underwent addition to decene in 40% yield when Et3B (0.1 equiv.)/air was used to initiate the reaction. Under these conditions, traces of alkane 1b (< 5%) were also isolated. When 2.5 equiv. of Et3B were used, compound 1a failed to yield adduct 1c but instead was reduced to the corresponding alkane 1b in 63% yield. No other identifiable product could be isolated. Zard and Nozaki mentioned similar reductions from α-acylxanthate[19] and α-acyliodide[20,21] respectively. These authors proposed the formation of an intermediate boron enolate that hydrolyses on work-up. A detailed report of our results concerning the addition process is given in a subsequent article.[13]

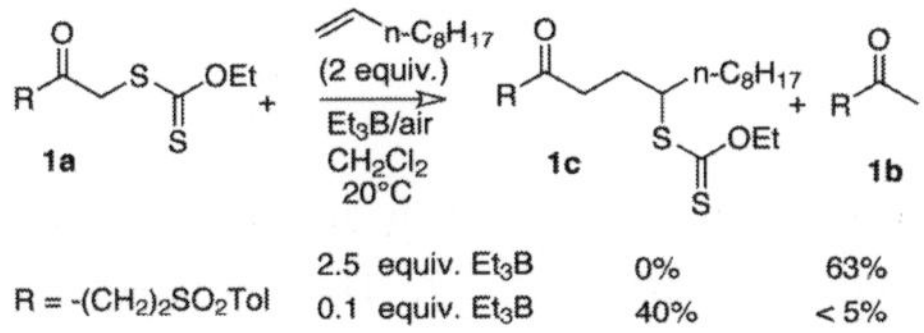

Figure 1. Reaction of α-acylxanthate 1a with 1-decene and Et_3B/air.

As the absence or the low yield of the addition product 1c might be due to its consumption by an unanticipated reaction, we decided to examine the reactivity of S-alkylxanthates that do not bear a carbonyl function in the adjacent position, under similar conditions, but without added olefin. In a first set of experiments [Figure 2, Table 1, Method A], we showed that the primary and secondary 2-oxoalkyl xanthates 1a and 2a behaved similarly (entries 1 and 2), as expected, giving high yields of compounds 1b and 2b, respectively. Interestingly, secondary S-alkylxanthates 3a–5a were cleanly reduced to the corresponding alkanes in excellent yields (up to 80%) (Table 1, entries 3–5). When the starting material was soluble enough, there was no need for an extra solvent other than the mixture of hexanes present in the commercial Et3B solution (entries 3 and 6). Experimentally, in method A, a solution of xanthate, Et3B (5 equiv., 1M solution in hexanes), in the given solvent (if needed) was simply stirred for 2 h in the presence of air under anhydrous conditions.

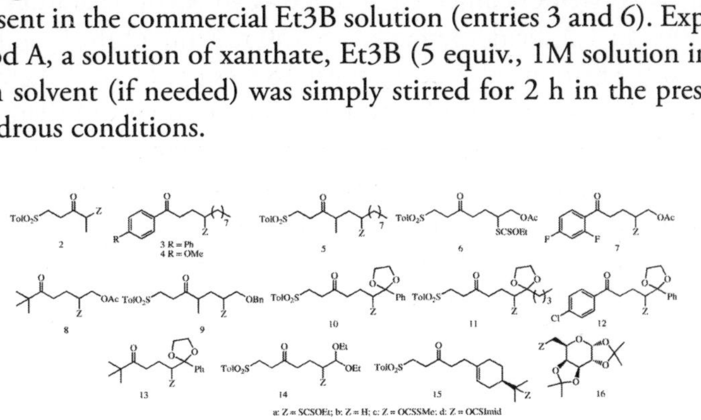

Figure 2. Xanthates and thionoimidazolides 2–16 and their reduced derivatives.

Table 1. Reduction of S-alkylxanthates with Et_3B/air at 20°C according to Method A.

Entry	Xanthate	Product	Et_3B (equiv.)	Solvent	Yield (%)
1	**1a**	**1b**	2.5	$(CH_2Cl)_2$	77
2	**2a**	**2b**	2.5	$(CH_2Cl)_2$	75
3	**3a**	**3b**	5	-	79
4	**4a**	**4b**	5	THF	79
5	**5a**	**5b**	7	Et_2O	80
6	**7a**	**7b**	5	-	50
7	**8a**	**8b**	5	Et_2O	63
8	**10a**	**10b**	5	$(CH_2Cl)_2$	51
9	**12a**	**12b**	5	$(CH_2Cl)_2$	45
10	**13a**	**13b**	5	Et_2O	67
11	**14a**	**14b**	5	$(CH_2Cl)_2$	60
12	**15a**	**15b**	5	CH_2Cl_2	52
13	**16a**	**16b**	5	-	0 (90) [a]

None of the reactions reported was fully optimised. For details on method A, see Experimental Section. [a] Yield of recovered starting material.

Secondary xanthates 7a and 8a, bearing an acetyl group, were also reduced, but in slightly lower yields (entries 6, 7). The reduction with Et3B/air is also adapted to S-alkylxanthates which are a part of fragile compounds such as ketals or carbonyl derivatives that contain a leaving group in the β-position. Thus, compounds 10a, 12a–14a gave the corresponding alkanes in fair to good yields (entries 8–11). The reduction of a tertiary xanthate, without risk of any pseudo-Tchugaev thermal elimination, was also feasible in good yield as shown by reaction of compound 15a (entry 12). Attempts to reduce a primary xanthate under these conditions failed. No trace of compound 16b was produced when xanthate 16a was subjected to the action of Et3B/air at 20°C, but, noteworthy, 90% of the starting material was recovered (entry 13).

In a second set of experiments, we tried to gain information concerning the rate of the reduction. In this context, the Surzur-Tanner/Giese rearrangement was particularly illustrative. [22-25] The 2 → 1 migration of the acyloxy group in glycosyl radicals is well documented and allows an easy preparation of 2-deoxy-sugars. At 20°C in 1,2-dichloroethane, a 50/50 mixture of the "regular" 1-deoxy sugar 17b and the rearranged 2-deoxy sugar 18 was produced (Scheme 1). When the temperature was raised to 60°C, the latter compound became largely predominant (ratio 17b/18 = 10/90). On the contrary, at low temperature (-20°C), only traces of the rearranged product (< 5%) could be detected by 1H NMR. Regardless of the temperature, the yield of the reduction exceeds 82%. It is important to note that, at 20°C, the rate of reduction equals the rate of the 1,2-acetyl group migration. The rate constant for the latter process has been estimated to be 4.0 x 102 s-1 in benzene at 75°C.[26]

Scheme 1. Reduction of xanthate 17a at different temperatures with Et_3B (5 equiv.)/air.

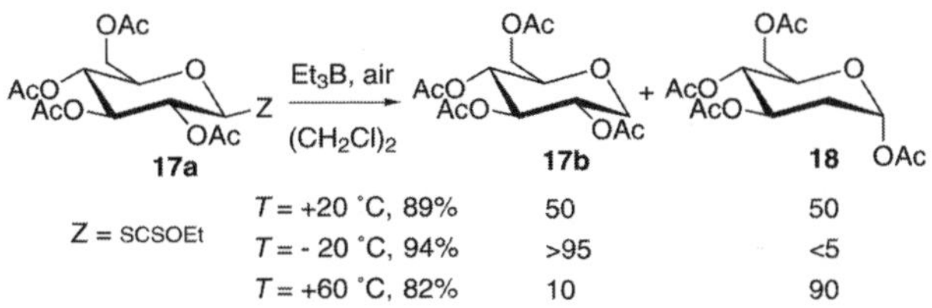

We then turned our attention to the reduction of O-alkylxanthates using the same technique (Method A). Despite our efforts, changing the solvent, the amount of Et3B (from 1 to 7 equiv.) or the concentrations (0.1 or 0.2 M), compound 19 led repeatedly to complex mixtures. Our interpretation is based on the following observations: – the attack of a carbon radical on the sulfur atom of a thiocarbonyl function is known to be a fast process; – the hydrogen atom transfer in the system Et3B/air is a relatively slow process as demonstrated above. In the case of S-alkylxanthates a, the production of radical R• results from the reaction of xanthate a with Et• radical generated from O2 and Et3B, as depicted by equation (1) (Scheme 2). When R• is a secondary radical, the reaction is pulled to the right side, i.e. to the formation of the secondary radical, more stable than the primary Et• . If the hydrogen atom transfer is slow, the radical may easily react with a or e [equation (2)], thus establishing a degenerate process that does not consume the starting material a, preserves the radical character and allows slow hydrogen atom transfer to occur. For O-alkylxanthates, the situation is different. The reaction of Et• with f gives rise to h and R• [equation (3)]. If the hydrogen atom transfer is slow, the latter reacts with f to afford the rearranged S-alkyl S-methyldithiocarbonate j. As R• radical cannot react with the carbonyl function of j, the starting material is progressively consumed by the unwanted formation of j [equation (4)]. Furthermore, as the reaction proceeds, the concentration of f decreases, giving radical R• the opportunity to undergo uncontrolled reactions.

Scheme 2. Reduction of S-alkylxanthates and O-alkylxanthates.

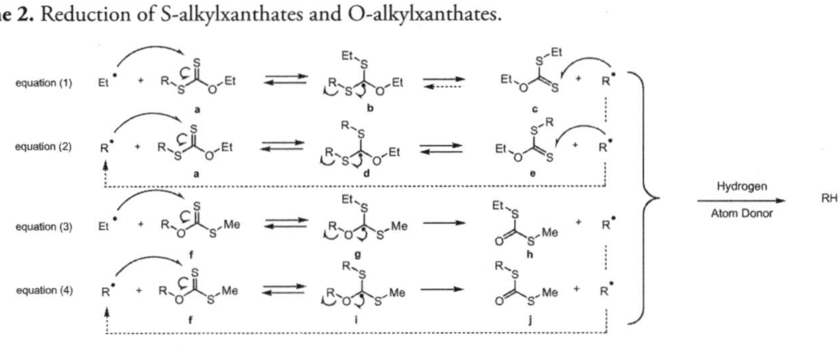

This rationale permitted reduction of O-alkylxanthates. To minimise the unwanted reactions, both the concentration of R• and the concentration of the substrate must be kept low. The former may be adjusted by controlling the amount of oxygen in the medium while the latter may be regulated either by high dilution technique or by slow addition of the substrate. Simply carrying out the reduction of xanthate 19 with Et3B (5 equiv.), under high dilution conditions (C = 0.04 M) combined with a slow addition of air with a syringe pump (40 mL/h for 2 h), led to a gratifying 27% yield of reduced material 20 (Table 2, entry 2). Increasing the amounts of Et3B to 10 equiv. or decreasing to 3 equiv. did not significantly improve the outcome of the reaction (Table 2, entries 3 and 4). When both air and xanthate were added slowly with two syringe pumps, the yield of reduced compound 20 was increased (Table 2, entries 6–8). The best result (57%) was obtained by the association of a highly diluted medium and slow additions of air and substrate.

Table 2. Reduction of O-alkylxanthate 19 to 20 under various conditions.

Entry	Solvent	Concentration (M)[a,b]	Et_3B (equiv.)	Rate of addition of xanthate **19**	Rate of addition of air	Yield (%) **20**
1	Various[c]	0.1 or 0.2	1 to 7	All at once	Open to air	<5
2	CH_2Cl_2	0.04	5	All at once	80 mL/2 h	27
3	$(CH_2Cl)_2$	0.02	10	All at once	80 mL/1.2 h	26
5	$(CH_2Cl)_2$	0.027	3	All at once	60 mL/0.8 h	31
6	$(CH_2Cl)_2$	0.03	10	30 min.	40 mL/2 h	44
7	$(CH_2Cl)_2$	0.02	5	15 min.	15 mL/1.5 h	57
8	$(CH_2Cl)_2$	0.12	5	15 min	15 mL/1.5 h	32

[a] Experiments carried out on 0.5 mmol of xanthate **19**. [b] Concentration (M) refers to the overall amount of substrate added. [c] CH_2Cl_2, $(CH_2Cl)_2$, THF, Et_2O

Thionoimidazolide 21 and iodide 22 were also converted to the corresponding 3-deoxy glucofuranose 20, according to method B, in 50 and 80% yields respectively (Scheme 3). The reduction of compound 21 according to method A gave a complex mixture mostly constituted of the undesired rearranged S-alkylimidazolide. Not unexpectedly, primary O-alkylxanthate 16c and thionoimidazolide 16d, under the same conditions, failed to produce the corresponding 6-deoxy compound but led to complex mixtures. Wood and coworkers also noticed that the deoxygenation of primary O-alkylxanthates under similar conditions is impractical (yield 3%, determined by GC).[14]

Scheme 3. Reduction of O-alkyl-S-methyl xanthate 19, thionoimidazolide 21 and iodide 22 by Et_3B/air at 20°C.

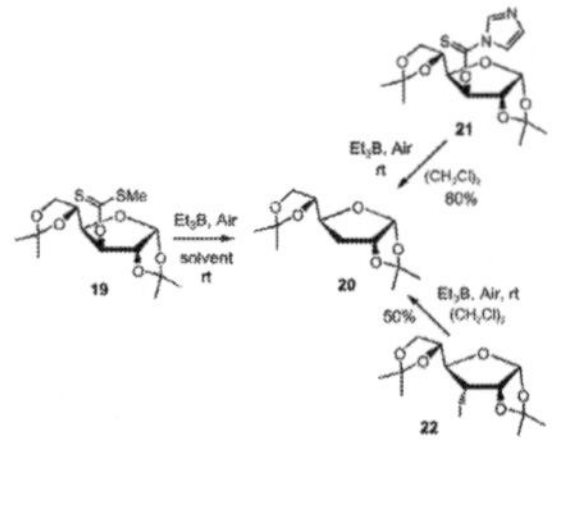

The comparison of entries 1 and 7 in Table 2 shows that an impressive improvement was obtained in the deoxygenation of O-alkylxanthates by a better understanding of the various steps involved and by subsequent modifications of the experimental procedure.

We therefore returned our attention to the reduction of S-alkylxanthates that sometimes led to poor or irreproducible results when Method A was employed. In particular, some dimerisation of the radical generated from the xanthate was sometimes observed by mass spectrometry. This is clearly an indication that the concentration of radicals in the medium is too high. Utilisation of Method B permitted high and reproducible yields of the corresponding alkanes to be obtained by suppressing most of the side reactions (Table 3). For example, the yield of the reduction of xanthate 8a increased from 63% (Method A, Table 1, entry 7) to 84% (Method B, Table 3, entry 2). Under these conditions, the amounts of Et3B may be lowered to 3 equiv. without altering the efficiency of the reaction (Table 3, entry 3). However, when 1.5 equiv. of Et3B was used, the reaction failed to reach completion (Table 3, entry 4). Bu3B/air is also an efficient reducing agent as exemplified by conversion of xanthate 8a to compound 8b in a gratifying 85% yield (Table 3, entry 5). With the noticeable exception of derivative 9a possessing a benzyloxy group in the β-position, yields of reduced compounds always exceeded 70% (Table 3, entries 1, 2–5, and 7–9).

Table 3. Reduction of S-alkylxanthates with Et_3B/air at 20°C according to Method B.

Entry	Xanthate	Product	Borane (equiv.)	Solvent	Yield (%)
1	**6a**	**6b**	5	CH_2Cl_2	71
2	**8a**	**8b**	5	$(CH_2Cl)_2$	84
3	**8a**	**8b**	3	$(CH_2Cl)_2$	83
4	**8a**	**8b**	1.5	$(CH_2Cl)_2$	55 (27) [a]
5	**8a**	**8b**	5	CH_2Cl_2	85 [b]
6	**9a**	**9b**	5	CH_2Cl_2	42 (10) [a]
7	**11a**	**11b**	5	Et_2O	77
8	**13a**	**13b**	5	Et_2O	80
9	**15a**	**15b**	5	$(CH_2Cl)_2$	70

None of the reactions reported was fully optimised. For details on method B, see Experimental Section. [a] Yield of recovered starting material. [b] Bu_3B was used instead of Et_3B.

To explain the relatively low yield of the reduction of 9a into 9b, the question of a putative 1,5-hydrogen atom transfer has been raised. This hypothesis calls for several remarks. Compounds 5a, 6a, 10a, 11a, and 14a that display the same type of skeleton led to satisfactory yields of the corresponding reduced compounds (Method A or B, Tables 1 and 3). If such a transfer occurred, radical B would be converted to radical C (represented as two mesomeric forms, Scheme 4). Radical C would certainly evolve rapidly to xanthate D or would be trapped by Et3B to afford a stable enolboronate F that would give G upon work-up, as postulated above for reduction of compounds 1a and 2a. This would thus give the same product as in the direct reduction from radical B, without lowering the yield of reduced compound. Alternatively, radical C might fragment to conjugated enone E and stabilised TolSO2• radical. Neither compounds D nor enones E were ever isolated in the present work. Similarly, these compounds were not seen in a precedent work in which a "fast" hydrogen atom donor (hypophosphorus acid) or a "slow" hydrogen donor (diethylphosphite) were used on the same type of substrate.[11] Of course the hypothesis of a putative 1,5-hydrogen shift that would operate to a minor extent is not fully ruled out, and should certainly deserve further experimentation. However, one should stress that the reduction with the system Et3B/O2 is not limited to compounds where such shifts are possible and that such hydrogen migration does not pertain to the method used to generate the radical but to the intrinsic reactivity of radical B, as it is generally admitted that the fate of a free radical is independent of the method used to produce it. But this is beyond the scope of the present study.

Scheme 4. Products formed through a putative 1,5-hydrogen atom transfer.

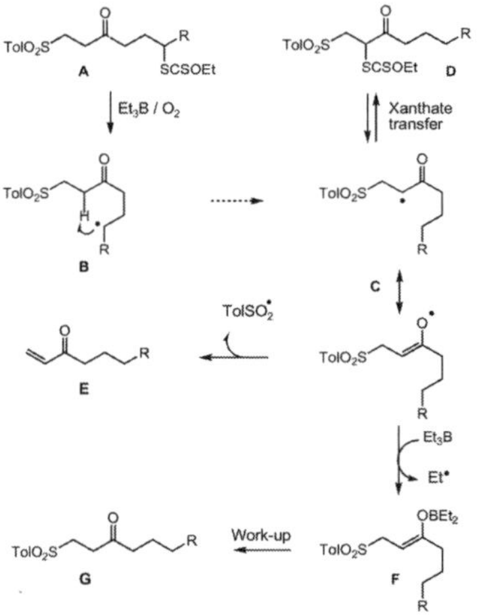

Conclusion

In this note, we have described a new, practical and efficient method to reduce S-alkylxanthates, iodides, and O-alkylxanthates, into the corresponding alkanes under very mild conditions using commercially available reagents and simple experimental procedures. We are currently exploring the scope of the reaction by extending this process to other radicophilic species. Undeniably, the question of the origin of the hydrogen atom is compelling and an in-depth mechanistic study is needed. This intriguing point is investigated in part 2 of this series.[12]

Acknowledgements

This research was supported by grants from the "Institut de Chimie des Substances Naturelles". We are grateful to Prof. J-Y Lallemand for much help and encouragement.

References

1. Barton DHR, Ferreira JA, Jaszberenyi JC: Preparative Carbohydrate Chemistry. Edited by: Hanessian S. New York: Marcel Dekker; 1997:151–172.
2. Renaud P, Sibi MP (Eds): Radicals in Organic Synthesis. Volume 1 and 2. Weinheim, Germany: Wiley-VCH; 2001.
3. Barton DHR, McCombie SW: J Chem Soc, Perkin Trans I. 1975, 1574–1585.
4. Zard SZ: Radical Reaction in Organic Synthesis. Oxford: Oxford University Press; 2003.
5. Zard SZ: Angew Chem, Int Ed. 1997, 36:673–685.
6. Boyer IJ: Toxicology. 1989, 55:253–298.
7. Crich D, Sun S: J Org Chem. 1996, 61:7200–7201.
8. Baguley PA: Angew Chem, Int Ed. 1998, 37:3072–3082.
9. Studer A, Amrein S: Synthesis. 2002, 835–849.
10. Walton JC, Studer A: Acc Chem Res. 2005, 38:794–802.
11. Boivin J, Jrad R, Juge S, Nguyen VT: Org Lett. 2003, 5:1645–1648.
12. Allais F, Boivin J, Nguyen VT: Part 2 : Beilstein Journal of Organic Chemistry 2007. [http://bjoc.beilstein-journals.org/content/3/1/46] BJOC 2007., 3(46):
13. Boivin J, Nguyen VT: Part 3 : Beilstein Journal of Organic Chemistry 2007. BJOC 2007., 3(47):
14. Spiegel DA, Wiberg KB, Schacherer LN, Medeiros MR, Wood JL: J Am Chem Soc. 2005, 127:12513–12514.
15. Pozzi D, Scanlan EM, Renaud P: J Am Chem Soc. 2005, 127:14204–14205.
16. Jin J, Newcomb M: J Org Chem. 2007, 27:5098–5103.
17. Olivier C, Renaud P: Chem Rev. 2001, 101:3415–3434.

18. Yorimitsu H, Oshima K: Radicals in Organic Synthesis. Volume 1. Edited by: Renaud P, Sibi MP. Weinheim, Germany: Wiley-VCH; 2001:11–27.
19. Legrand N, Quiclet-Sire B, Zard SZ: Tetrahedron Lett. 2000, 41:9815–9818.
20. Nozaki N, Oshima K, Utimoto K: Tetrahedron Lett. 1988, 29:1041–1044.
21. Nozaki N, Oshima K, Utimoto K: Bull Chem Soc Jpn. 1991, 64:403–409.
22. Surzur J-M, Teissier PCR: Acad Sci Fr Ser C. 1967, 264:1981–1984.
23. Tanner DD, Law FC: J Am Chem Soc. 1969, 91:7535–7537.
24. Giese B, Groninger KS, Witzel T, Korth H-G, Sustmann R: Angew Chem, Int Ed. 1987, 26:233–236.
25. Crich D: Can J Chem. 2004, 82:75–79.
26. Korth H-G, Sustmann R, Groninger KS, Leisung M, Giese B: J Org Chem. 1988, 53:4364–4369.

Part 2. Mechanistic Aspects of the Reduction of S-Alkyl-Thionocarbonates in the Presence of Triethylborane and Air

Florent Allais, Jean Boivin and Van Tai Nguyen

ABSTRACT

Experiments conducted with deuterated compounds demonstrated that during the reduction of S-alkylxanthates with triethylborane, the hydrogen atom transferred has several competing origins. Hydrogen abstraction from water (or an alcohol) is the most favourable route.

Background

In the first part of this series,[1] we showed that trialkylboranes and especially commercial solutions of Et_3B are efficient reducing agents that permit the conversion, at room temperature, of S-alkylxanthates, iodides and O-alkylxanthates into the corresponding alkanes with good to excellent yields. Such a process complies with the long-standing pursuit of an environmentally acceptable process for desulfurisation, dehalogenation or deoxygenation that operates under mild reaction conditions. In this context, special attention must be paid to a paper published by Jaszberenyi and Barton in 1990.[2] The authors described a reduction process of O-alkylxanthate and related compounds with Bu_3SnH/Et_3B/air at room temperature. They mentioned very briefly that when Bu3SnH was omitted, the reduction still occurred. The O-cyclododecyl S-methylxanthate derived from cyclododecanol afforded cyclododecane in a remarkable 62% yield. No hypothesis about the origin of the hydrogen atom that replaced the original xanthate function was proposed. Recently, as the work reported here was largely completed as already mentioned in the first part of this series,[3,4] Wood and coworkers also observed an "anomalous" reduction of a closely related tertiary O-alkylxanthate into the corresponding alkane instead of an expected rearrangement.[5] In a subsequent report on the reduction of B-alkylcatecholboranes, Renaud and coworkers showed that the O-H bond in methanol may also be the source of hydrogen. However, the deuterium incorporations in experiments aimed at elucidating the mechanism are relatively low,[6] and obviously the experimental results do not support the hypothesis that the O-H group is the sole source of the hydrogen transferred.

Results and Discussion

In this Note, we wish to report our findings concerning the intriguing question of the origin of the hydrogen atom that replaces the radicophilic group in the reduction of S-alkyl-thionocarbonates. Several hypotheses may reasonably be proposed. The first possibility is a hydrogen transfer from the solvent. However, dichloromethane, 1,2-dichloroethane and hexane (from the commercial solution of Et_3B) utilised in the preceding article,[3] are considered to be poor hydrogen atom donors, especially at low temperature.[7] There are only rare reports concerning the ability of an alkane (cyclohexane) to cleanly transfer a hydrogen atom to a specific type of carbon radical.[8,9] Disproportionation between the ethyl radical and the carbon-centred radical derived from the xanthate would have probably given the corresponding olefins, especially in the case of a tertiary radical. Such olefins were never found. A more seemly hypothesis is a hydrogen abstraction from the α-position of the boron atom. This reaction is well documented and energetically acceptable (BDE (C-H) = 80 ± 3 Kcal/mol). For example, the methyl radical abstracts α-boronyl hydrogen of Et3B 124 times faster than hydrogen in the methyl group of toluene.[10] We also envisioned the possible intermediacy of a transient organoborane species that would undergo protonolysis during work-up. However, a "simple" trialkylborane intermediate is not a plausible route since the attack of a carbon radical on the boron atom is not favourable. [10]

In an attempt to gain information about the mechanism, we performed deuteration experiments (Figure 1 and Table 1). Most of the results reported herein concern the reduction of xanthate 1a. This substrate was chosen because of its low molecular weight and its simple structure that ensure easy spectral analyses (NMR, MS, GC-MS) and also because a putative 1,5-hydrogen shift between the intermediate radical and a hydrogen atom in the α-position to the ketone cannot intervene. This point was discussed at the end of the first part of this series. [3] On the other hand, we proved that a similar 1,5-hydrogen shift in which the acetyl group would be implicated does not occur either (see below). The standard experiment (entry 1) performed without any source of deuterium serves to evaluate the natural abundance of 13C in compound 1c and to calibrate the deuterium measurements.

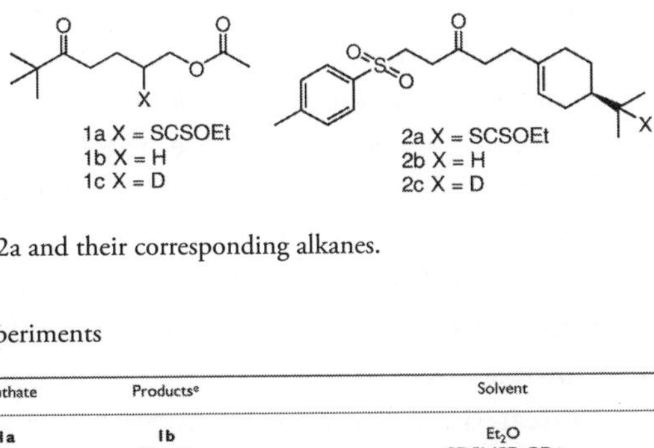

Figure 1. Xanthates 1a, 2a and their corresponding alkanes.

Table 1. Deuteration experiments

Entry	Xanthate	Products[e]	Solvent	D %
1	**1a**	**1b**	Et_2O	0
2	**1a**	**1b + 1c**	$CDCl_3/CD_3OD$ [a]	88
3	**1a**	**1b + 1c**	$(CH_2Cl)_2/CD_3OD$ [a]	85
4	**1a**	**1b + 1c**	$(CH_2Cl)_2/CH_3OD$ [a]	76
5	**1a**	**1b + 1c**	$(CH_2Cl)_2/D_2O$ [a]	83
6	**1a**	**1b + 1c**	THF/D_2O [a]	83
7	**1a**	**1b + 1c**	THF 3 h. then D_2O [a]	<1
8	**1a**	**1b + 1c**	3 h then CH_3OD [a]	6
9	**1a**	**1b + 1c**	20 min. then CH_3OD [a]	17
10	**1a**	**1b + 1c**	THF; D_2O/H_2O = 25 equiv/25 equiv	6
11	**1a**	**1b + 1c**	H_2O/D_2O = 5 equiv/100 equiv[b]	66
12	**1a**	**1b + 1c**	H_2O/D_2O = 20 equiv/80 equiv[b]	32
13	**2a**	**2b + 2c**	$(CH_2Cl)_2/CH_3OD$[a]	93[f]
14	**1a**	**1b + 1c**	$CDCl_3$[c]	57[g]
15	**1a**	**1b + 1c**	$CDCl_3$[c]/D_2O[d]	83
16	**1a**	**1b + 1c**	C_6D_6[c]	<1
17	**1a**	**1b + 1c**	C_6D_6[c]/D_2O[d]	70
18	**1a**	**1b + 1c**	$CDCl_3$[c]/H_2O[d]	3
19	**1a**	**1b + 1c**	THF-d_8	<1[h]

Reduction of S-alkylxanthates (0.3–0.4 mmol) with Et_3B (5 equiv from a commercial solution in hexanes)/dry air for 3 h, unless otherwise stated. Xanthates were reduced according to method A except in experiment 13 where method B was used. [a] 50 equiv of deuterated methanol or water were used. [b] No other organic solvent than hexanes from the commercial Et_3B solution. [c] Reactions performed with freshly prepared 1M solutions of pure Et_3B in $CDCl_3$ or C_6D_6. [d] 5 equiv of D_2O or H_2O. [e] All the yields in pure **1b + 1c** are superior to 61% except for experiment 16 (yield 40%). [f] Yield **2b + 2c** = 70%. [g] When the reaction was performed with $CDCl_3$ dried over K_2CO_3, the deuterium incorporation fell to 15.8%. [h] This experiment was carried out twice.

A possible source of hydrogen could be an alcohol present as a contaminant in various solvents, therefore we chose to examine the effect of addition of methanol to the reaction mixture. When xanthate 1a was treated with Et3B/dry air, according to method A,[3] in a mixture of chloroform-d and methanol-d4, the incorporation of deuterium was determined to be 88% (entry 2). The analysis of the spectral data (1H, 2H and 13C NMR and mass) unambiguously shows that the incorporated deuterium is located at the same position as the parent xanthate. This proves that no 1,5-hydrogen shift occurred. The replacement of chloroform-d by 1,2-dichloroethane gave similar results (85%, entry 3). More interestingly, the incorporation was still high when methanol-d4 was replaced by CH3OD (76%, entry 4). Methanol-d4 or CH3OD can be in turn replaced by D2O without significant change in the level of deuterium incorporation (entries 5 and 6), even in a solvent known to be a good hydrogen donor such as THF (entry 6). Experiments 2–6 established that, in the presence of deuterated methanol or water, the incorporated deuterium may come from rupture of an O-D bond. The results given in entries 7 and 8 show that there is no noticeable incorporation of deuterium when D2O or CH3OD are added after 3 h, a period after which all the starting material 1a has been consumed. On the other hand, experiment 9, where CH3OD was added after 20 min. before consumption of all the starting material, indicates that the deuterated compound is formed pro-rata with time of contact with CH3OD (compare entries 9 and 4). Under the same conditions, we observed that there is no more incorporation of deuterium when CH3OD was added after 30 min. (results not shown). Therefore, there is no accumulation of an intermediate species responsible for the conversion of the xanthate group into the alkane. Experiments 10, 11 and 12 prove that the incorporation of deuterium does not parallel the ratio H2O/D2O. Even small amounts of H2O led to a significant decrease of deuterium incorporation. This may be explained as the consequence of a significant isotopic effect or of an alternative mechanism in which the O-H(D) bond is not implicated. We showed that the incorporation of deuterium was uppermost in the case of tertiary xanthate 2a (entry13).

Eventually, we performed several experiments using solutions of pure Et3B in deuterated solvents, with or without added H2O or D2O (entries 14–19). When chloroform-d was used as solvent, the percentage of deuterium incorporation is still high (57%, entry 14). By contrast, when the reaction is performed in benzene-d6, no deuterium is incorporated (entry 16). These two experiments indicate that, depending on its ability to transfer hydrogen, the solvent may also be a source of hydrogen. Finally, experiments 6, 15 and 17 show that when D2O is present the incorporation of deuterium is very high and thus transfer of D from D2O surpasses that from the other sources. Experiment 18 corroborates this conclusion: deuterium incorporation from CDCl3 is suppressed when H2O is present. Noteworthy is the observation that when the reaction is performed in THF-d8, no deuterium incorporation occurs (entry 19).

The above experiments shed light on the role of water or alcohols and solvent's in the reduction of S-alkylxanthates and related compounds. Hydrogen transfer from the O-H bond present in water or in an alcohol is not an obvious hypothesis in radical chemistry because of the high BDE of the O-H bond. Our results corroborate Wood's findings concerning reduction of O-alkylxanthates.[5] This author also came to the conclusion that water is the source of hydrogen and proposed two routes. The first one invokes an innovative concept in which water, complexed to a trialkylborane, is the hydrogen atom donor. Gaussian 3G calculations indicate that such a complexation would lower the BDE of the O-H bond by 30 Kcal/mol when compared to free H2O, thus rendering hydrogen abstraction from water a plausible process. The second route implies hydrogen abstraction from •O-O-H radical (O-H BDE 47.6 Kcal/mol). The latter could possibly be formed by interaction of dialkylboron peroxy radical R2BOO• with water and should depend on the amount of dioxygen. It is worth noting that both routes necessitate at least stoechiometric amounts of water. Wood and his colleagues[5] serendipitously observed this anomalous reduction of O-alkylxanthates, working on very small scales (0.006 mmol.). Conspicuously, in further experiments aimed at elucidating the origin of the hydrogen atom, the authors deliberately added high quantities (5 to 20 equiv.) of H2O or D2O.

This mechanism is different from the process involved in the case of α-acyl xanthates where an intermediate enol boronate is a likely intermediate that can be easily protonolysed during work-up. When xanthate 1a was reduced with Et3B/air in methanol-d4, a mixture of compound 1b and its mono-deuterated analogue (55% deuterium incorporation, determined by 1H NMR) was obtained. However, because of a possible deuterium-proton exchange through enolisation, this observation cannot be used to ascertain the mechanism.

The results reported in Table 1 prove that competitive radical mechanisms are operative: when water (or an alcohol) is present, the abstraction of the hydrogen atom obviously takes place by breaking an O-H bond, whereas in the absence of water, this hydrogen transfer could happen by hydrogen abstraction from the solvent (provided that it is a reasonably good hydrogen atom donor). However, despite all our efforts to operate under strictly anhydrous conditions, we observed good yield of hydrogen incorporation in experiments carried out in completely deuterated solvents (Table 1, entries 14 and 16). This suggests that no water is needed and that a third source of hydrogen atom, different from water and solvent, might be involved.

We first focused our attention on O-ethyl-S-ethyl dithiocarbonate as a possible source of hydrogen atom. This by-product is produced by reaction of ethyl radical, generated from Et3B, with the xanthate function. Deuterated compounds 3–5 were easily prepared from ethanol, carbon disulfide and ethylbromide (Figure 2). The results reported in Table 2 clearly show that no deuterium transfer is observed from either compounds 3, 4, or 5 when xanthate 1a was subjected to reduction (Method A) in the presence of one equivalent of O-ethyl-S-ethyl dithiocarbonate, using benzene as solvent to suppress any hydrogen atom transfer from the solvent.

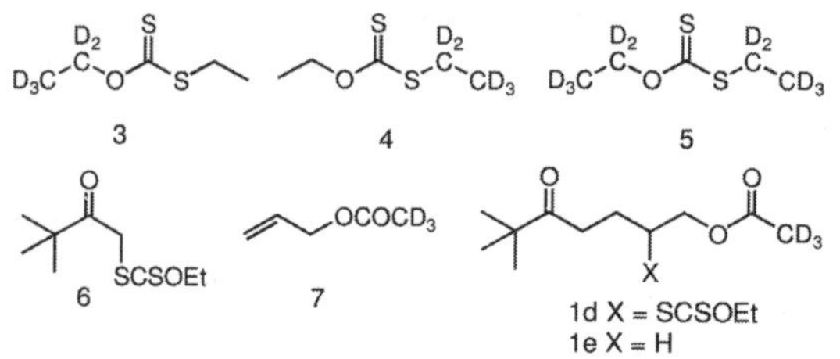

Figure 2. Deuterated O-ethyl-S-ethyl dithiocarbonates 3–5, deuterated adduct 1d and alkane 1e.

Table 2. Reduction of xanthate 1a in the presence of deuterated O-ethyl S-ethyl dithiocarbonates 3–5 in C6H6.

Entry[a]	Xanthate	Additive(1equiv)	Et_3B (eq)[b]	Yield % **1b** + **1c**	Yield % **1a**	D%
1	**1a**	**3**	5	40	55	<1
2	**1a**	**4**	5	36	58	<1
3	**1a**	**5**	5	37	58	<1

[a] Reactions performed in Teflon "glassware" using 0.312 mmol. of xanthate **1a.** [b] Freshly prepared 1 M solutions of pure Et_3B in C_6H_6were used.

Along the same lines, we prepared compound 1d by addition of xanthate 6 to deuterated allylacetate 7 (Figure 2). The reaction of xanthate 1d with Et3B (5 equiv) in benzene (Method A) afforded the corresponding reduced compound 1e. Examination of NMR and mass spectra of compounds 1d and 1e shows that during both the addition and the reduction processes no deuterium loss or scrambling occurred. In particular, a putative 1,5-hydrogen transfer between the transient radical and the CD3 group is clearly ruled out.

Incidentally, we noticed that when the reduction of xanthate 1d (or 1a) with 5 equiv of Et3B was performed in benzene, the reaction was not complete even after 18 h, and much starting material was recovered (see Table 2). We therefore performed a kinetic monitoring of the reaction using a GC technique. The reduction was carried out according to Method A (0.2 M solution of xanthate 1a in a freshly prepared solution of 1 M Et3B in benzene, i.e. 5 equiv). The experimental graph (Figure 3) shows that after 100 min. the formation of the reduced compound 1b slows down rapidly to reach a plateau. Upon addition of a further three equivalents of Et3B (1 M solution in benzene, t = 1700 min.), the reduction started again and rapidly attained another plateau. Plotting of Ln [1a] vs time indicates that, when BEt3 is used in excess (0 < t < 120 min), the reduction follows a pseudo-first order kinetic equation relative to the concentration [1a]. Linear regression calculations gave Ln [1a] = 10.7 + 4.9 10-3 t with a correlation factor of 0.988. GCMS analysis at 1800 min. showed that all the starting material was consumed. The yield of compound 1b then reached 82% (isolated yield). At this time, we cannot explain why the reduction is more sluggish in benzene than in other solvents. Noteworthy, Wood and colleagues [5] used a huge amount of Et3B (20–50 equiv.) when performing the reaction in benzene.

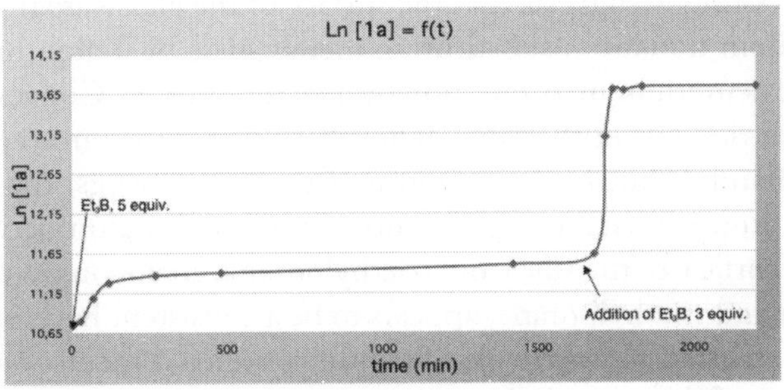

Figure 3. Kinetics of the reduction of compound 1a in C_6H_6 at 20°C.

As put forth above, a hydrogen abstraction from Et3B is an energetically credible hypothesis. To verify this, we undertook the synthesis of Et3B-d15. After testing an improvement of a procedure published by H.C. Brown and co-workers in which Et3B is prepared from ethylbromide, we succeeded in obtaining Et3B-d15 8 in high purity from commercially available ethylbromide-d5 (Scheme 1).[11] The selectively deuterated compounds (CH3CD2)3B 9 and (CD3CH2)3B 10 were prepared in a similar fashion from CH3CD2Br and CD3CH2Br, respectively.

Scheme 1. Preparation of Et3B-d15.

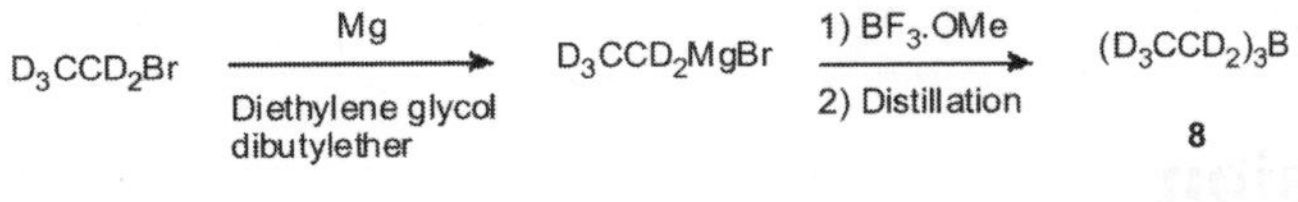

When xanthate 1a was reacted with perdeuterated triethylborane (5 equiv), an 8.3% incorporation of deuterium was observed in the resulting reduced compounds 1b +1c (Table 3, entry 1). This result is perfectly reproducible as shown by comparison of entries 1 and 2. When the reaction was performed in a dried Teflon round bottom flask, the incorporation increased to 26.7%. This result is also quite reproducible (entry 3 vs entry 4). We have shown above that no deuterium transfer was seen from dithiocarbonate 4, a by-product formed during the reduction of compound 1a in the presence of deuterated triethylborane. Therefore, one must assume that the deuterium transfer occurs directly from Et3B-d15 to the transient carbon radical formed by fragmentation of the xanthate function in compound 1a. Interestingly, the deuterium abstraction occurs not only from the methylene group as anticipated, but also from the methyl group (entries 5 and 6). However, the abstraction from the methylene group, after statistical correction, is approximately 5.5 times more efficient than from the methyl group. Examination of Table 3 also strongly suggests that the surface of the glassware might play a role in hydrogen atom donation: substituting merely glass by Teflon brings about an increase of the deuterium incorporation by a factor of three. Conversely, when the reaction was carried out in the same Teflon flask under the same conditions as in experiments 3 and 4, but in the presence of small glass rings, the incorporation of deuterium dropped back to 8.3% (entry 7). Complexation between the O-H groups on the surface of the silica and triethylborane, analogous to that postulated between water and triethylborane, appears to be a consistent hypothesis to explain the hydrogen transfer. However, this hypothesis seems difficult to test directly as the preparation of deuterated glassware is not a straightforward process: the replacement of hydrogen by deuterium on the surface of silica requires very harsh conditions.[12,13]

Table 3. Reduction of xanthate 1a with Et3B-d15, Et3B-d6 and Et3B-d9 in C6H6.

Entry[b]	Xanthate **1a** (mmol.)	Deuterated Et_3B	**1b** + **1c** Yield %	**1a** (recovered) Yield %	D %
1	0.300	Et_3B-d_{15} **8**	38	52	8.3
2	0.300	Et_3B-d_{15} **8**	39	60	8.3
3	0.300[a]	Et_3B-d_{15} **8**	37	61	26.7
4	0.300[a]	Et_3B-d_{15} **8**	38	55	25.9
5	0.300[a]	Et_3B-d_6 **9**[d]	43	48	12.9
6	0.300[a]	Et_3B-d_9 **10**[d]	40	52	3.5
7[c]	0.300[a]	Et_3B-d_{15} **8**	35	57	8.3

[a] The reaction was performed in a dry Teflon 25 mL round bottom flask.[b] Experiments 1–2 were carried out using the same 25 mL round bottom flask. [c] Same reaction conditions as in experiments 3 and 4, except that small glass rings (1 g) were introduced into the flask. All the reactions were performed using 5 equiv of triethylborane. [d] $(CH_3CD_2)_3B$ **9** and $(CD_3CH_2)_3B$ **10** were prepared from commercially available CH_3CD_2Br and CD_3CH_2Br respectively, according to the same procedure as the one described for the synthesis of perdeuterated triethylborane.

Conclusion

Undeniably, the reduction of S-alkyl-thionocarbonates conceals several subtleties. In our quest for the origin of the transferred hydrogen, we demonstrated by unambiguous deuteration experiments that three types of sources are implicated.

The first one relies on hydrogen donation from an O-H group (presumably complexed to the trialkylborane). This O-H group may belong to water, an alcohol, and, in all likelihood, to the surface of the glassware. We showed that, when present, water or an alcohol (methanol) is the preferred hydrogen donor. The second involves a hydrogen transfer from the solvent, as far as it constitutes an acceptable hydrogen donor. The last source is triethylborane itself. The hydrogen abstraction occurs at the α position to boron and also, to a minor extent, at the β position. It is important to note that the reduction gave good yields even under anhydrous conditions (or without alcohol). This allows extensions of this method to water-sensitive compounds to be envisioned. This will be the subject of a forthcoming paper. In fact, the qualitative results reported in this article indicate that the overall reduction process seems to be the result of a delicate balance between three different elementary mechanisms.

Very recently, Newcomb,[14] following Wood's and Renaud's reports, carried out kinetic estimations of the rate of hydrogen abstraction by an alkyl radical from the complex between Et3B and water using the routine radical clock method. This author also came to the conclusion that hydrogen abstraction from the α position of Et3B must be invoked. No spectral or chemical evidence was furnished either to ascertain the formation of a complex between Et3B and water or to demonstrate that hydrogen abstraction from triethylborane is effective. However, it is very gratifying for us to note that this author found the same order of magnitude for the rate of the hydrogen transfer as we estimated in this paper and in the preceding part of this series.[3]

Despite a lack of experimental proof at this time, one may postulate that the reduction of O-alkylxanthates and iodides with triethylborane/air also embodies the same plurality of mechanisms.[3] The reactivity of the combination alkylborane/air/water touches many other domains and certainly deserves further study.

Acknowledgements

This research was supported by grants from the "Institut de Chimie des Substances Naturelles." We are grateful to Prof. J-Y Lallemand for much help and encouragement. Many thanks also to Jean-François Gallard for hetero nuclear NMR experiments.

References

1. Boivin J, Nguyen VT: Part 3 : Beilstein Journal of Organic Chemistry 2007. [http://bjoc.beilstein-journals.org/content/3/1/47] BJOC 2007., 3(47):

2. Barton DHR, Jang DO, Jaszberenyi JC: Tetrahedron Lett. 1990, 31:3991–3994.
3. Boivin J, Nguyen VT: Part 1: Beilstein Journal of Organic Chemistry 2007. [http://bjoc.beilstein-journals.org/content/3/1/45] BJOC 2007., 3(45):
4. Nguyen VT: PhD thesis. University Paris XI, UFR Scientifique d'Orsay 2006 January 19th.
5. Spiegel DA, Wiberg KB, Schacherer LN, Medeiros MR, Wood JL: J Am Chem Soc. 2005, 127:12513–12514.
6. Pozzi D, Scanlan EM, Renaud P: J Am Chem Soc. 2005, 127:14204–14205.
7. Fossey J, Lefort D, Sorba S: Les Radicaux Libres en Chimie Organique. Paris: Masson; 1993.
8. Quiclet-Sire B, Zard SZ: J Am Chem Soc. 1996, 118:9190–9191.
9. Boivin J, Quiclet-Sire B, Ramos L, Zard SZ: J Chem Soc, Chem Commun. 1997, 353–354.
10. Grotewild J, Lissi EA, Scaiano JC: J Chem Soc, B. 1971, 1187–1191.
11. Brown HC, Racherla US: J Org Chem. 1986, 51:427–432.
12. Boccuzzi F, Coluccia S, Ghiotti G, Morterra C, Zecchina A: J Phys Chem. 1978, 82:1298–1303.
13. Tsuchiya I: J Phys Chem. 1982, 86:4107–4112.
14. Jin J, Newcomb M: J Org Chem. 2007, 27:5098–5103.

Part 3. Triethylborane-Air: A Suitable Initiator for Intermolecular Radical Additions of S-2-Oxoalkyl-Thionocarbonates (S-Xanthates) to Olefins

Jean Boivin and Van Tai Nguyen

ABSTRACT

Under carefully controlled conditions, the triethylborane-air combination proves to be an efficient radical initiator that allows intermolecular radical additions of S-2-oxoalkyl-thionocarbonates (S-xanthates) to olefins. Depending on both the structures of the xanthate and the olefin, the addition process can be achieved at room temperature or slightly higher.

Background

Alkylboranes, mainly triethylborane, have became more and more popular as radical initiators because of their ability to generate alkyl radicals by reaction with dioxygen (or air) even at very low temperature (-78°C). [1-4] To the best of our knowledge, only one attempt to use Et3B as a radical initiator at 0°C for the intermolecular addition of an S-alkylxanthate onto 1,1-dimethoxy-2-cyclopropene, has been mentioned in the literature but without success.[5,6] In part 1 of this series, we reported that trialkylboranes are convenient reagents, when used in excess, to reduce S-alkyl-thionocarbonates (S-xanthates), O-alkyl-thionocarbonates (O-xanthates) and related compounds to the corresponding alkanes at room temperature.[7] In the present article, we wish to report that a more comprehensive understanding of the different routes involved permits the premature reduction of the starting 2-oxoalkylxanthate to be avoided. Then, by carefully choosing the modus operandi, the transient α-acyl carbon radical can then be trapped by a suitable olefin, thus offering a mild and efficient method to achieve intermolecular radical additions. In a recent paper, Zard described additions of various S-alkylxanthates to vinyl epoxides and related derivatives using an excess of triethylborane (2 equiv. vs xanthate) at room temperature. The mechanism is different from that reported in this note as the radical chain is maintained by the ring opening of the oxirane that produces an alkoxy radical. The latter reacts rapidly with Et_3B to afford a borinate and ethyl radical.[8]

Results and Discussion

The pivotal experiments at the origin of this article are depicted in Scheme 1. In the first experiment, 2.5 equiv. of Et_3B were added to a mixture of xanthate 1a and 1-decene in dichloromethane under argon at 20°C. The stopper was then removed and air was allowed to enter the flask. After 1 h, purification afforded reduced starting material 1b as the only isolated compound (63%). Such a reactivity was not surprising in view of previous observations [3] and from the literature data. [9-14] Trapping with benzaldehyde gave aldol 1d (48%, Scheme 1) and thus confirmed that a boron enolate is a plausible intermediate in the reduction of compound 1a into 1b. One cannot put aside the possibility that the reduction of the transient α-acyl radical may also occur, to a minor extent, via a direct transfer from a hydrogen donor.[4,15]

Scheme 1. Reactivity of 2-oxoalkylxanthates toward 1-decene in the presence of Et_3B/O_2: competition between addition and reduction.

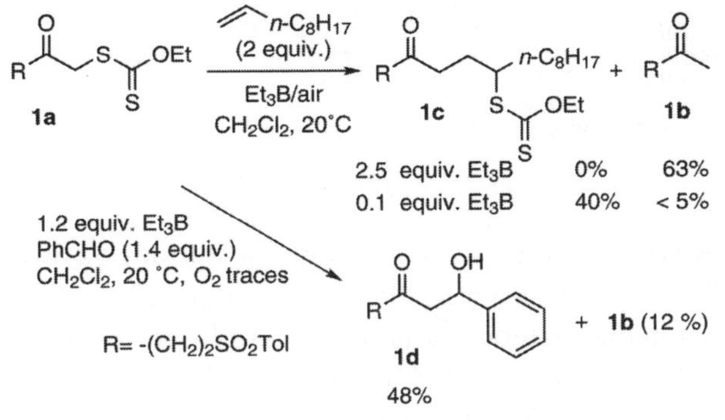

Scheme 1 shows that xanthate 1a undergoes two main types of reactions. The group transfer reaction operates through a radical chain mechanism and affords the adduct 1c.[16] The reduction of compound 1a into 1b results from a bimolecular process in which Et3B is implicated not only in the generation of the α-acyl radical but also in the reaction with the latter, in a stoichiometric manner, to afford an intermediate boron enolate. Lowering the amount of Et3B would therefore minimise the premature unwanted reduction. This hypothesis was then tested. When Et3B (0.1 equiv.) was used in catalytic amounts, adduct 1c was isolated in a modest but remarkable 40% yield (Scheme 1), together with traces of compound 1b (<5%). We reasoned that a slow addition of Et3B would diminish more efficiently the unwanted reduction into 1b. On the other hand, slow addition of air would maintain a low concentration of the radical species and hence minimise the usual unwanted side reactions (dimerisations, abstractions...) that could hamper a clean addition process. Accordingly, as Et3B was added slowly with a syringe pump to a 0.4 M solution of xanthate 1a (0.6 mmol) and 1-decene (2 equiv.) in dichloromethane at 20°C, air was injected (10 mL/h) at the same time in the reaction medium with a second syringe pump. The data reported in Table 1 show that, with this technique, diminishing the rate of addition of Et3B from 0.15 mmol/h to 0.03 mmol/h and increasing simultaneously the total amounts of Et3B (from 0.2 to 0.4 equiv.) resulted in a marked improvement. The yield of adduct 1c increased from 35 to 64%. At the same time, the amount of recovered 1a dropped from 30 to 11% and the reduction into compound 1b was totally suppressed.

Table 1. Addition of xanthate 1a to decene at r.t., catalysed by Et_3B/air

entry	Et_3B mmol/h (equiv.)	Time (h)	Decene (equiv.)	**1c** (%)[a]	**1b** (%)
1	0.15 (0.2)	1.1	2	35 (30)	11
2	0.06 (0.3)	3.15	2.5	47 (21)	6
3	0.03 (0.4)	7.3	2.5	64 (11)	-

[a] in parentheses, percentage of recovered starting material **1a**

Using this procedure, xanthate 1a was added to various olefins (Figures 1 and 2, Table 2). Addition to allylacetate 7 furnished adduct 17a in 51% yield accompanied by some starting material 1a (entry 1). Addition to pinene 8 gave compound 18a (44%) and some reduced adduct 18b (12%, entry 2). Interestingly, addition of xanthate 1a to allylsilane 9 gave adduct 19a in a high yield (71%, entry 3), while addition of xanthate 1a to vinylsilane 10 afforded adduct 20a (57%, entry 4). Reaction of xanthate 1a with allylboronate 11 gave compound 21a in a modest yield (41%, entry 5). Similarly, addition to acrolein diethylacetal 12, at 20°C, led to the desired compound 22a in a low yield (25%), and much starting material 1a (60%) was recovered. A small amount of reduced compound 1b was also isolated (5%, entry 6).

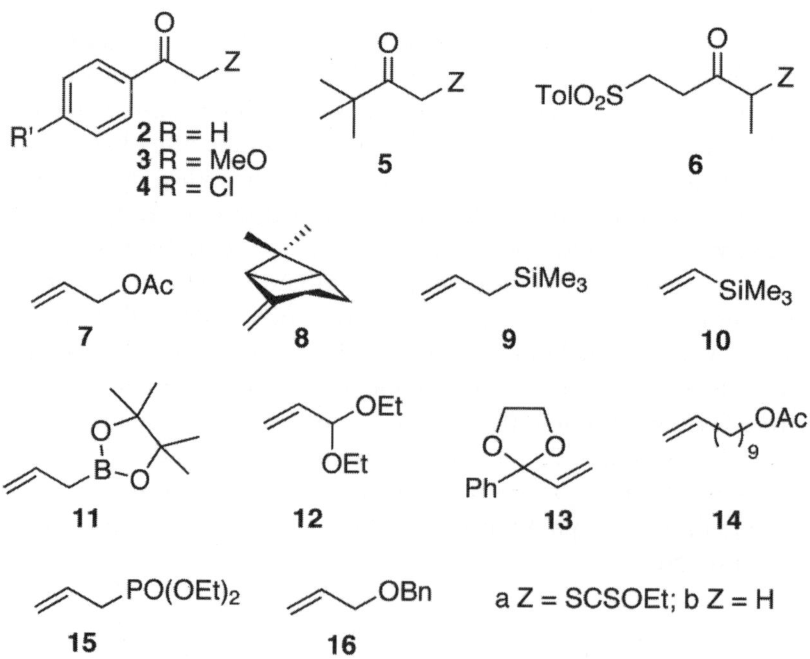

Figure 1. Starting xanthates and olefins.

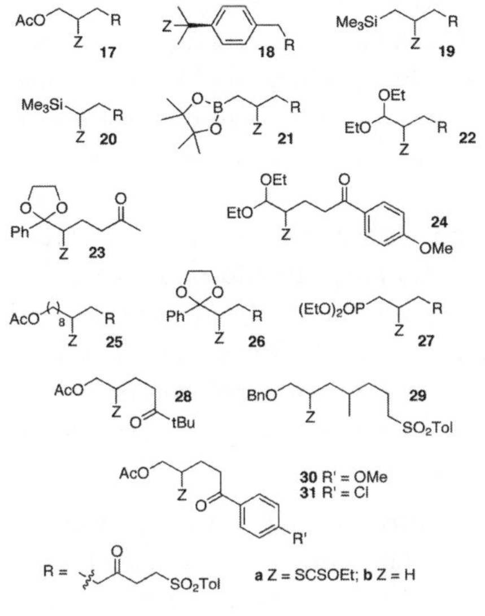

Figure 2. Adducts between xanthates and olefins.

Table 2. Et_3B/air catalysed intermolecular radical additions to olefins

Entry	Xanthate	Olefin (equiv.)	Et_3B (equiv.)	Solvent	T (°C)	Time (h)	Products (yields %)
1	**1a**	**7** (2.5)	0.5	CH_2Cl_2	20	8	**17a** (51); **1a** (20)
2	**1a**	**8** (2.5)	0.5	CH_2Cl_2	20	8	**18a** (44); **18b** (12)
3	**1a**	**9** (2.5)	0.5	CH_2Cl_2	20	8	**19a** (71)
4	**1a**	**10** (2.5)	0.5	CH_2Cl_2	20	8	**20a** (57)
5	**1a**	**11** (2.5)	0.5	CH_2Cl_2	20	8	**21a** (41)
6	**1a**	**12** (2.5)	0.5	CH_2Cl_2	20	8	**22a** (25); **1b** (5); **1a** (60)
7	**2a**	**13** (2.5)	0.5	CH_2Cl_2	20	8	**23a** (0); **2a** (41)
8	**3a**	**12** (2.5)	0.5	CH_2Cl_2	20	8	**24a** (0); **3a** (54); **3b** (12)
9	**2a**	**13** (2)	0.3	CH_2Cl_2	40	8	**23a** (51); **2b** (12)
10	**3a**	**12** (3)	0.5	CH_2Cl_2	40	8	**24a** (47); **3b** (26)
11	**3a**	**7** (2.5)	0.5	CH_2Cl_2	40	8	**30a** (49); **3b** (39)
12	**4a**	**7** (5)	0.4	CH_2Cl_2	40	7	**31a** (45); **4b** (34)
13	**1a**	**12** (2.5)	0.3	CH_2Cl_2	40	6	**22a** (74)
14	**1a**	**14** (2)	0.3	CH_2Cl_2	40	8	**25a** (52); **25b** (20)
15	**1a**	**13** (2)	0.3	CH_2Cl_2	40	8	**26a** (66)
16	**1a**	**9** (2.5)	0.3	CH_2Cl_2	40	4	**19a** (77) (71)[c]
17	**1a**	**10** (2.5)	0.3	CH_2Cl_2	40	4	**20a** (42)
18	**1a**	**11** (2.5)	0.3	CH_2Cl_2	40	4	**21a** (54)
19	**1a**	**15** (2.5)	0.3	CH_2Cl_2	40	4	**27a** (59)
20	**1a**	**10** (2.5x2)	0.3	CH_2Cl_2	40	4	**20a** (73)[d]
21	**5a**	**7** (2)	0.1[b]	$(CH_2Cl)_2$	83	4	**28a** (77)[a]
22	**6a**	**16** (2)	0.4[b]	$(CH_2Cl)_2$	83	14	**29a** (54)
23	**1a**	**7** (2)	0.1[b]	$(CH_2Cl)_2$	83	4	**17a** (78)
24	**1a**	**8** (2)	0.15[b]	$(CH_2Cl)_2$	83	6	**18a** (79)
25	**3a**	**7** (2)	0.1[b]	$(CH_2Cl)_2$	83	4	**30a** (71)
26	**4a**	**7** (2)	0.1[b]	$(CH_2Cl)_2$	83	4	**31a** (73)
27	**3a**	**12** (2)	0.15[b]	$(CH_2Cl)_2$	83	4	**24a** (60)

[a] When the reaction was performed at 40°C, no adduct was formed. The starting material **7a** was recovered. [b] Et_3B (1 M solution in hexanes) was added with a syringe pump (0.03 mmol/h). [c] reaction flask equipped with a condenser opened to air. [d] The second portion of vinyltrimethylsilane (2.5 equiv.) was added after 90 min.

We then turned our attention to the highly delocalised radicals derived from aromatic ketones 2a, 3a, and 4a. As anticipated, this represented one of the worst situations, as premature reduction to methyl ketone should be relatively fast when compared to intermolecular addition to an olefin. Experiments 7 and 8 validated this hypothesis. When xanthate 2a was reacted at 20°C with phenylvinyl dioxolane 13 in the presence of Et3B, no trace of the adduct 23a could be isolated, and only starting material 2a was recovered (Table 2, entry 7). Under the same conditions, xanthate 3a also failed to add to acrolein diethylacetal 12 (entry 8). The only compounds that could be isolated were the starting material 3a (54%) and the reduced product 3b (12%). Obviously, the reaction conditions needed to be adjusted in order to favour the addition process with regard to both premature reduction of the starting material and useless degenerate reaction of the α-acyl carbon radical with its precursor.[8]

We were delighted to observe that gently warming the reaction in refluxing dichloromethane (40°C) totally turned the course of the reaction. Thus, xanthate 2a added to olefin 13 in a fair 51% yield (entry 9). All the starting material was consumed and only 12% of acetophenone 2b were formed. Similarly, xanthate 3a, in the presence of olefin 12, succeeded in giving adduct 24a (47%, entry 10) accompanied by some p-methoxyacetophenone 3b (26%). For these two substrates, comparison between experiments 7–10 showed a striking effect of the temperature on the outcome of the reaction: at 20°C no addition was observed but simply warming the reaction medium to 40°C ensured a clean intermolecular addition process. Under the same conditions, α-phenacyl xanthates 3a and 4a were also reacted with allylacetate (entries 11 and 12). In both cases the corresponding adducts were the major products (30a and 31a, 49% and 45% yield, respectively), accompanied by some reduced starting materials (39% and 34% respectively). The astonishing effect of the temperature increase from 20 to 40°C noticed with the aromatic ketones also held for "normal" ketones, albeit to a less dramatic extent. Thus, xanthate 1a condensed with olefin 12 with a much higher yield (74%, entry 13) than at 20°C (25%, entry 6). Xanthate 1a also reacted with olefins 14 (entry 14) and 13 (entry 15) to afford adducts 25a and 26a in satisfactory 52% and 66% yields, respectively. Similarly, 1a condensed with olefins 9, 10, 11 and 15 to furnish adducts 19a, 20a, 21a, and 27a in 77%, 42%, 54% and 59% yield respectively (entries 16–19). In the case of addition to vinyltrimethylsilane 10, the yield obtained at 40°C, lower than the one observed at 20°C, is clearly due to the volatility of vinyltrimethylsilane, as demonstrated by experiment 20 where the addition of more olefin (2.5 equiv.) during the reaction resulted in a marked increase of the yield (73%). It is important to note that, at 40°C, no trace of prematurely reduced starting material 1b could be isolated.

Nevertheless, some substrates were still refractory. Thus, at 40°C in dichloromethane, xanthate 5a failed to add to allylacetate and was recovered unchanged.

However, when the reaction was performed in refluxing 1,2-dichloroethane (83°C), adduct 28a was isolated in an excellent 77% yield (Table 2, entry 21). Under the same conditions, secondary xanthate 6a reacted cleanly with allyl benzyl ether 16 to give compound 29a in 54% yield (Table 2, entry 22). We reexamined reactions that gave moderate yields at 20 or 40°C. In all cases, the yields were markedly improved (compare entries 1 vs 23, and entries 2 vs 24) even for less reactive aromatic ketones (compare entries 8 vs 10, entries 11 vs 25, and entries 12 vs 26). When the addition was carried out at 83°C, the reaction time was shorter and the amounts of Et3B could be lowered to only 0.10–0.15 equiv. vs the starting xanthate (entries 21, 23, 25, 26, and 27).

From a mechanistic viewpoint, the results reported herein may be rationalised as follows (Scheme 2). The initiation of the process is governed by interaction of dioxygen with Et3B to give Et•. This reaction occurs within a wide range of temperatures. The reaction of ethyl radical with the highly radicophilic species A leads to stabilised radical B. The latter fragments either to xanthate A and Et• or, more easily, to stabilised α-acyl radical C and dithiocarbonate D. From the intermediate radical C, three possible routes determine the outcome of the reaction. Route a1 represents the xanthate group transfer between radical C and any O-ethyl dithiocarbonate (A, D, or E) present in the reaction mixture. The xanthate group transfer (route a2) leads to the formation of C. Routes a1, a2 (and routes a'1 and a'2, see below) constitute a body of fast but useless processes (degenerate reactions)[16] that preserve the radical character but do not let the system evolve. Route c is a relatively slow reaction when compared to the degenerate reactions or to reaction of Et• with A. Routes b and b' depend on the concentration of Et3B that can be controlled by maintaining a low concentration of Et3B. The addition to olefins (route c) is practically irreversible because of the formation of a strong C-C bond. However, the efficiency of route c, compared to routes b, b' and a1, is strongly linked to the structures of both xanthate A and olefin R3-CH=CH2, and can be dramatically modified by varying the reaction temperature. Fortunately, such an increase of reaction temperature enhances route c much more than route b.

Scheme 2. Postulated mechanism for the reaction of 2-oxoalkyl xanthates with olefins in the presence of Et$_3$B.

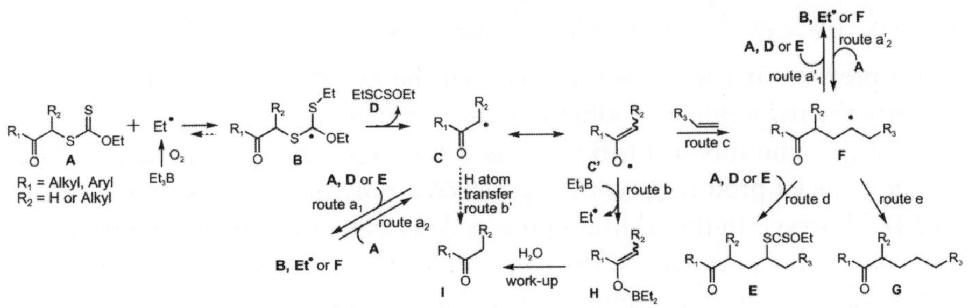

For planar radicals, Rûchardt and Beckwith established that the C-H bond dissociation energy (BDE) for H-CXYZ compounds displays a good linear correlation with the measured α and β-proton ESR hyperfine splitting constants. [17-19] When Y = Me, Z = H, the H-C BDE follows the order for X: CH=CH2 < Ph < PhCO = MeCO < CN < CO2Et < Me (Table 3, entries 1–7) The BDE for compounds where Y = Z = H follows the same order, albeit the value is of course slightly higher when compared to their methylated counterparts (entries 8–11). On the other hand, trialkylboranes react much faster with an oxygen centered radical [20-22] than with a carbon radical.[23,24] Therefore, when R1 = aryl, the highly stabilised and delocalised radical C <–> C' (R2CH•-COAr <–> R2CH=CO•Ar) has a strong propensity to react with Et3B on the oxygen part where a high electron density is located, thus affording the enolboronate H (Scheme 2). Therefore it is not surprising that the rare literature reports of successful intermolecular radical additions of α-oxo carbon radicals are limited to esters[2,3] that correspond to less stabilised radicals more likely to react on the carbon centre.

Table 3. Selected values of H-C BDE for compounds H-CXYZ from references 17–19.

Entry	X	Y	Z	C-H BDE (kcal mol^{-1})
1	Me	Me	H	95.7
2	CO_2Et	Me	H	95.6
3	CN	Me	H	94.9
4	PhCO	Me	H	92.9 (91)[a]
5	EtCO	Me	H	91.2[a]
6	Ph	Me	H	90.3
7	$CH=CH_2$	Me	H	86.1
8	COMe	H	H	97
9	PhCO	H	H	96[a]
10	Ph	H	H	91[a]
11	$CH=CH_2$	H	H	88.8[a]

[a] Determined according to Bordwell's method.

Nevertheless, we have shown in this article that an efficient control of the various reaction parameters (slow additions, temperature) permitted us to elude this problem. For α-oxo carbon radicals derived from aliphatic ketone derivatives, we succeeded in reducing this impediment, and the usual intermolecular addition could take place readily, even at low temperature.

In a previous article, [7] we showed that the Et3B/air combination efficiently promotes the reduction of S-alkylxanthates. It is an apparent paradox that, even when "large" amounts of Et3B were used (i.e. 0.3–0.5 equiv, entries 1–20 and 22), the reduced product G could be detected only in a few instances (entries 2 and 14). However, in the addition process described in this paper, contrary to the reduction method (see ref. 7), the concentration of Et3B is maintained very low by slow addition with a syringe pump, thus minimising route e. Moreover, we demonstrated that the reduction process is relatively slow. As a consequence, the "degenerate" route a'1 is much more efficient than the route e.

Conclusion

We have described a new, efficient, and extremely mild method for performing radical additions of 2-oxoalkylxanthates to various olefins. The efficiency of the addition process vs the premature reduction depends on the reactivity of a particular substrate toward a specific olefin for given reaction conditions. This approach can be extended to cyclisations that should operate even at low temperature.

Acknowledgements

This research was supported by grants from the "Institut de Chimie des Substances Naturelles." We are grateful to Prof. J-Y Lallemand for much help and encouragement.

References

1. Miura K, Ichinose Y, Nozaki K, Fugami K, Oshima K, Utimoto K: Bull Chem Soc Jpn. 1989, 62:143–147.
2. Yorimitsu H, Oshima K: Radicals in Organic Synthesis. Volume 1. Edited by: Renaud P, Sibi MP. Weinheim, Germany: Wiley-VCH; 2001:11–27.
3. Ollivier C, Renaud P: Chem Rev. 2001, 101:3415–3434.
4. Li Liu, Xing Wang, Chaozhong Li: Org Lett. 2003, 5:361–364.
5. Legrand N, Quiclet-Sire B, Zard SZ: Tetrahedron Lett. 2000, 41:9815–9818.
6. Legrand N: PhD thesis. École Polytechnique; 2001.
7. Boivin J, Nguyen VT: Part 2 : Beilstein Journal of Organic Chemistry 2007. [http://bjoc.beilstein-journals.org/content/3/1/46] BJOC 2007., 3(46).
8. Charrier N, Gravestock D, Zard SZ: Angew Chem, Int Ed. 2006, 45:6520–6523.
9. Nozaki N, Oshima K, Utimoto K: Tetrahedron Lett. 1988, 29:1041–1044.
10. Nozaki N, Oshima K, Utimoto K: Bull Chem Soc Jpn. 1991, 64:403–409.
11. Horiuchi Y, Taniguchi M, Oshima K, Utimoto K: Tetrahedron Lett. 1995, 36:5353–5356.

12. Kabalka GW, Brown HC, Suzuki A, Honma S, Arase A, Itoh M: J Am Chem Soc. 1970, 92:710–712.
13. Beraud V, Gnanou Y, Walton JC, Maillard B: Tetrahedron Lett. 2000, 41:1195–1198.
14. Olivier C, Renaud P: Chem Eur J. 1999, 5:1468–1473.
15. Spiegel DA, Wiberg KB, Schacherer LN, Medeiros MR, Wood JL: J Am Chem Soc. 2005, 127:12513–12514.
16. Zard SZ: Radicals in Organic Synthesis. Volume 1. Edited by: Renaud P, Sibi MP. Weinheim, Germany: Wiley-VCH; 2001:90–108.
17. Brocks JJ, Beckhaus H-D, Beckwith ALJ, Rûchardt C: J Org Chem. 1998, 63:1935–1943.
18. Viehe HG, Janousek Z, Merényi H: Substituent Effects in Radical Chemistry. Reidel D Publishing Co.: New York; 1986.
19. Welle FM, Beckhaus H-D, Rüchardt C: J Org Chem. 1997, 62:552–558.
20. Davies AG, Roberts BP: Free Radicals. Volume 1. Edited by: Kochi JK. Wiley: New York; 1973:547.
21. Krusic PJ, Kochi JK: J Am Chem Soc. 1969, 91:3942–3944.
22. Davies AG, Roberts BP, Scaiano JC: J Chem Soc B. 1971, 2171–2176.
23. Grotewold J, Lissi EA: J Chem Soc, Chem Commun. 1965, 21–22.
24. Grotewold J, Lissi EA, Scaiano JC: J Chem Soc B. 1971, 1187–1191.

Catalytic Oxidative Cleavage of C=N Bond in the Presence of Zeolite H-NaX Supported $Cu(NO_3)_2$, as a Green Reagent

A. Lalitha, K. Sivakumar, K. Parameswaran, K. Pitchumani and C. Srinivasan

ABSTRACT

Copper(II) nitrate supported on faujasite zeolites such as H-NaX is employed as solid acid catalysts for the clean and less hazardous catalytic oxidative cleavage of C=N bond under mild conditions. The reactions proceed very smoothly, and the yields are excellent.

Introduction

Increasing awareness of environmental hazards forces chemists to look for "eco-friendly" reaction conditions. In this connection, the use of heterogeneous catalysts involving solid reagents supported on high surface area materials, obtained by introduction of the reagent onto or into an organic polymeric or an inorganic porous or layered support material meets the fundamental challenges in the protection of the environment. These supported reagents have advantages such as easy handling, good dispersion of active sites leading to improved reactivity, safer and milder reaction conditions and minimal pollution [1, 2]. Oximes, hydrazones, and semicarbazones are useful as preferred derivatives for the identification and characterization of carbonyl compounds [3]. Several reagents have been reported for the regeneration of the carbonyl groups from the mentioned derivatives [4–9]. Although some of the methods involve mild reaction conditions, most of them require strong acidic media, long reaction time, a strong oxidizing agent (which causes over oxidation), and expensive and not readily available reagents.

Anhydrous metallic nitrates with a bidentate covalent coordination and with the available lower intermediate oxidation state for the metal, find maximum reactivity, and wider applicability [10, 11]. Their use, however, in organic syntheses is limited by solubility problems.

Laszlo et al. have used the K10-montmorillonite clay-supported copper(II) nitrate (claycop) and iron(III) nitrate (clayfen) for many of the organic reactions like oxidation of alcohols, oxidative coupling of thiols, hydrolytic cleavage of imine derivatives of carbonyl compounds, cleavage of tosylhydrazones, phenyl hydrazones, 2,4-dinitrophenylhydrazones and semicarbazones [12]. K10-montmorillonite supported Thallium nitrate is used for the oxidative rearrangement of alkyl aryl ketones into alkyl aryl carboxylates [13]. Pyridinium chlorochromate [14], potassium dichromate [15] and sodium metaperiodate [16] are some alumina-supported oxidants used for the chemoselective oxidation of alcohols and sulfides. In this paper we report a general method for the oxidative cleavage of C=N to their carbonyl derivatives in excellent yields without using any microwave or ultrasonic irradiation. To the best of our knowledge, this is the first report using a zeolite-supported cupric nitrate as the reagent for the regeneration of carbonyl group from oximes, hydrazones and semicarbazones. Moreover, copper-salts are inexpensive, easy to handle and environmentally friendly.

Experimental

Preparation of Zeolite H-Nax Supported Copper(II) Nitrate

To a solution of $Cu(NO_3)_2{\cdot}3H_2O$ (0.483 g, 2 mmoL) in acetone (15 mL), activated zeolite H-NaX (1 g) (obtained by activating NH_4-NaX form of zeolite which has been prepared by partial exchange of sodium with ammonium salt in commercially available NaX) was added at once with stirring over a magnetic stirrer for 2 hours. Then the solvent was removed in a rotary evaporator. The blue powder formed was dried further at 130°C under reduced pressure. 1 g of H-NaX zeolite supported copper(II) nitrate reagent contains about 0.326 g of Cu(NO3)2 (1.35 mmoL).

General Procedure for the Oxidative Cleavage of C=N

0.05 g of substrate was ground with 0.25 g of activated supported reagent using mortar and pestle and refluxed using 10 mL of dichloromethane as solvent for specified time. The reaction mass was cooled to room temperature, filtered the catalyst and washed with dichloromethane twice. The filtrate was washed with distilled water thrice. After drying over anhydrous sodium sulphate, the solvent was evaporated to give the product. Percentage conversion of deprotection was checked by GC analysis.

Results and Discussions

The results illustrated in Table 1 indicate that the reaction is successful for a variety of aliphatic and aromatic oximes, phenylhydrazones, p-nitrophenylhydrazones and semicarbazones (Scheme 1). All these carbonyl derivatives were converted back to their corresponding aldehydes and ketones in dichloromethane as the optimal solvent, among the tested solvents including: methanol, ethanol, and acetonitrile taking benzaldehyde oxime as a representative example where the yields are found to be 79%, 81%, 72%, respectively. The reaction is found to be general with compounds having variety of functional groups like chloro, nitro, phenolic hydroxyl, and methoxy groups.

Table 1. Conversion of oximes, phenyl hydrazones, 2,4-dinitrophenylhydrazones and semicarbazones to corresponding carbonyl compounds.

Entry	Substrate	Product	Time (h)	Yield %
1	Benzaldehyde oxime	Benzaldehyde	0.5	92
2	2-chlorobenzaldehyde oxime	2-chlorobenzaldehyde	0.25	78
3	4-chlorobenzaldehyde oxime	4-chlorobenzaldehyde	0.25	89
4	4-nitrobenzaldehyde oxime	4-nitrobenzaldehyde	0.25	88
5	Furfural oxime	Furfural	0.5	88
6	2-hydroxy 4-methoxy benzaldehyde oxime	2-hydroxy4-methoxy benzaldehyde	0.5	87
7	Acetophenone oxime	Acetophenone	1	89
8	Benzophenone oxime	Benzophenone	1	89
9	4-hydroxyacetophenone oxime	4-hydroxyacetophenone	1	79
10	Cyclohexanone oxime	Cyclohexanone	0.5	90
11	Benzaldehyde phenylhydrazone	Benzaldehyde	1	66
12	Acetophenone phenylhydrazone	Acetophenone	2	79
13	Benzophenone phenylhydrazone	Benzophenone	2	69
14	Benzaldehydes 2,4-dinitrophenylhydrazone	Benzaldehyde	1	65
15	Acetophenone 2,4-dinitrophenylhydrazone	Acetophenone	3	86
16	Benzophenone 2,4-dinitrophenylhydrazone	Benzophenone	3	79
17	Furfural 2,4-dinitrophenylhydrazone	Furfural	1.5	75
18	Cyclohexanone 2,4-dinitrophenylhydrazone	Cyclohexanone	2	87
19	4-chlorobenzalde 2,4-dinitrophenylhydrazone	4-chlorobenzalde	3	66
20	2-chlorobenzalde 2,4-dinitrophenylhydrazone	2-chlorobenzaldehyde	2	78
21	2-hydroxy -4-methoxy benzaldehyde 2,4-dinitrophenylhydrazone	2-hydroxy -4-methoxy benzaldehyde	0.5	69
22	2-hydroxy -4-methoxy benzaldehyde semicarbazone	2-hydroxy -4-methoxy benzaldehyde	1	78
23	Benzaldehyde semicarbazone	Benzaldehyde	2	91
24	Acetophenone semicarbazone	Acetophenone	2	78
25	Benzophenone semicarbazone	Benzophenone	2	76

Table 2. The effect of various solid supports on the oxidative cleavage of benzaldehyde semicarbazone (a mixture of benzaldehyde semicarbazide (0.05 g) and supported reagent (0.25 g) was refluxed in dichloromethane for 2 hour).

Run	Solid acid support	Yield of Benzaldehyde %
1	H-NaX	91
2	NaX	20
3	NaY	24
4	ZSM-5	34
5	MCM-41	12
6	SiO_2	14
7	None	—

Scheme 1. Catalytic oxidative cleavage of C=N bond in the presence of Zeolite H-NaX supported $Cu(NO_3)_2$.

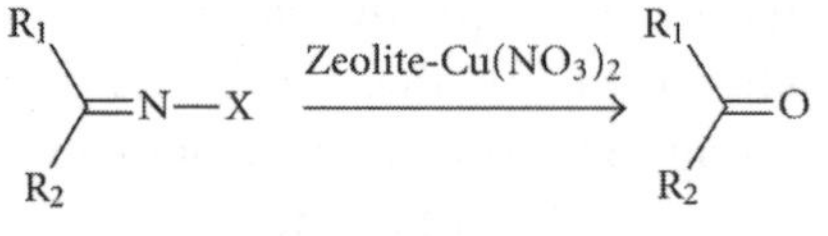

R_1, R_2 = alkyl, aryl or H
X = OH, NH-Ph, $NHCONH_2$

As evident from the results, aldehyde derivatives were generally deprotected relatively faster than keto derivatives. It was also interesting to note that by controlling the amounts of the reagent, it was possible to avoid further oxidation of the liberated aldehydes to the corresponding carboxylic acids (entries 1–6), while we have demonstrated a facile aromatic nitration reaction with cupric nitrate in the presence of solid support [17], we did not observe any nitration of aromatic substrates during the cleavage reaction when we used optimum ratio of substrates and oxidants. Prolonged reaction time as well as excess of supported reagent in the case of aldehydes leads to further oxidation to the respective acids. The reaction failed to produce the ketones without cupric nitrate or H-NaX zeolite.

The superiority of H-NaX supported cupric nitrate as catalyst has been proved by the inefficiency in the regeneration of carbonyl groups from benzaldehyde semicarbazones by simple NaX supported Cu(NO3)2 or with unsupported Cu(NO3)2, revealing the involvement of acidic sites present in the solid support. The optimum ratio of substrate to oxidant (1:5) was determined for complete conversion of oximes, semicarbazones and phenyl hydrazones to the corresponding carbonyl compounds. The recovered catalyst was verified for three times to catalyze the deprotection of benzaldehyde oxime. The efficiency of catalyst decreases considerably during the successive reusability tests.

Regarding the mechanism of the deprotection, it is proposed that the diffusion of the cupric ions into the zeolite H-NaX surface lattices and the subsequent formation of N2O4 finally lead to the generation of NO3- and NO+ ions. The presence of these ions has been confirmed by comparison of IR spectra of Cu(NO3)2 and Cu(NO3)2-loaded zeolite, which shows strong peaks around 1030–1070 cm-1 due to symmetric stretching of nitrate ion, 1374 cm-1 due to asymmetric stretching of nitrate ion, 810 cm-1 due to in-plane deformation of nitrate ion and 710 cm-1 due to out of-plane deformation of nitrate ion, and peak around 2332–2415 cm-1 due to NO+. The nitrosonium ion may act as an electrophile, giving the corresponding carbonyl compounds. Based on the observations, the proposed mechanism for cleavage of the carbon–nitrogen double bond in oximes is presented in Scheme 2.

Scheme 2. Proposed mechanistic pathway.

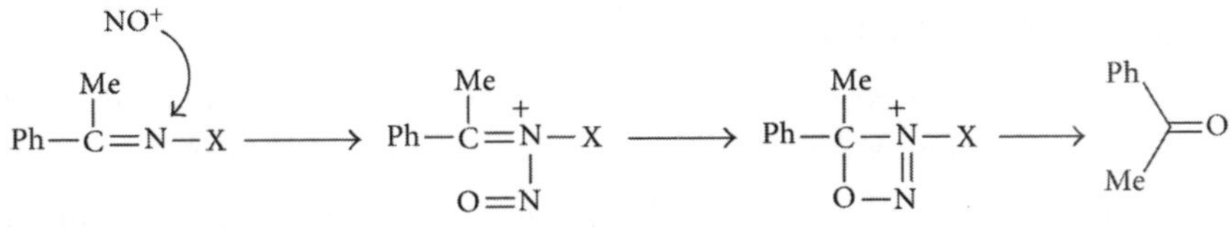

X = OH, NHC_6H_5, 2, 4-$(NO_2)_2C_6H_5NH$, $NHCONH_2$

Conclusions

In conclusion, from commercially available NaX zeolite, a facile heterogeneous catalytic method, involving the more acidic form, namely, H-NaX with cupric nitrate has been employed for oxidative cleavage of C=N for the regeneration of carbonyl compound. It will be obvious that advantages of heterogeneous catalysis in terms of easy separation; and consistent yields are noteworthy. The operational simplicity, selectivity and cheapness, and good yields in very short times make this procedure a useful, attractive alternative to previously available methods.

Acknowledgement

K. Sivakumar thanks Jawaharlal Nehru Memorial Fund (JNMF) for the award of fellowship.

References

1. P. Lazslo, Preparative Chemistry Using Supported Reagents, Academic Press, San Diego, Calif, USA, 1987.
2. J. H. Clark, A. P. Kybett, and D. J. Macquarrie, Supported Reagents: Preparation, Analysis and Applications, VCH, New York, NY, USA, 1992.
3. N. D. Cheronis and J. B. Entrikin, Identification of Organic Compounds, Interscience, New York, NY, USA, 1963.
4. M. M. Heravi, L. Ranjbar, F. Derikvand, H. A. Oskooie, and F. F. Bamoharram, "Catalytic oxidative cleavage of C=N bond in the presence of mixed-addenda vanadomolybdophosphate, H6PMo9V3O40 as a green and reusable catalyst," Journal of Molecular Catalysis A, vol. 265, no. 1-2, pp. 186–188, 2007.
5. S. B. Shim, K. Kim, and Y. H. Kim, "Direct conversion of oximes and hydrazones into their ketones with dinitrogen tetroxide," Tetrahedron Letters, vol. 28, no. 6, pp. 645–648, 1987.
6. B. P. Bandgar, L. B. Kunde, and J. L. Thote, "Deoximation with N-haloamides," Synthetic Communications, vol. 27, no. 7, pp. 1149–1152, 1997.
7. M. Giurg and J. Młochowski, "Regeneration of carbonyl compounds from azines with cerium(IV) ammonium nitrate," Synthetic Communications, vol. 29, no. 24, pp. 4307–4313, 1999.
8. M. M. Heravi, D. Ajami, M. Tajbakhsh, and M. Ghassemzadeh, "Clay supported bis-(trimethylsilyl)-chromate: an efficient reagent for oxidative deoximation," Monatshefte für Chemie, vol. 131, no. 10, pp. 1109–1113, 2000.
9. M. M. Heravi, D. Ajami, and M. M. Mojtahedi, "Regeneration of carbonyl compounds from oximes on clayfen under conventional heating and microwave irradiation," Journal of Chemical Research, vol. 2000, no. 3, pp. 126–127, 2000.
10. C. C. Addison, N. Logan, S. C. Wallwork, and C. D. Garner, "Structural aspects of co-ordinated nitrate groups," Quarterly Reviews, Chemical Society, vol. 25, no. 2, pp. 289–322, 1971.
11. C. C. Addison, "The relation between chemical reactivity of ligands and the nature of the metal-ligand bond: nitrato-complexes," Coordination Chemistry Reviews, vol. 1, no. 1-2, pp. 58–65, 1966.
12. A. Cornelis and P. Laszlo, "Clay-supported copper(II) and iron(III) nitrates: novel multi-purpose reagents for organic synthesis," Synthesis, vol. 10, p. 909, 1985.

13. E. C. Taylor, C.-S. Chiang, A. McKillop, and J. F. White, "Oxidative rearrangements via oxythallation with thallium(III) nitrate supported on clay," Journal of the American Chemical Society, vol. 98, no. 21, pp. 6750–6752, 1976.

14. D. Savoia, C. Trombini, and A. Umani-Ronchi, "Synthesis of 2-(6-carboxyhexyl)cyclopent-2-en-1-one, an intermediate in prostaglandin synthesis," Journal of Organic Chemistry, vol. 47, no. 3, pp. 564–566, 1982.

15. J. H. Clark, A. P. Kybett, P. Landon, D. J. Macquarrie, and K. Martin, "Catalytic Oxidation of Organic Substrates using Alumina Supported Chromium and Manganese," Journal of Chemical Society., Chemical Communication, p. 1355, 1989.

16. G. W. Kabalka and M. Richard, "Organic reactions on alumina," Tetrahedron, vol. 53, pp. 7999–8065, 1997.

17. A. Lalitha and K. Sivakumar, "Zeolite H-Y-supported copper(II) nitrate: a simple and effective solid-supported reagent for nitration of phenols and their derivatives," Synthetic Communications, vol. 38, no. 11, pp. 1745–1752, 2008.

A Divergent Asymmetric Approach to Aza-Spiropyran Derivative and (1S,8ar)-1-Hydroxyindolizidine

Jian-Feng Zheng, Wen Chen, Su-Yu Huang, Jian-Liang Ye and Pei-Qiang Huang

ABSTRACT

Background

Spiroketals and the corresponding aza-spiroketals are the structural features found in a number of bioactive natural products, and in compounds possessing photochromic properties for use in the area of photochemical erasable memory, self-development photography, actinometry, displays, filters, lenses of variable optical density, and photomechanical biomaterials etc. And (1R,8aS)-1-hydroxyindolizidine (3) has been postulated to be a biosynthetic precursor of hydroxylated indolizidines such as (+)-lentiginosine 1,

(-)-2-epilentiginosine 2 and (-)-swainsonine, which are potentially useful antimetastasis drugs for the treatment of cancer. In continuation of a project aimed at the development of enantiomeric malimide-based synthetic methodology, we now report a divergent, concise and highly diastereoselective approach for the asymmetric syntheses of an aza-spiropyran derivative 7 and (1S,8aR)-1-hydroxyindolizidine (ent-3).

Results

The synthesis of aza-spiropyran 7 started from the Grignard addition of malimide 4. Treatment of the THP-protected 4-hydroxybutyl magnesium bromide with malimide 4 at -20°C afforded N,O-acetal 5a as an epimeric mixture in a combined yield of 89%. Subjection of the diastereomeric mixture of N,O-acetal 5a to acidic conditions for 0.5 h resulted in the formation of the desired functionalized aza-spiropyran 7 as a single diastereomer in quantitative yield. The stereochemistry of the aza-spiropyran 7 was determined by NOESY experiment. For the synthesis of ent-3, aza-spiropyran 7, or more conveniently, N,O-acetal 5a, was converted to lactam 6a under standard reductive dehydroxylation conditions in 78% or 77% yield. Reduction of lactam 6a with borane-dimethylsulfide provided pyrrolidine 8 in 95% yield. Compound 8 was then converted to 1-hydroxyindolizidine ent-3 via a four-step procedure, namely, N-debenzylation/O-mesylation/Boc-cleavage/cyclization, and O-debenzylation. Alternatively, amino alcohol 8 was mesylated and the resultant mesylate 12 was subjected to hydrogenolytic conditions, which gave (1S,8aR)-1-hydroxyindolizidine (ent-3) in 60% overall yield from 8.

Conclusion

By the reaction of functionalized Grignard reagent with protected (S)-malimide, either aza-spiropyran or (1S,8aR)-1-hydroxyindolizidine skeleton could be constructed in a concise and selective manner. The results presented herein constitute an important extension of our malimide-based synthetic methodology.

Background

Spiroketals of general structure A (Scheme 1) constitute key structural features of a number of bioactive natural products isolated from insects, microbes, fungi, plants or marine organisms. [1-3] The corresponding aza-spiroketal (cf: general structure B) containing natural products, while less common, are also found in plants, shellfish and microbes.[4,5] For example, pandamarilactone-1 and pandamarine were isolated from the leaves of Pandanus amaryllifolius;[6] solasodine and its derivatives were isolated from Solanum umbelliferum, which exhibited significant activity toward DNA repair-deficient yeast mutants;[7] azaspiracids are marine phycotoxins isolated from cultivated mussels in Killary harbor, Ireland;[8] and chlorofusin A is a novel fungal metabolite showing the potential as a lead in cancer therapy.[9] In addition, aza-spiropyrans C, being able to equilibrate with the corresponding non-spiro analogue D, is a well known class of compounds possessing photochromic properties for use in the area of photochemical erasable memory,[10] and also found applications as self-development photography, actinometry, displays, filters, lenses of variable optical density,[11] and photomechanical biomaterials etc.[12]

Scheme 1. The skeletons of useful aza-spiroketals and some naturally occurring hydroxylated indolizidines.

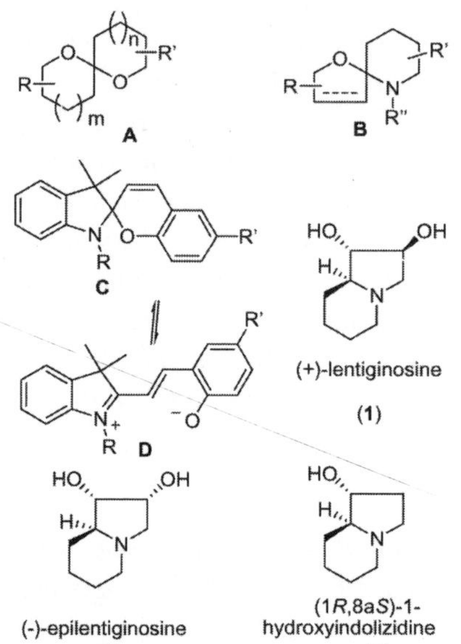

On the other hand, hydroxylated indolizidines [13-20] such as castanospermine, (-)-swainsonine, (+)-lentiginosine [21-23] (1) and (-)-2-epilentiginosine [21-26] (2) constitute a class of azasugars showing potent and selective glycosidase inhibitory activities. [13-20] (1R,8aS)-1-Hydroxyindolizidine 3 has been postulated as a biosynthetic precursor [21-26] of (+)-lentiginosine (1), (-)-2-epilentiginosine (2) and (-)-swainsonine, a potentially useful antimetastasis drug for the treatment of cancer.[15] In addition, these molecules serve as platforms for testing synthetic strategies, and several asymmetric syntheses of both enantiomers of 1-hydroxyindolizidine (3) have been reported. [27-34]In continuation of our efforts in the development of enantiomeric malimide-based synthetic methodologies, [35-38] we now report concise and highly diastereoselective syntheses of an aza-spiropyran derivative 7 and (1S,8aR)-1-hydroxyindolizidine (ent-3).

Results and Discussion

Previously, we have shown that the addition of Grignard reagents to N,O-dibenzyl malimide 4 leads to N,O-acetals 5 in high regioselectivity (Scheme 2), and the subsequent reductive dehydroxylation gives 6 in high trans-diastereoselectivity. [35] On the other hand, treatment of N,O-acteals 5 with an acid furnished enamides E, which can be transformed stereoselectively to either hydroxylactams F or G under appropriate conditions. [36-38] It was envisioned that if a C_4-bifunctional Grignard reagent was used, both aza-spiroketal H (such as aza-spiropyran, n = 1, path a) and indolizidine ring systems I (path b) could be obtained.

Scheme 2. Synthetic strategy based on N,O-dibenzylmalimide (4).

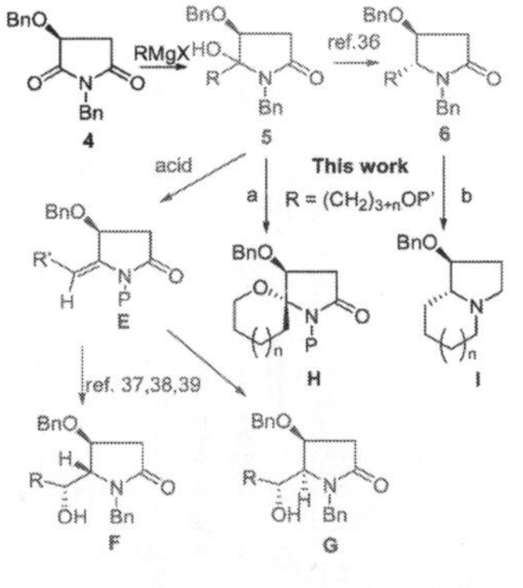

The synthesis of aza-spiropyran 7 started from the Grignard addition of malimide 4. Treatment of the THP-protected 4-hydroxybutyl magnesium bromide with malimide 4 at -20°C for 2.5 h afforded N,O-acetal 5a as an epimeric mixture in 7:1 ratio and with a combined yield of 89% (Scheme 3). If the reaction was allowed to stir at room temperature overnight, the diastereomeric ratio was inversed to 1: 1.8. Subjection of the diastereomeric mixture of the N,O-acetal 5a to acidic conditions [TsOH (cat.)/CH2Cl2, r.t.] for 0.5 h resulted in the formation of the desired functionalized aza-spiropyran derivative 7 as a single diastereomer in quantitative yield. The result means that a tandem dehydration-THP cleavage-intramolecular nucleophilic addition occurred. When the stirring was prolonged to 2 h, about 5% of another epimer (no shown) was also formed according to the 1H NMR analysis.

Scheme 3. Stereoselectivity synthesis of aza-spiropyran 7.

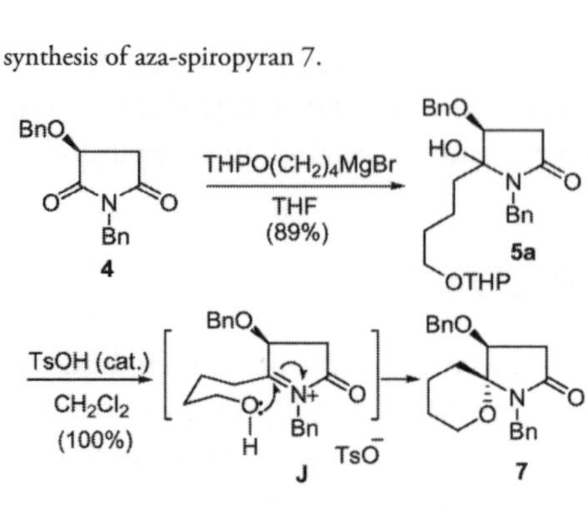

The stereochemistry of the aza-spiropyran 7 was determined on the basis of the NMR analysis. This was done firstly by a 1H-1H COSY experiment to confirm the proton assignments, and then by NOESY experiment. As shown in Figure 1, the strong NOE correlation of H-9a (δH 3.59) and H-4 (δH 4.22) indicates clearly O4/O5-trans relationship in compound 7.

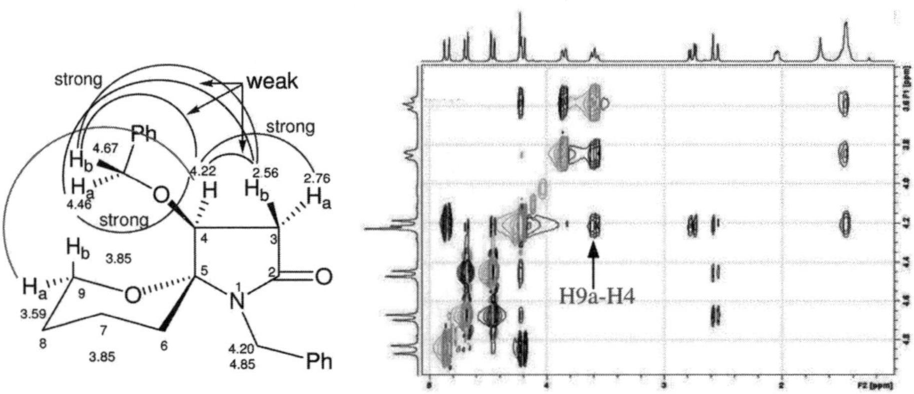

Figure 1. The observed NOE correlations (in part) and the region expanded NOESY spectrum of compound 7.

These findings are surprising comparing with our recent observations. In our previous investigations, it was observed that the treatment of N,O-acetals 5 with an acid leads to the dehydration products E (Scheme 1), and the two diastereomers of 5 shows different reactivities towards the acid-promoted dehydration. [36-38] The trans-diastereomer reacts much more slower than the cis-diastereomer, and some un-reacted trans-epimer was always recovered even starting with a pure cis-diastereomer. In the present study, not only both two diastereomers have been completely converted to the aza-spiropyran 7, what is equally surprising is that no dehydration product was observed under acidic conditions!

For the synthesis of ent-3, aza-spiropyran 7, a cyclic N,O-acetal, was converted to lactam 6a under standard reductive dehydroxylation conditions (Et3SiH, BF3·OEt2, -78°C, 6 h; warm-up, yield: 78%) (Scheme 4). Under the same conditions, N,O-acetal 5a was converted to lactam 6a in 77% yield. It was observed that during the reaction of 5a, 7 was first formed as an intermediate after the addition of Et3SiH and BF3·OEt2, and stirring for 1 hour.

Scheme 4. Stereoselective synthesis of (1S,8aR)-1-hydroxyindolizidine (ent-3).

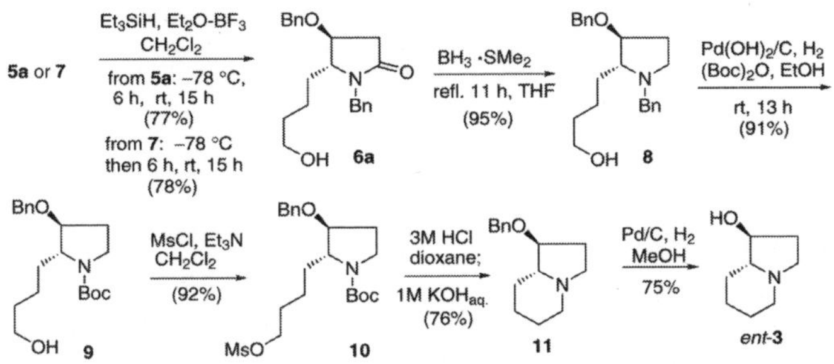

Reduction of lactam 6a with borane-dimethylsulfide provided pyrrolidine derivative 8 in 95% yield. Compound 8 was then converted to (1S,8aR)-1-hydroxyindolizidine (ent-3) {[α]D27 +50 (c 0.90, EtOH); lit.[29] [α]D +51.0 (c 0.54, EtOH); lit.[32] -49.7 (c 0.95, EtOH) for the antipode} via a four-step procedure, namely, one-pot N-debenzylation-N-Boc formation/O-mesylation/Boc-cleavage/cyclication, and O-debenzylation.

In searching for a more concise method, amino alcohol 8 was mesylated (MsCl, NEt3, 0°C) and the resultant labile mesylate 12 was subjected to catalytic hydrogenolysis (H2, l atm, 10% Pd/C, r.t.), which gave (1S,8aR)-1-hydroxyindolizidine (ent-3) in 60% overall yield from 8 (Scheme 5).[39,40] The one-pot N,O-bis-debenzylation and cyclization of mesylate 12 deserves comment. Because the N-debenzylation generally required longer reaction time,[41] or using of Pearlman's catalyst (cf. Scheme 4). The easy debenzylation of 12 allows assuming that an intramolecular substitution occurred firstly, and the formation of the quaternary ammonium salt K [40] then favors the reductive debenzylation. This mechanism is supported by the following observations. First, in a similar case, Thompson et al. observed that the formation of a mesylate resulted in spontaneous quarternization leading to the bicyclic indolizidine.[40] Second, we have also observed that the tosylate of 8 is too labile to be isolated, and mesylate 12 decomposed upon flash column chromatography on silica gel, which are due to the spontaneous formation of a polar quaternary ammonium salt. In addition, the presence of the

O-benzyl group in K is an assumption based on our previous observation on a similar case.[42]

Scheme 5. One-pot synthesis of ent-3 from amino alcohol 8.

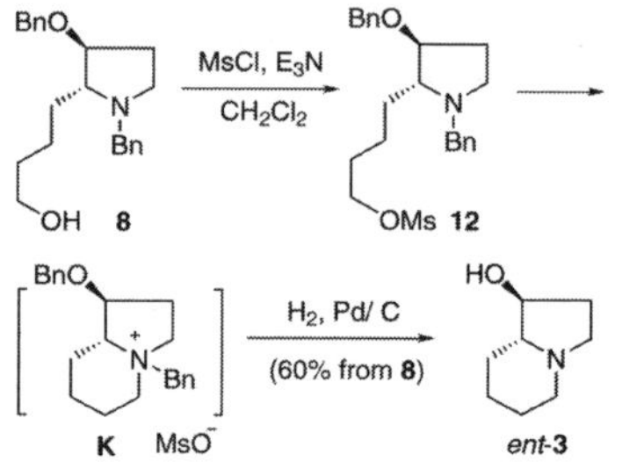

Conclusion

In summary, we have demonstrated that by the reaction of functionalized Grignard reagent with the protected (S)-malimide 4, either aza-spiropyran derivative 7 or (1S,8aR)-1-hydroxyindolizidine skeleton (ent-3) can be constructed in a concise and selective manner. It is worthy of mention that in addition to the reductive dehydroxylation leading to 2-pyrrolidinones 6, and the acid-promoted dehydration leading to (E)-enamides E (and then F, G), acid treatment of the N,O-acetal 5a could provide, chemoselectively and quantitatively, the aza-spiropyran ring system 7. The results presented herein constitute a valuable extension of our malimides-based synthetic methodology.

Acknowledgements

The authors are grateful to the NSFC (20572088), NSF of Fujian Province and Xiamen City (2006J0268; 3502z20055019) and the program for Innovative Research Team in Science & Technology (University) in Fujian Province for financial support. We thank Professor Y. F. Zhao for the use of her Bruker Dalton Esquire 3000 plus LC-MS apparatus.

References

1. Perron F, Albizati KM: Chem Rev. 1989, 89:1617–1661.
2. Boivin TLB: Tetrahedron. 1987, 43:3309–3362.
3. Brimble MA, Farès FA: Tetrahedron. 1999, 55:7661–7706.

4. Nonato MG, Garson MJ, Truscott RJW, Carver JA: Phytochemistry. 1993, 34:1159–1163.
5. Byrne LT, Guevara BQ, Patalinghug WC, Recio BV, Ualat CR, White AH: Aust J Chem. 1992, 45:1903–1908.
6. Pradhan R, Patra M, Behera AK, Mishra BK, Behera RK: Tetrahedron. 2006, 62:779–828.
7. Kim YC, Che QM, Gunatilake AAL, Kingston DGI: J Nat Prod. 1996, 59:283–285.
8. Satake M, Ofuji K, Naoki H, James KJ, Furey A, McMahon T, Silke J, Yasumoto T: J Am Chem Soc. 1998, 120:9967–9968.
9. Duncan SJ, Gruschow S, Williams DH, McNicholas C, Purewal R, Hajek M, Gerlitz M, Martin S, Wrigley S, Moore M: J Am Chem Soc. 2001, 123:554–560.
10. Fisher E, Hirshberg Y: J Chem Soc. 1952, 4522–4524.
11. Berkovic G, Krongauz V, Weiss V: Chem Rev. 2000, 100:1741–1754.
12. McCoy CP, Donnelly L, Jones DS, Gorman SP: Tetrahedron Lett. 2007, 48:657–661.
13. Asano N, Nash RJ, Molyneux RJ, Fleet GWJ: Tetrahedron: Asymmetry. 2000, 11:1645–1680.
14. Ahmed EN: Synthetic Methods for the Stereoisomers of Swainsonine and its Analogues. Tetrahedron 2000, 56:8579–8629.
15. Watson AA, Fleet GWJ, Asano N, Molyneux RJ, Nash RJ: Phytochemistry. 2001, 56:265–295.
16. Michael JP: Nat Prod Rep. 2000, 17:579–602.
17. Michael JP: Nat Prod Rep. 2001, 18:520–542.
18. Michael JP: Nat Prod Rep. 2003, 20:458–475.
19. Michael JP: Nat Prod Rep. 2004, 21:625–649.
20. Michael JP: Nat Prod Rep. 2005, 22:603–626.
21. Pastuszak I, Molyneux RJ, James LF, Elbein AD: Biochemistry. 1990, 29:1886–1891.
22. Rasmussen MO, Delair P, Greene AE: J Org Chem. 2001, 66:5438–5443. For recent asymmetric syntheses of lentiginosine, see: references 22 and 23.
23. Ha D-C, Yun C-S, Lee Y: J Org Chem. 2000, 65:621–623.
24. Harris TM, Harris CM, Hill JE, Ungemach FS, Broquist HP, Wickwire BM: J Org Chem. 1987, 52:3094–3098.

25. Harris CM, Campbell BC, Molyneux RJ, Harris TM: Tetrahedron Lett. 1988, 29:4815–4818.
26. Harris CM, Schneider MJ, Ungemach FS, Hill JE, Harris TM: J Am Chem Soc. 1988, 110:940–949.
27. Aaron HS, Pader CP, Wicks GE Jr: J Org Chem. 1966, 31:3502–3505. For the synthesis of racemic 1-hydroxyindolizidine, see: references 27 and 28.
28. Clevenstine EC, Walter P, Harris TM, Broquist HP: Biochemistry. 1979, 18:3663–3667.
29. Harris CM, Harris TM: Tetrahedron Lett. 1987, 28:2559–2562. For the asymmetric synthesis of (1S,8aR)-1-hydroxyindolizidine, see: references 29 and 30.
30. Klitzke CF, Pilli RA: Tetrahedron Lett. 2001, 42:5605–5608.
31. Shono T, Kise N, Tanabe T: J Org Chem. 1988, 53:1364–1367. For the asymmetric synthesis of (1R,8aS)-1-hydroxyindolizidine, see: references 29–34.
32. Takahata H, Banba Y, Momose T: Tetrahedron: Asymmetry. 1990, 1:763–764.
33. Guerreiro P, Ratovelomanana-Vidal V, Genêt JP: Chirality. 2000, 12:408–410.
34. Rasmussen MO, Delair P, Greene AE: J Org Chem. 2001, 66:5438–5443.
35. Huang P-Q: Synlett. 2006, 1133–1149.
36. Zhou X, Huang P-Q: Synlett. 2006, 1235–1239.
37. Zhou X, Zhang P-Y, Ye J-L, Huang P-Q: Comptes Rendus Chimie. 2008., 11: doi:10.1016/j.crci.2007.02.018.
38. Zhou X, Liu W-J, Ye J-L, Huang P-Q: J Org Chem. 2007, 72:8904–8909.
39. Ikota N, Hanaki A: Heterocycles. 1987, 26:2369–2370.
40. Gren DLC, Kiddle JJ, Thompson CM: Tetrahedron. 1995, 51:2865–2874.
41. Liu L-X, Ruan Y-P, Guo Z-Q, Huang P-Q: J Org Chem. 2004, 69:6001–6009.
42. Huang P-Q, Meng W-H: Asymmetric syntheses of protected (2S,3S,4S)-3-hydroxy-4-methylproline and 4′-tert-butoxyamido-2′-deoxythymidine. Tetrahedron: Asymmetry 2004, 15:3899–3910.

Flexible Synthesis of Poison-Frog Alkaloids of the 5,8-Disubstituted Indolizidine-Class. II: Synthesis of (-)-209B, (-)-231C, (-)-233D, (-)-235B″, (-)-221I, and an Epimer of 193E and Pharmacological Effects at Neuronal Nicotinic Acetylcholine Receptors

Soushi Kobayashi, Naoki Toyooka, Dejun Zhou, Hiroshi Tsuneki, Tsutomu Wada, Toshiyasu Sasaoka, Hideki Sakai, Hideo Nemoto, H. Martin Garraffo, Thomas F. Spande and John W. Daly

ABSTRACT

Background

The 5,8-disubstituted indolizidines constitute the largest class of poison-frog alkaloids. Some alkaloids have been shown to act as noncompetitive blockers at nicotinic acetylcholine receptors but the proposed structures and the biological activities of most of the 5,8-disubstituted indolizidines have not been determined because of limited supplies of the natural products. We have therefore conducted experiments to confirm proposed structures and determine biological activities using synthetic compounds. Recently, we reported that one of this class of alkaloids, (-)-235B', acts as a noncompetitive antagonist for α4β2 nicotinic receptors, and its sensitivity is comparable to that of the classical competitive antagonist for this receptor, dihydro-β-erythroidine.

Results

The enantioselective syntheses of (-)-209B, (-)-231C, (-)-233D, (-)-235B", (-)-221I, and what proved to be an epimer of natural 193E, starting from common chiral lactams have been achieved. When we performed electrophysiological recordings to examine the effects of the synthetic alkaloids on two major subtypes of nicotinic receptors (α4β2 and α7) expressed in Xenopus laevis oocytes, (-)-231C effectively blocked α4β2 receptor responses (IC_{50} value, 1.5 μM) with a 7.0-fold higher potency than for blockade of α7 receptor responses. In contrast, synthetic (-)-221I and (-)-epi-193E were more potent in blocking α7 receptor responses (IC_{50} value, 4.4 μM and 9.1 μM, respectively) than α4β2 receptor responses (5.3-fold and 2.0-fold, respectively).

Conclusion

We achieved the total synthesis of (-)-209B, (-)-231C, (-)-233D, (-)-235B", (-)-221I, and an epimer of 193E starting from common chiral lactams, and the absolute stereochemistry of natural (-)-233D was determined. Furthermore, the relative stereochemistry of (-)-231C and (-)-221I was also determined. The present asymmetric synthesis of the proposed structure for 193E revealed that the C-8 configuration of natural 193E should be revised. The selectivity for α4β2 and α7 nicotinic receptors differed markedly for the 5,8-disubstituted indolizidines tested, and thus it appears that the nature of the side chains in these indolizidines is crucial with regard to subtype-selectivity.

Introduction

In the preceding article [1], we have reported the synthesis of the chiral lactam building blocks (1, 2, Schemes 1, 2) for the flexible synthesis of poison-frog alkaloids of the 5,8-disubstituted indolizidine class. The utility of these chiral building blocks was demonstrated by the synthesis of alkaloids (-)-203A, (-)-205A from 1, and of (-)-219F from 2. Although the biological activity of most of the 5,8-disubstituted indolizidines has not been investigated, certain 5,8-disubstituted indolizidines have been shown to act as noncompetitive blockers of nicotinic acetylcholine receptors. [2,3]

Scheme 1. Syntheses of (-)-209B, (-)-231C, (-)-233D, and (-)-235B.

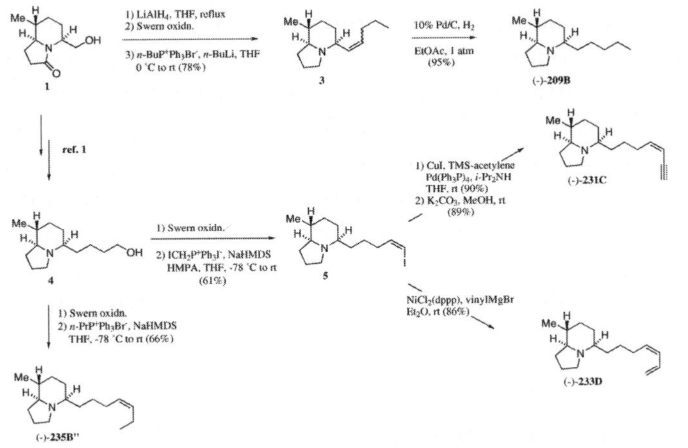

Scheme 2. Syntheses of (-)-221I and (-)-7 (an epimer of 193E).

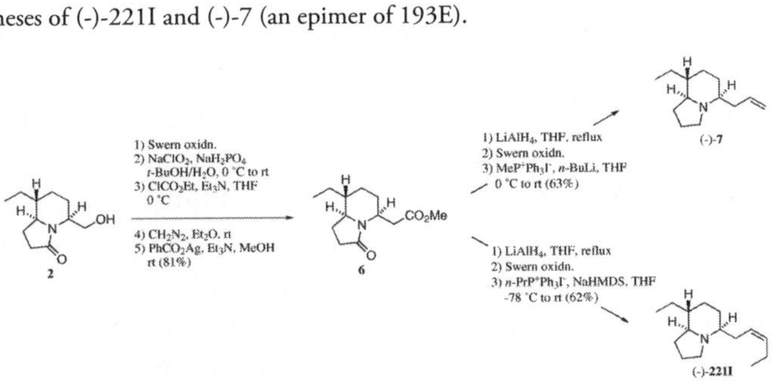

Nicotinic receptors are ligand-gated ion channels composed of five subunits. [4] To date, 12 nicotinic receptor subunits (α2-α10, β2-β4) have been identified. Subtypes of neuronal nicotinic receptors are constructed from numerous subunit combinations, which confer varied functional and pharmacological characteristics. [5] Nicotinic receptors have been implicated in a wide range of neuronal dysfunctions and mental illness, such as epilepsy, Tourette's syndrome, Alzheimer's disease, Parkinson's disease, and schizophrenia. [5,6] Since different subtypes of nicotinic receptors are involved in different neurological disorders, subtype-selective nicotinic ligands would be valuable for investigation and potentially for treatment of cholinergic disorders of the central nervous system. However, there are only a limited number of compounds that elicit subtype-selective blockade of nicotinic receptors because of the similarity of receptor-channel structure among the subtypes. Recently, we have investigated the effect of synthetic (-)-235B', one of the 5,8-disubstituted indolizidine class of poison-frog alkaloids, on several subtypes of nicotinic receptors, and found that this alkaloid exhibits selective and potent blocking effects at the α4β2 nicotinic receptor. [3] The potency of (-)-235B' for this receptor is comparable to that of the classical α4β2 competitive antagonist, dihydro-β-erythroidine. In this study, we have synthesized 5,8-disubstituted indolizidines (-)-209B, (-)-231C, (-)-233D, (-)-235B", (-)-221I, and an alkaloid that proved to be an epimer of natural indolizidine 193E. The alkaloids (-)-209B and (-)-235B" are known to be noncompetitive nicotinic blockers [2], but effects of the other compounds have not yet been tested. To explore possible subtype selectivity, we examined the effects of (-)-231C, (-)-221I and (-)-epi-193E on α4β2 and α7 nicotinic receptors, the most abundant subtypes in the mammalian brain. [4]

Results and Discussion

Reduction of the lactam 1 [1] with $LiAlH_4$ followed by Swern oxidation of the resulting alcohol and Wittig reaction gave the olefin 3 in 78% overall yield (Scheme 1). Hydrogenation of the double bond in 3 with 10% Pd/C provided (-)-209B, whose spectral data were identical with reported values. [7] The lactam 1 was also converted to the alcohol 4, [1] which was transformed into (-)-235B" by Swern oxidation followed by Wittig reaction under high dilution and 'salt free' conditions (Scheme 1). The spectral data of synthetic (-)-235B" were identical with reported values. [8,9] Indolizidines (-)-231C [10] and (-)-233D [10] were synthesized from common intermediate 5 prepared from the alcohol 4. Thus, the Swern oxidation of 4 and then the Wittig reaction of the resulting aldehyde under Stork's conditions [11] provided the Z-iodoolefin 5 in a highly stereoselective manner. The Sonogashira coupling reaction [12] of 5 with TMS-acetylene followed by cleavage of the trimethylsilyl group with K_2CO_3 afforded (-)-231C. Although the rotation of the natural alkaloid is unknown, the relative stereochemistry was determined to be 5,8-E and 5,9-Z by GC-MS and GC-FTIR comparison with natural 231C in extracts from a Panamanian dendrobatid frog, Dendrolbates pumilio. A similar, Ni-catalyzed cross coupling [13] reaction of 5 with vinylmagnesium bromide provided the (-)-233D, whose spectral data were identical with values reported for the natural alkaloid isolated from the Panamanian dendrobatid frog. [10] Although differing in magnitude, the HCl salts of both synthetic (-) 233D and the natural alkaloid had negative optical rotations.

Indolizidine (-)-7 with the relative stereochemistry proposed for 193E [14] and indolizidine (-)-221I [14] were synthesized from the lactam 2 [1] via the ester 6 (Scheme 2). The two-step oxidation of 2 followed by Arndt-Eistert homologation of the resulting carboxylic acid provided the ester 6. Reduction of both lactam and ester moieties of 6 with LiAlH4 followed by Swern oxidation and Wittig reaction of the resulting aldehyde furnished the indolizidine (-)-7. Coinjections of synthetic material with an alkaloid fraction from a Madagascan mantellid frog, Mantella viridis that contained natural 193E, revealed that the synthetic material had a slightly longer GC retention time than the natural product. The GC-mass spectra of (-)-7 and natural product were virtually identical and their GC-FTIR spectra were very similar in the Bohlmann band region (indicating 5,9-Z configurations in both), although differing slightly in their fingerprint regions. These results indicate that the natural 193E is most likely the 8-epimer of (-)-7 and that the proposed configuration [14] of the ethyl substituent at C-8 was in error. The indolizidine (-)-221I was also synthesized from 6 following a procedure similar to that used for the synthesis of (-)-7 as shown in Scheme 2.

The relative stereochemistry of natural 221I was determined to be the same as that of synthetic (-)-221I by GC-MS and GC-FTIR comparison with natural 221I, in the alkaloid fraction from the Madagascan mantellid frog, Mantella viridis.

We then conducted electrophysiological experiments to examine the effect of three of the synthetic alkaloids on nicotinic receptors. When Xenopus laevis oocytes expressing the $\alpha4\beta2$ nicotinic receptor were treated with 3 μM (-)-231C, the peak amplitude of the acetylcholine (ACh)-elicited currents was greatly decreased, whereas the responses elicited in oocytes expressing the $\alpha7$ nicotinic receptor were not strongly affected (Figure 1A). When the concentration-response curves were compared between these receptor subtypes, (-)-231C blocked the $\alpha4\beta2$ receptor-mediated currents [50% inhibitory concentration (IC50) = 1.5 μM, 95% confidence intervals (CI): 1.1 to 2.1 μM] with 7.0-fold higher sensitivity than blockade of the $\alpha7$ receptor-mediated currents (IC50 = 10.7 μM, 95% CI: 8.6 to 13.3 μM) (Figure 1B). These results indicate that (-)-231C selectively blocked the responses mediated by the $\alpha4\beta2$ receptor.

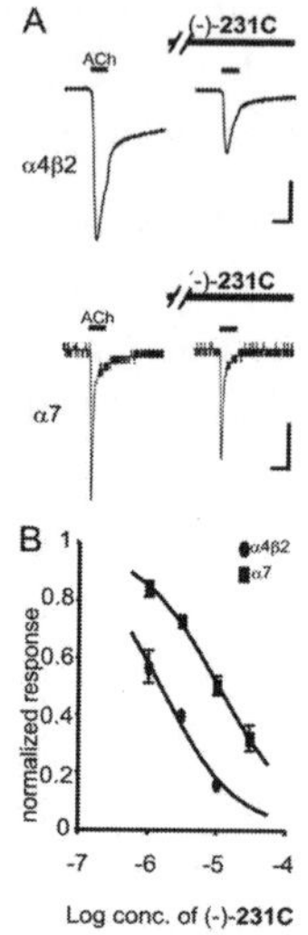

Figure 1. Inhibitory effect of (-)-231C on ACh-induced currents in X. laevis oocytes expressing recombinant nicotinic receptors. Currents were recorded in the voltage-clamp mode at -60 mV. Concentrations of ACh used were 1 μM for the α4β2 receptor and 100 μM for the α7 receptor. For test responses, oocytes were preincubated with (-)-231C for 3 min and then exposed to ACh with (-)-231C. A, representative traces showing the ACh-elicited currents in the absence and presence of (-)-231C (3 μM). Horizontal bars indicate the period of perfusion with ACh for 5 s. Vertical scale bars represent 0.5 μA on the α4β2 receptor, and 0.1 μA on the α7 receptor. B, concentration-response curves for (-)-231C on recombinant nicotinic receptors. Current responses to ACh in the presence of (-)-231C in each oocyte were normalized to the ACh responses (control responses) recorded in the same oocytes. Values represent the mean ± S.E.M. for five to six separate experiments.

The 5,8-disubstituted indolizidine (-)-231C is an analogue of (-)-235B', both of which have a seven-carbon unsaturated side-chain at C-5 and a methyl at C-8. Both synthetic compounds have the same absolute stereochemistry (5R, 8R, 9S). Our previous [3] and present data demonstrate that both (-)-235B' and (-)-231C produce potent blockade of the α4β2 nicotinic receptor with a similar selectivity of 6- to 7-fold over blockade of the α7 receptor. However, the potency of (-)-235B' in blocking the α4β2 receptor is approximately 20-fold greater than that of (-)-231C. These results suggest that either flexibility or degree of unsaturation of the seven-carbon side-chain at C-5 in these 5,8-disubstituted indolizidines is crucial for potent interaction with the α4β2 receptor.

The synthetic (-)-221I and (-)-epi-193E are 5,8-disubstituted indolizidines with an ethyl rather than a methyl at C-8 and a five-carbon or three-carbon side-chain, respectively, at C-5. The alkaloid (-)-221I blocked α7 receptor responses (IC50 = 4.4 μM, 95% CI: 3.1 to 6.1 μM) with 5.3-fold higher potency than for blockade of the α4β2 receptor responses (IC50 = 23.1 μM, 95% CI: 18.5 to 28.9 μM) (Figure 2). Synthetic (-)-epi-193E was more potent in blocking the α7 receptor response (IC50 = 9.1 μM, 95% CI: 7.5 to 11.1 μM) compared to blockade of the α4β2 receptor (IC50 = 18.0 μM, 95% CI: 12.2 to 26.7 μM) (Figure 3).

Previously, we examined the effects of three synthetic 5,8-disubstituted indolizidines with an n-butyl group at C-8 and an n-propyl group at C-5 in blocking different subtypes of nicotinic receptors. [3] Two of these compounds, namely (+)-8,9-diepi-223V and (-)-9-epi-223V were 6.7-fold and 11.2-fold more potent in blocking α7 receptor compared to blockade of α4β2 receptor, while the third, (-) 223V, was only slightly more potent at blocking the responses mediated by the α7 receptor. [3,15] These results suggest that the α4β2 receptor does not interact well with indolizidines having substituents larger than methyl at C-8, while the α7 receptor is more accepting of larger side-chains at C-8. Further analogous synthetic alkaloids need to be tested. Overall, the side chains of 5,8-disubstituted indolizidines appear to be of critical importance in determining selectivity and potency in blocking responses mediated by subtypes of neuronal nicotinic receptors. Further study of structure-activity relationships of synthetic 5,8-disubstituted indolizidines at nicotinic subtypes could lead to even more subtype-selective ligands as research probes and as potentially useful drugs.

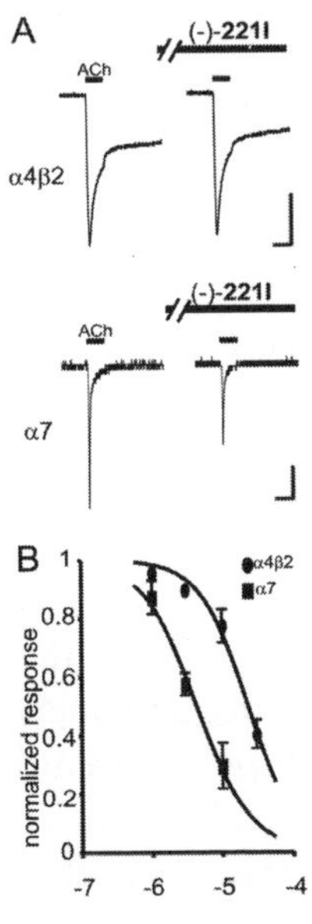

Figure 2. Inhibitory effect of (-)-221I on ACh-induced currents in X. laevis oocytes expressing recombinant nicotinic receptors. Currents were recorded in the voltage-clamp mode at -60 mV. Concentrations of ACh used were 1 μM for the α4β2 receptor and 100 μM for the α7 receptor. For test responses, oocytes were preincubated with (-)-221I for 3 min and then exposed to ACh with (-)-221I. A, representative traces showing the ACh-elicited currents in the absence and presence of (-)-221I (3 μM). Horizontal bars indicate the period of perfusion with ACh for 5 s. Vertical scale bars represent 0.5 μA on the α4β2 receptor, and 0.1 μA on the α7 receptor. B, concentration-response curves for (-)-221I on recombinant nicotinic receptors. Current responses to ACh in the presence of (-)-221I in each oocyte were normalized to the ACh responses (control responses) recorded in the same oocytes. Values represent the mean ± S.E.M. for five separate experiments.

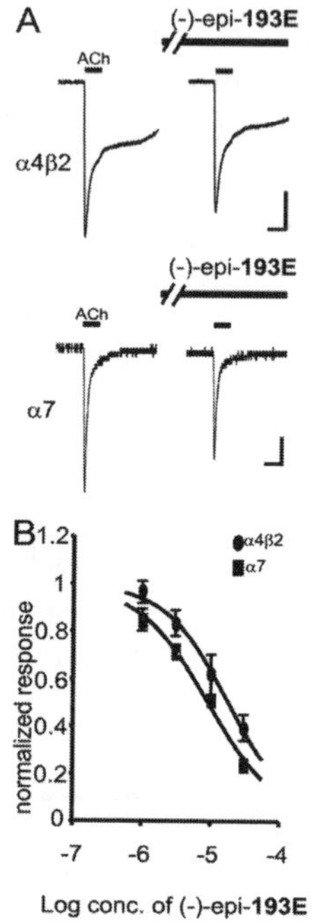

Figure 3. Inhibitory effect of (-)-epi-193E on ACh-induced currents in X. laevis oocytes expressing recombinant nicotinic receptors. Currents were recorded in the voltage-clamp mode at -60 mV. Concentrations of ACh used were 1 μM for the α4β2 receptor and 100 μM for the α7 receptor. For test responses, oocytes were preincubated with (-)-epi-193E for 3 min and then exposed to ACh with (-)-epi-193E. A, representative traces showing the ACh-elicited currents in the absence and presence of (-)-epi-193E (3 μM). Horizontal bars indicate the period of perfusion with ACh for 5 s. Vertical scale bars represent 0.5 μA on the α4β2 receptor, and 0.1 μA on the α7 receptor. B, concentration-response curves for (-)-epi-193E on recombinant nicotinic receptors. Current responses to ACh in the presence of (-)-epi-193E in each oocyte were normalized to the ACh responses (control responses) recorded in the same oocytes. Values represent the mean ± S.E.M. for five separate experiments.

Neuronal nicotinic receptors have been implicated in the physiological processes of reward, cognition, learning and memory. [5,6] Some ligand-binding and autoradiography studies with postmortem human brain suggest that loss of neuronal nicotinic receptors is related to central cholinergic disorders such as Alzheimer's disease, Parkinson's disease and schizophrenia. [4,6] For instance, in schizophrenic patients, decrease in binding of α-bungarotoxin (α-Bgt), a major specific ligand for α7 nicotinic receptors, has been detected in hippocampus, thalamus and frontal cortex [16,17]. Therefore, loss of α7 nicotinic ligand-binding appears to be an early presymptomatic diagnostic marker for schizophrenia. For in vivo mapping of brain receptors, positron emission tomography (PET) and single photon emission computed tomography (SPECT) using specific ligands are powerful, non-invasive techniques. Although 125I-methyllycaconitine has been used for α7-selective binding in rat brain, [18] neither PET nor SPECT ligand of α7 nicotinic receptors has been available so far. Radiolabeled α-Bgt could not be used for in vivo mapping because of the large molecular weight, high toxicity and poor blood-brain barrier permeability. [19] Indolizidines are low molecular weight, lipophilic compounds that should penetrate well into brain and, as shown in our research, some exhibit high affinity and selectivity for either α4β2 or α7 nicotinic receptors. Further structure-activity relationship studies of synthetic indolizidines may lead to the development of radioactive α4β2-selective or α7-selective ligands useful for in vivo mapping of these important central nicotinic receptors.

Acknowledgements

We are grateful to Dr. John A. Dani (Baylor College of Medicine, Houston, TX, USA) for his support with electrophysiological data acquisition, and to Dr. Jerry A. Stitzel (University of Colorado) for providing us with plasmid DNA. This work was supported in part by a grant-in-aid for Scientific Research (C, No. 17590004, and No. 16590435) by the Japan Society for the Promotion of Science (JSPS). Work at NIH was supported by the intramural research program of NIDDK.

References

1. Toyooka N, Zhou D, Nemoto H, Garraffo HM, Spande TF, Daly JW: Beilstein J Org Chem. 2007, 3:30.
2. Daly JW, Nishizawa Y, Padgett WL, Tokuyama T, Smith AL, Holmes AB, Kibayashi C, Aronstam RS: Neurochem Res. 1991, 16:1213–1218.
3. Tsuneki H, You Y, Toyooka N, Kagawa S, Kobayashi S, Sasaoka T, Nemoto H, Kimura I, Dani JA: Mol Pharmacol. 2004, 66:1061–1069.
4. Gotti C, Zoli M, Clementi F: Trends Pharmacol Sci. 2006, 27:482–491.
5. Dani JA, Bertrand D: Annu Rev Pharmacol Toxicol. 2007, 47:699–729.
6. Gotti C, Clementi F: Prog Neurobiol. 2004, 74:363–396.
7. Holmes AB, Smith AL, Williams SF, Hughes LR, Lidert Z, Swithenbank C: J Org Chem. 1991, 56:1393–1405.
8. Tokuyama T, Nishimori N, Shimada A, Edwards MW, Daly JW: Tetrahedron. 1987, 43:643–652.
9. Comins DL, Lamunyon DH, Chen X: J Org Chem. 1997, 62:8182–8187.
10. Tokuyama T, Tsujita T, Shimada A, Garraffo HM, Spande TF, Daly JW: Tetrahedron. 1991, 47:5401–5414.

11. Stork G, Zhao K: Tetrahedron Lett. 1989, 30:2173–2174.
12. Takahashi S, Kuroyama Y, Sonogashira K, Hagihara N: Synthesis. 1980, 627–630.
13. Tamao K, Sumitani K, Kumada M: J Am Chem Soc. 1972, 94:4374–4376.
14. Daly JW, Spande TF, Garraffo HM: J Nat Prod. 2005, 68:1556–1575.
15. Toyooka N, Tsuneki H, Kobayashi S, Dejun Z, Kawasaki M, Kimura I, Sasaoka T, Nemoto H: Current Chemical Biology. 2007, 1:97–114.
16. Martin-Ruiz CM, Haroutunian VH, Long P, Young AH, Davis KL, Perry EK, Court JA: Biol Psychiatry. 2003, 54:1222–1233.
17. Leonard S, Adams C, Breese CR, Adler LE, Bickford P, Byerley W, Coon H, Griffith JM, Miller C, Myles-Worsley M, Nagamoto HT, Rollins Y, Stevens KE, Waldo M, Freedman R: Schizophr Bull. 1996, 22:431–445.
18. Navarro HA, Xu H, Zhong D, Abraham P, Carroll FI: Synapse. 2002, 44:117–123.
19. Navarro HA, Zhong D, Abraham P, Xu H, Carroll FI: J Med Chem. 2000, 43:142–145.

A Novel, One-Step Palladium and Phenylsilane Activated Amidation from Allyl Ester on Solid Support

Zheming Ruan, Katy Van Kirk, Christopher B. Cooper and R. Michael Lawrence

ABSTRACT

The direct conversion of solid-supported carboxylic acid allyl esters to carboxamides through the use of phenylsilane and catalytic $Pd(PPh_3)_4$ *under mild reaction conditions is reported. The use of this methodology for the generation of a 48 compound solid-phase array is described herein.*

Introduction

The synthesis of large combinatorial libraries of low molecular weight, drug-like molecules requires robust chemistry employing a wide variety of diversity elements on solid support. The carboxylic acid functional group has been widely used in solid-phase chemistry, especially when protected as an ester. An allyl ester is a commonly used protecting group that can be used with many acid or base-labile linkers, and is removed easily with $Pd(PPh_3)_4$ in various solvent systems with the aid of a scavenger reagent [1–3]. Phenylsilane, acting as a hydride donor, has been reported to be an excellent scavenger when used in conjunction with $Pd(PPh_3)_4$ in the removal of the allyl ester group [4]. Recently, it was also reported from our laboratory that phenylsilane could be directly used as an active amidation reagent of carboxylic acids like other coupling agents [5]. Based on this finding, a novel, one-step to convert allyl ester to amide, using palladium and phenylsilane as activating agents, has been developed on solid support (Scheme 1). In an effort to explore the generality and scope of this method, this reaction was examined using various structurally diverse carboxylic acids and amines (primary, secondary, or anilines) on solid support. A general method to protect carboxylic acids with an allyl ester group on solid support is also reported.

Scheme 1

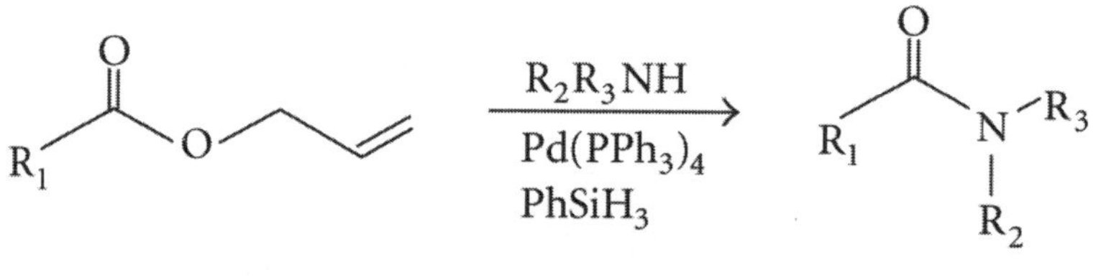

SCHEME 1

Results and Discussion

(1) Three representative allyl ester resins (Scheme 2, 4–6) were initially prepared from the resin-bound N-benzoyl carboxylic acids 1–3, which could be easily prepared through standard procedures. Resin bound allylation proceeds well under optimized reaction conditions (5 equivalents of allylbromide, 5 equivalents of CsF, DMF, room temperature).

Scheme 2

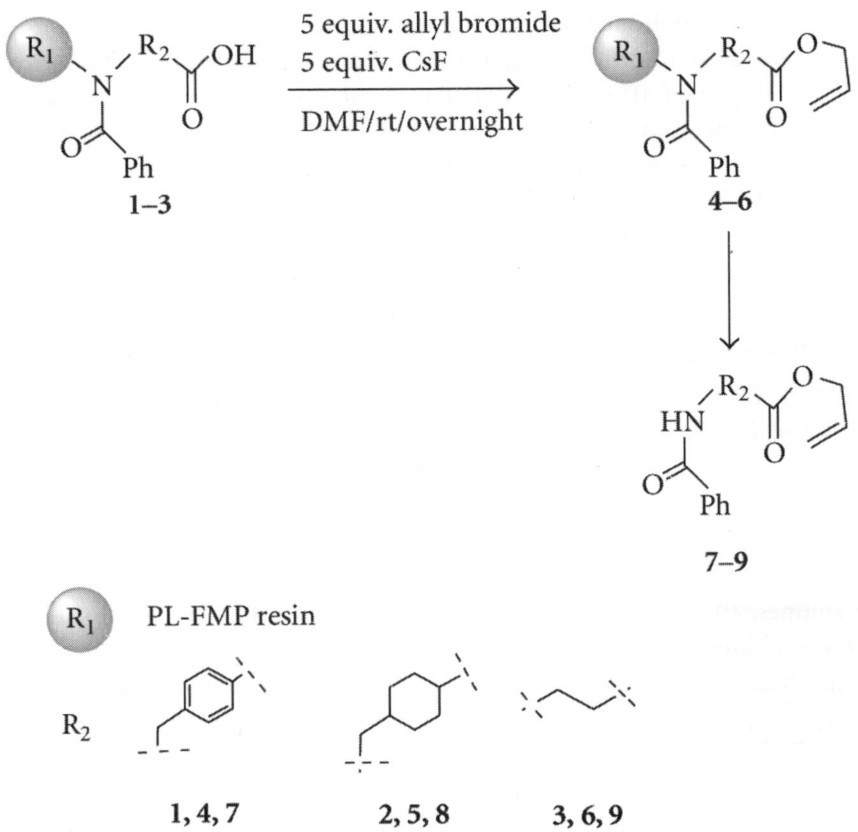

Typical Procedure

(a) The acid resin 1–3 (2 g, ~1.0 mmol/g loading) was swollen with ~15 mL DMF at room temperature and then 5 equiv of allybromide and 5 equiv of CsF were added into the resin mixture. The reaction was allowed to agitate overnight. After that, the reaction was washed with DMF(2x), THF(3x), DCM(4x). The resin 4–6 was dried under high vacuum pump. The resin was checked by TFA cleavage, and the product 7–9 had 90~96% of purity and 96~100% of yield.

(b) A 48-compound array was constructed from a matrix of these three R2 allyl esters crossed with 16 R3 amines and anilines. The process was car-

> ried out using IRORI MicroKans, which is Rf-encoded split pool synthesis technology. The reactions were performed on a 0.02 mmol scale (0.02 mmol/one microkan) in anhydrous CH2Cl2 at room temperature for 24 hours with 10 equivalents of amines or anilines, 20 equiv of PhSiH3 and 0.05 equiv of Pd(PPh3)4. After that, all microkans were pooled together and then washed with DMF(2x), THF(3x), DCM(4x). All microkans were dried under high vacuum pump and then sorted into IRORI cleavage station. The final products 10 were cleaved into 96 well-format plate with 30% of TFA solution in DCM. The solvent was removed and the products were directly analyzed with flow NMR and LC-MS. Overall yield was calculated based on the initial loading of resins.

Following cleavage from the resin with TFA, nearly quantitative yields of the allyl ester products 7–9 (based on the initial loading of resin 1–3) were obtained with high purity (>90%) (Scheme 2).

Using resins 4–6, a 48 compound array was constructed from a matrix of these three R2 allyl esters crossed with 16 R3R4NH amines and anilines (see Typical Procedures above). The reactions were performed on a 0.02 mmol scale in anhydrous CH2Cl2 at room temperature for 24 hours with 10 equivalents of amines or anilines, 20 equiv of PhSiH3 and 0.05 equiv of Pd(PPh3)4. The final products 10 were obtained after cleavage from the resins. This entire process was carried out using commercially available IRORI MicroKans [6] (Scheme 3).

Solid-phase array results are provided in Table 1 next page. The general reaction conditions employed worked equally well for both primary and secondary amines with the various allyl esters. Additional experiments have indicated that the best solvents for this method are DMF or NMP and the reaction could be carried out with 5 equivalents of amines and 10 equiv of PhSiH3.

Direct analyses of the cleaved products (HPLC and LC/MS) indicated high purity and yields in most cases. All calculations were based on the initial loading of resin 1–3. All reactions were clean with the major side product being the unprotected carboxylic acid. With an increase in steric congestion of the amine component (entries 3, 10, 19, 26, and 35), more severe conditions were required to drive reactions to completion (i.e., higher reaction temperatures and the use of additional equivalents of amines). Reactions with anilines were disappointing with few amide products obtained and high recovery of the unprotected acids observed even under more forcing reaction conditions. No product was formed in the reaction without addition of both PhSiH3 and catalytic Pd(Ph3P)4 reagents. It was shown in our controlling experiments, entry 38.

Table 1. Purity and yield data for 48-compound automation library.

Entry	Resin 4-5	Amines or anilines	Product characterization (**10**)[c]		
			MH$^+$	Purity (%)	Yield (%)
1	**4**	Pyrrolidine	309.20	90	73
2	**4**	Morpholine	325.17	99	72
3	**4**	Diethylamine	311.25	27(53[a])	74
4	**4**	Tetrahydrofurylamine	339.24	90	68
5	**4**	N-Acetylethylenediamine	340.20	72	68
6	**4**	Phenethylamine	359.26	95	64
7	**4**	1-(3-aminopropyl)imidazole	363.30	77	65
8	**4**	4-(2-aminoethyl)morpholine	368.30	84	49
9	**4**	Tyramine	375.30	86	61
10	**4**	Cyclohexylamine	337.25	20(67[a])	67
11	**4**	Benzylamine	345.22	85	50
12	**4**	4-(Aminomethyl)pyridine	346.23	87	62
13	**4**	4-Bromobenzylamine	424.14	95	55
14	**4**	Aniline		Only acid	0
15	**4**	p-Anisidine		Only acid	0
16	**4**	4-Nitroaniline		Only acid	0
17	**5**	Pyrroline	315.20	86	71
18	**5**	Morpholine	331.23	94	47
19	**5**	Diethylamine	317.37	7 (14[a])	56
20	**5**	Tetrahydrofurylamine	345.40	97	43
21	**5**	N-Acetylethylenediamine	346.10	98	42
22	**5**	Phenethylamine	365.33	86	61
23	**5**	1-(3-Aminopropyl)imidazole	369.35	94	54
24	**5**	4-(2-Aminoethyl)morpholine	374.10	97	55
25	**5**	Tyramine	381.33	86	35
26	**5**	Cyclohexylamine	343.32	52(73[a])	64
27	**5**	Benzylamine	351.29	85	64
28	**5**	4-(Aminomethyl)pyridine	352.29	89	38
29	**5**	4-Bromobenzylamine	430.13	85	51
30	**5**	Aniline		Only acid	0
31	**5**	p-Anisidine		Only acid	0
32	**5**	4-Nitroaniline		Only acid	0
33	**6**	4-Nitroaniline	247.10	100	56
34	**6**	Pyrroline	263.20	97	83
35	**6**	Morpholine	249.24	37(67[a])	71
36	**6**	Diethylamine	277.29	99	69
37	**6**	Tetrahydrofurylamine	278.10	100	54
38	6	N-Acetylethylenediamine	297.10	98 (0[b])	92 (0[b])
39	**6**	Phenethylamine	301.26	99	91
40	**6**	1-(3-Aminopropyl)imidazole	306.10	99	80
41	**6**	4-(2-Aminoethyl)morpholine	313.10	98	58
42	**6**	Tyramine	275.32	92	56
43	**6**	Cyclohexylamine	283.26	99	72
44	**6**	Benzylamine	284.27	100	66
45	**6**	4-(Aminomethyl)pyridine	362.17	99	52
46	**6**	4-Bromobenzylamine		Only acid	0
47	**6**	Aniline		Only acid	0
48	**6**	p-Anisidine		Only acid	0

[a]60°C in DMF for 24 hours. [b]Without the addition of both $PhSiH_3$ and $Pd(Ph_3P)_4$, or one of them. [c]Overall isolated yield after cleavage. All products were analyzed by LC-MS and flow ^1H NMR spectroscopy.

Scheme 3

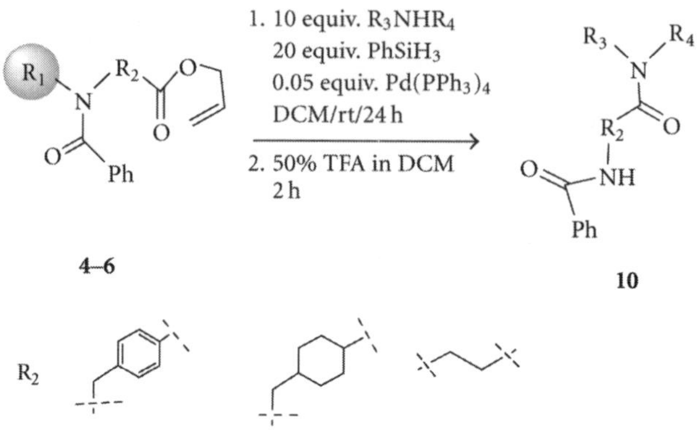

Conclusion

In summary, a Pd(0)/$PhSiH_3$ system has been successfully applied to convert resin-bond allyl esters to amides. The reaction can typically be carried out in a single step at room temperature. A systematic investigation of amine and aniline inputs has demonstrated that in general, primary amines and unhindered secondary amines give excellent yields of amides with high purity. Higher reaction temperatures and additional equivalents of amines can be used to push reactions to completion. This methodology has been recently used in a solid-phase sequence to prepare a 10000-compound library directed at the identification of protease inhibitors.

References

1. D. G. Hall, S. Manku, and F. Wang, "Solution- and solid-phase strategies for the design, synthesis, and screening of libraries based on natural product templates: a comprehensive survey," Journal of Combinatorial Chemistry, vol. 3, no. 2, pp. 125–150, 2001.
2. P. Seneci, Solid-Phase Synthesis and Combinatorial Technologies, Wiley-Interscience, New York, NY, USA, 2000.
3. R. E. Dolle, "Comprehensive survey of combinatorial library synthesis: 2000," Journal of Combinatorial Chemistry, vol. 3, no. 6, pp. 477–517, 2001.

4. F. Guibé, "Allylic protecting groups and their use in a complex environment—part II: allylic protecting groups and their removal through catalytic palladium π-allyl methodology," Tetrahedron, vol. 54, no. 13, pp. 2967–3042, 1998.
5. M. Dessolin, M.-G. Guillerez, N. Thieriet, F. Guibé, and A. Loffet, "New allyl group acceptors for palladium catalyzed removal of allylic protections and transacylation of allyl carbamates," Tetrahedron Letters, vol. 36, no. 32, pp. 5741–5744, 1995.
6. H. Lee and C. R. Sarko, "Analysis of a combinatorial library synthesized using a split-and-pool Irori MicroKan method for development and production," in High Throughput Analysis for Early Drug Discovery, pp. 37–56, Elsevier, San Diego, Calif, USA, 2004.

The First Organocatalytic Carbonyl-Ene Reaction: Isomerisation-Free C-C Bond Formations Catalysed By H-Bonding Thio-Ureas

Matthew L. Clarke, Charlotte E. S. Jones and Marcia B. France

ABSTRACT

Intramolecular carbonyl ene reactions of highly activated enophiles can be catalysed by H-bonding thio-ureas to give tertiary alcohols in high yields without extensive isomerisation side products. An asymmetric variant of this reaction was realised using a chiral thiourea but was limited by low enantioselectivity (up to 33% e.e.) and low turnover frequencies.

Background

The intermolecular carbonyl-ene reaction is a useful and completely atom-efficient C-C bond forming reaction. These reactions take place without a catalyst at relatively high temperatures (>150°C), but are more often carried out with either a stoichiometric or catalytic Lewis acid catalyst. [1-4] One of the drawbacks of the intermolecular carbonyl ene reaction is substrate scope. The majority of all successful catalytic ene reactions have utilised the highly activated glyoxylate esters as enophile. Extending asymmetric intramolecular carbonyl ene reactions to include ketone enophiles is a highly desired process that can potentially deliver enantioenriched tertiary alcohols. The most promising results in this regard come from Evans and co-workers who demonstrated that 1,1-disubstituted alkenes would react with the activated ethyl pyruvate to give the desired tertiary alcohols in good yield. However, this reaction suffered from low turnover frequencies and was limited to certain types of alkene nucleophiles which were required in a large excess for the reaction to proceed.[5] However, a range of ketone ene reactions have been promoted by stoichiometric Et_2AlCl, suggesting some potential for this reaction. [6] We have screened catalysts from across the periodic table with a wide range of ligands in an attempt to develop catalytic asymmetric ketone ene reactions of ketones of type $RC(O)CO_2Et$ with scant success. The ketones were unreactive, and the alkenes used often polymerised under the reaction conditions. An exception was the ene reaction of ethyl trifluoropyruvate 2 with alkenes which actually took place very readily, although sometimes accompanied by isomerisation and other side products. This type of ene reaction has recently been explored by other groups using Pd, Ni and Pt catalysts.[7,8] The importance of such reactions is derived from the ever-increasing occurrence of trifluoromethyl substituents in drugs and biologically active compounds.[9,10] Given that trifluoropyruvate ene reactions seemingly have a low activation barrier, and as part of our ongoing interest in hydrogen bond mediated catalysis,[11] we have investigated thioureas as H-bonding additives for organocatalytic carbonyl ene reactions and report these results here. Despite the explosive growth in organocatalytic reactions in recent years, [12-16] this represents, to the best of our knowledge, the first example of an organocatalytic carbonyl ene reaction. [17]

Findings

We first carried out a thermal ene reaction of α-methyl styrene 1 with ethyl trifluoropyruvate under microwave heating to produce racemic product 3 (Scheme 1). This reaction was accompanied by several side products. The major one of these was clearly an (E + Z) alkene containing product. This was identified as isomerisation product 4 on the basis of ^{19}F, 1H NMR spectroscopy and most informatively, GCMS of the isolated product mixture, which showed three species with the same mass but different retention times. Other minor side products were formed from the decomposition of ethyl trifluoropyruvate, but were readily separated from compounds 3 and 4. This decomposition becomes more significant as the reaction is left longer.

Scheme 1. Microwave promoted ene reaction of ethyl trifluoropyruvate with α-methyl styrene.

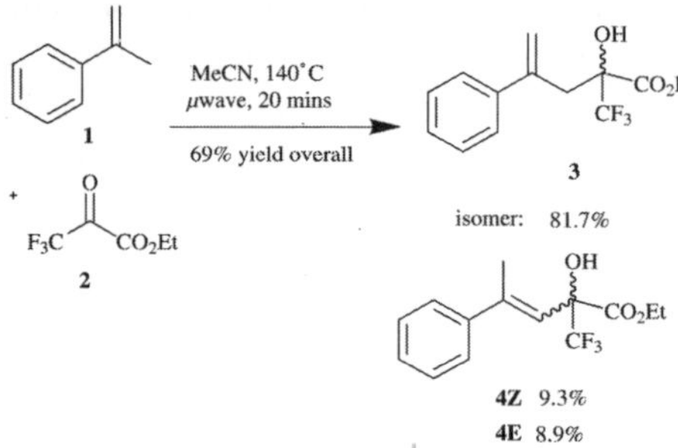

An investigation into the effect of various H-bonding additives on this reaction was carried out (Scheme 2 and Table 1). N,N'-di [3,5-bis(trifluoromethyl) phenyl]thiourea 5,[18], emerged as by far the most active promoter for this reaction. The catalyst was used at 20 mol % loading and compared against a control reaction with no catalyst. Products were isolated by chromatography after 3 hours reaction time. A 96% yield of the desired isomerically pure tertiary alcohol was obtained in the presence of 5. In contrast, the uncatalysed background reaction yields just 19%. Reactions monitored by 19F NMR reveal that good conversion (70%) was also possible after 1 hour using 10 mol % catalyst. Lower catalyst loadings required longer reaction times. Thus, in common with most organocatalytic procedures, turnover frequency is low.

Scheme 2. Thiourea catalysed ene reaction.

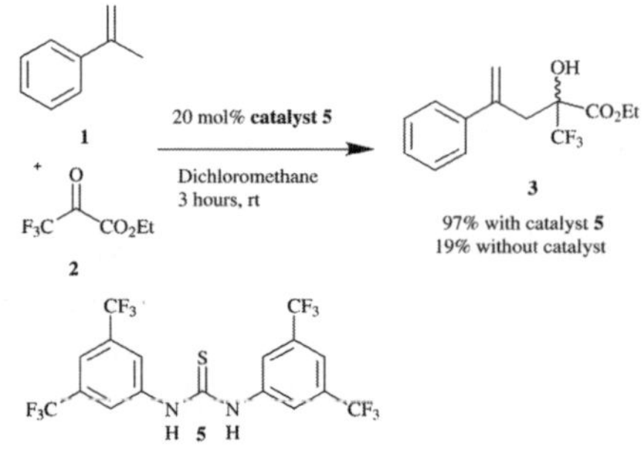

Table 1. Effect of different thiourea on ene reaction.

Thiourea	Time (hrs)	% Conv.[a]
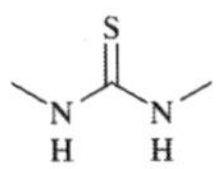	1	3
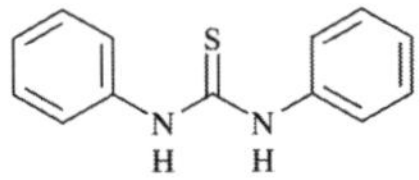	1	15
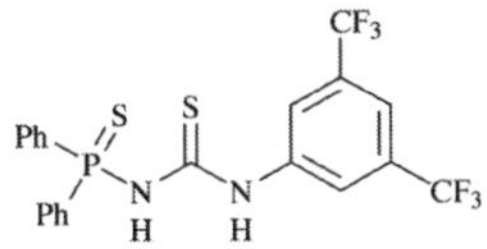	2	2
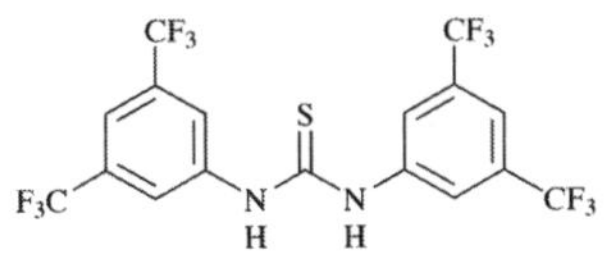	1	95

a: % Conversion determined by $^{19}F\ \{^{1}H\}$ NMR.
Ethyl trifluoropyruvate (0.098 ml, 0.75 mmol) was added to a solution of the thiourea (20 mol%) and the α-methyl styrene (0.12 ml, 0.9 mmol, ~1.2 eq) in dichloromethane (2 ml).

In order to explore if an asymmetric variant of this reaction was possible, we elected to employ a chiral thio-urea. It is almost certain that these reactions proceed by H-bonding of the ketone to the thio-urea moiety, lowering the LUMO for attack by uncomplexed alkene. An asymmetric catalyst would therefore need to be a diaryl thio-urea in order to maximise the acidity of the NH functions. It would also need to possess chirality that can potentially close off one face of the co-ordinated ketone that would also have to hydrogen bond in a defined manner. The most readily available thio-urea that could hopefully fulfil these two roles is thio-urea 6, which was prepared by reaction of (R)-(+)-1,1'-Bi(2-naphthylamine) with 3,5-bis(trifluoromethyl)phenyl isothiocyanate in 52% yield and has recently been reported by another group.[19] This catalyst was tested in the α-methyl styrene ene reaction to ethyl trifluoropyruvate with ultimately disappointing results. This catalyst did allow us to demonstrate the first asymmetric organocatalytic carbonyl ene reaction, but reactions run at both -20 and 0°C using 10 and 25 mol % catalyst respectively gave good yields and only ~30% e.e (Scheme 3 and Table 2). A reaction carried out with stoichiometric amounts of chiral thiourea also gave ~30% e.e. suggesting this is the true selectivity for this reaction with this catalyst (and not complicated by an unselective background reaction).

Scheme 3. Asymmetric carbonyl ene reaction mediated by chiral thiourea.

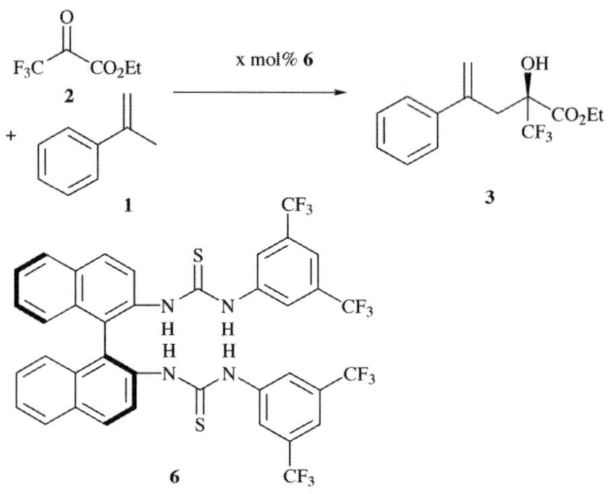

Table 2. Asymmetric ene reaction.

X mol% 6	Temp (°C)	Time (hrs)	Yield	e.e.*
10	0	16	32	23
10	-20	210	89	33
25	-20	345	65	30
25	0	90	70	26
100	-20	46	67	28

*Enantioselectivity determined using $Eu(hfc)_3$ and ^{19}F NMR
Ethyl trifluoropyruvate (0.098 ml, 0.75 mmol) was added to a solution of the thiourea (x mol%) and the α-methyl styrene (0.12 ml, 0.9 mmol, ~1.2 eq) in dichloromethane (2 ml) and left for the time shown before purification by column chromatography.

Before investigating any other asymmetric catalysts, the scope of the organocatalytic ene reaction was examined briefly. The reactions was found to proceed readily with 1,1'disubstituted alkenes, but was less effective with terminal alkenes which gave lower yields (Scheme 4 and Table 3). We note here that the ene reactions of the highly activated substrates proceed well without catalyst at room temperature: testament to trifluoropyruvate being a highly activated enophile. Catalyst 5 also did not catalyse ene reactions between α-methylstyrene and ethyl glyoxylate, trifluoro-acetophenone, or α-bromopyruvate. However, we have not had success with transition metal catalysed ketone ene reactions using these latter two substrates either. These studies reveal that the scope of the organocatalytic reactions is severely limited. A step change in catalytic performance is required for this reaction to truly reach its potential. It remains to be seen whether the more entropically favoured intramolecular carbonyl ene cyclisation reaction holds more promise.

Scheme 4. Reaction between ethyl trifluoropyruvate and various alkenes.

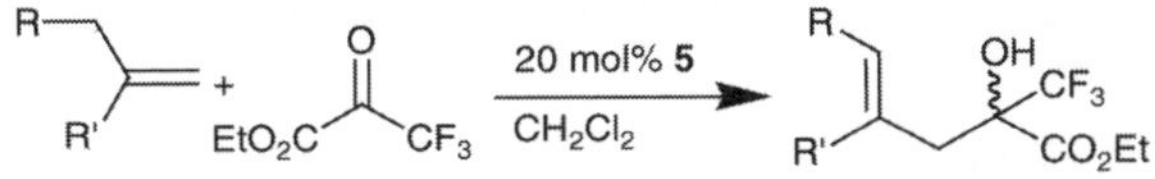

Table 3. Investigation of different alkenes in the ene reaction of trifluoropyruvate

	Alkene[a]	Time (hrs)	% conversion catalysed reaction	% conversion uncatalysed reaction
1		16	18	0
2	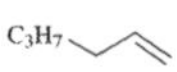	18	22	3
3	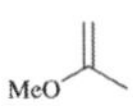	1	94	95
4	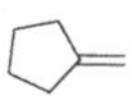	1	95	93

A: Ethyl trifluoropyruvate (0.098 ml, 0.75 mmol) was added to a solution of the thiourea **5** (75 mg, 20 mol%) and the α-methyl styrene (0.12 ml, 0.9 mmol, ~1.2 eq) in dichloromethane (2 ml).

In summary, we have developed the first organocatalytic carbonyl ene reaction and shown that some asymmetric induction is possible if a chiral thiourea catalyst is employed. However, turnover frequency, enantioselectivity and substrate scope are all very modest, suggesting that for intermolecular ene reactions, organocatalysis may not be a promising approach. Further work on intramolecular organocatalytic ene reactions is required, which may provide synthetically useful reactions.

Acknowledgements

The authors thank EPSRC for an equipment grant, the Donors of the American Chemical Society Petroleum Research Fund and Washington and Lee University (Glenn Grant) for partial support of this work (MBF). The authors would also like to thank Mrs M. Smith, S. Williamson and C. Horsburgh for technical assistance.

References

1. Snider BB: Acc Chem Res. 1980, 13:426.
2. Mikami K, Shimizu M: Chem Rev. 1992, 92:1021.
3. Dias LC: Curr Org Chem. 2000, 4:305.

4. Berrisford DJ, Bolm C: Angew Chem Int Ed Engl. 1995, 34:1717.
5. Evans DA, Tregay SW, Burgey CS, Paras NA, Vojkovsky T: J Am Chem Soc. 2000, 122:7936.
6. Jackson AC, Goldman BE, Snider BB: [http://mrw.interscience.wiley.com/cochrane/ clsysrev/ articles/ CD000284/ frame.html] J Org Chem. 1984, 49:3988.
7. Aikawa K, Kainuma K, Hatano M, Mikami K: Tetrahedron Lett. 2004, 45:183.
8. Doherty S, Knight JG, Smyth CH, Harrington RW, Clegg WL: J Org Chem. 2006, 71:9751.
9. Isanbor I, O'Hagan D: J Fluorine Chem. 2006, 127:303.
10. Mizuta S, Shibata N, Hibino M, Nagano S, Nakamura S, Toru T: Tetrahedron. 2007, 63:8521 and refs therein.
11. Clarke ML, Fuentes JA: Angew Chem Int Ed. 2007, 46:930.
12. Clarke ML: Lett Org Chem. 2004, 1:292.
13. Dalko PI, Moisan L: Angew Chem Int Ed. 2004, 43:4138.
14. Seayad J, List B: Org Biomol Chem. 2005, 3:719.
15. Taylor MS, Jacobsen EN: Angew Chem Int Ed. 2006, 45:1520.
16. Connon SJ: Angew Chem Int Ed. 2006, 45:3909.
17. Gotoh H, Masui R, Ogino H, Shoji M, Hayshi Y: Angew Chem Int Ed. 2006, 45:6853.
18. Witkopp A, Schreiner PR: Chem Eur J. 2003, 9:407.
19. Fleming EM, McCabe T, Connon SJ: Tetrahedron Lett. 2006, 47:7037.

Synthesis and Biological Evaluation of 7-O-Modified Formononetin Derivatives

Ying Yang, Wen-Jun Mao, Huan-Qiu Li, Tao-Tao Zhu, Lei Shi, Peng-Cheng Lv and Hai-Liang Zhu

ABSTRACT

Three series of novel formononetin derivatives were synthesized, in which formononetin and heterocyclic moieties were separated by 2-carbon, 3-carbon, and 4-carbon spacers. The chemical structures of these compounds were confirmed. All the derivatives were screened for antiproliferative activities against Jurkat cell line and HepG-2 cell line. In this paper, compounds prepared were also screened for their antibacterial activity of six bacterial strains. Compound 3b exihibited promising antibacterial activity against B. subtilis with minimal inhibitory concentration (MIC) value of 0.78 μg/mL, and compound 5e showed significant antiproliferative activities against Jurkat cell growth with IC_{50} *of* 1.35×10^{-4} *μg/mL. The preliminary investigation of structure-activity relationships (SARs) was also discussed based on the obtained experimental data.*

Introduction

Isoflavonoids are a broad class of polyphenolic secondary metabolites that are abundant in plants [1, 2] and in various common foods such as apples, onions, tea, and red wine [3, 4]. Isoflavonoids also have a potent activity against protein tyrosine kinase (PTK) [5]. Because of such a broad range of pharmacological properties, they receive considerable therapeutic importance. Protein tyrosine kinases (PTKs) have been intensively investigated because of their role in the transduction of proliferative signals in mammalian cells. Many transmembrane growth factor receptors possess intracellular PTK activity, with initiation of this activity following external binding of a growth factor, being the first step in the cellular signal transduction pathway which controls mitogenesis and cell proliferation [6, 7]. Therefore, selective interruption of signal transduction in tumor cells by specific inhibitors of PTK activity has recently emerged as a major new approach for the design of tumor-specific drugs [8, 9].

Formononetin (1, shown in Scheme 1), a kind of isoflavonoid, is reported to have many biological activities including antiproliferative, antioxidant, antidiabetic, antiestrogenic, antibacterial, antiangiogenic effects, and so on [10–14]. Formononetin is also a potent aryl hydrocarbon receptor agonist in vitro [14]. The versatile biological activities of formononetin prompt us to prepare a new series of its derivatives and evaluate their biological significance. Herein, we describe the synthesis of formononetin derivatives in which formononetin and heterocyclic moieties were linked by spacers,and investigate the effects of the size of the spacers and substitution patterns of the heterocyclic moieties. All of the compounds were assayed for their antiproliferative activities against a panel of two human tumor cell lines (Jurkat and HepG-2) by applying the MTT colorimetric assay. The results of this study may be useful to researchers attempting to gain more understanding of the PTK inhibitory activity of isoflavonoid derivatives.

Scheme 1. Synthesis of 7-O-heterocycle derivatives of formononetin. Reagents and conditions: (i) $BrCH_2CH_2Br$, $BrCH_2CH_2CH_2Br$ or BrCH$_2$CH$_2$CH2CH$_2$Br, K_2CO_3, DMF; (ii) R, K_2CO_3, dioxane, DMF, heating.

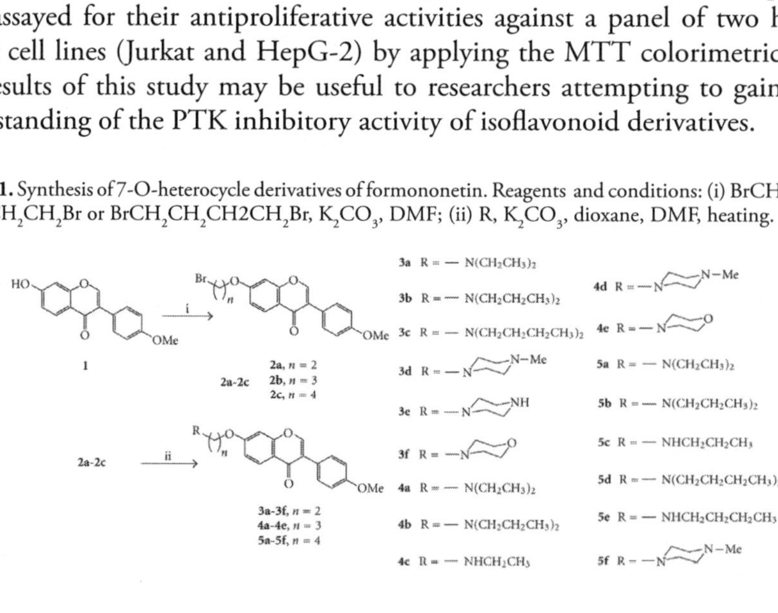

To our knowledge, this is the first report on the screening of 7-O-modified formononetin derivatives for their antimicrobial and antiproliferative activities.

Results and Discussions

Chemical Synthesis

Compounds 2a–c were the key intermedi ates for the synthesis of the compounds investigated. They were prepared from alkylation of 7-OH group by using 1,2-, 1,3-, or 1,4-dihaloalkanes in the presence of K_2CO_3 in anhydrous DMF [15]. The synthesis of compounds 3a–f, 4a–e, and 5a–f was accomplished according to the general pathway illustrated in Scheme 1. To increase the antimicrobial properties of formononetin, formononetin derivatives in which the formononetin ring system was linked to the alkylamines by different spacers at C-7 position were investigated, with a view to modify their lipophilicity. Literature survey revealed that the compounds containing alkyl amino side chains showed better activities against the test bacteria than those containing aromatic ring amino side chains [15]. To further optimize this activity, seventeen compounds reported in this article contain alkyl amino groups. Reaction of 2a–c with different cyclic and noncyclic alkylamines yielded 3a–f, 4a–e, and 5a–f, respectively, which were all first reported. All of the synthesized compounds gave satisfactory analytical and spectroscopic data, which were in full accordance with their depicted structures.

Biological Evaluation and Discussion

Antiproliferative Activities [16]

The antiproliferative activities of these compounds were evaluated against a panel of two human tumor cell lines (Jurkat and HepG-2) by applying the MTT colorimetric assay. The observed IC50's are listed in Table 1. From the results of the in vitro antiproliferative MTT tests of the prepared compounds, it followed that in series 1, most of the prepared compounds showed good antiproliferative activities. Among them compounds 3b and 3d exhibited strong activities on Jurkat cell growth. Also, compounds 3b and 3c had stronger activities on HepG-2 cell line than the positive control 5-UF. In series 2, compounds 4a, 4b, and 4e displayed remarkable antiproliferative activities on Jurkat cell, while they just showed moderate activities against HepG-2 cell. In series 3, compound 5e displayed significant antiproliferative activities on Jurkat cell growth, and compound 5d showed promising inhibitory activities on HepG-2 cell growth.

Table 1. Antiproliferative activity of the synthesized compounds.

Compounds	IC_{50} (μg/mL)	
	Jurkat[1]	HepG-2[2]
1	23.52	73.03
2a	48.04	>100.00
2b	0.21	4.14
2c	9.50	7.18
3a	13.93	5.57
3b	1.82	3.24
3c	>100.00	2.67
3d	3.70	8.14
3e	12.21	7.15
3f	16.49	7.54
4a	0.41	6.14
4b	5.32	8.58
4c	16.44	52.21
4d	18.08	5.90
4e	3.49	57.82
5a	1.94	6.54
5b	>100.00	7.52
5c	0.64	4.94
5d	1.27	1.49
5e	1.35×10^{-4}	13.14
5f	7.11	13.53
5-fluorouracil	18.41	2.50

[1]Jurkat: Human T cell lymphoblast-like cell line.
[2]HepG-2: Human hepatocellular liver carcinoma cell line.

Antibacterial Activities [17]

The antibacterial activities of the synthetic compounds were tested against B. subtilis, S. aureus and S. faecalis (Gram-positive bacteria), E. coli, P. aeruginosa, and E. cloacae (Gram-negative bacteria) by broth dilution method recommended by National Committee for Clinical Laboratory Standards (NCCLS) [18, 19]. Standard antimicrobial agents like penicillin and kanamycin were also screened under identical conditions for comparison. The minimal inhibitory concentration (MIC) values for the bacteria are listed in Table 2.

Table 2. Antimicrobial activity of the synthesized compounds.

Compounds	Minimum inhibitory concentrations MICs (μg/mL)					
	Gram positive			Gram negative		
	S. faecalis	*S. aureus*	*B. subtilis*	*E. coli*	*P. aeruginosa*	*E. cloacae*
1	3.12	25.00	50.00	50.00	25.00	50.00
2a	3.12	6.25	3.12	6.25	1.56	50.00
2b	3.12	50.00	25.00	6.25	25.00	1.56
2c	3.12	3.12	50.00	12.50	25.00	12.50
3a	3.12	6.25	50.00	3.12	3.12	25.00
3b	1.56	12.50	0.78	3.12	50.00	3.12
3c	3.12	6.25	6.25	25.00	6.25	3.12
3d	3.12	6.25	6.25	12.50	3.12	6.25
3e	3.12	1.56	50.00	25.00	25.00	6.25
3f	3.12	12.50	25.00	6.25	50.00	25.00
4a	3.12	3.12	50.00	6.25	50.00	50.00
4b	3.12	25.00	3.12	1.56	6.25	12.50
4c	3.12	12.50	25.00	6.25	6.25	12.50
4d	3.12	6.25	3.12	12.50	3.12	25.00
4e	3.12	3.12	50.00	1.56	50.00	50.00
5a	3.12	1.56	6.25	25.00	6.25	3.12
5b	3.12	12.50	1.56	25.00	50.00	50.00
5c	3.12	50.00	1.56	6.25	6.25	50.00
5d	3.12	50.00	25.00	6.25	3.12	6.25
5e	3.12	0.78	12.50	6.25	25.00	25.00
5f	3.12	25.00	50.00	50.00	3.12	1.56
Penicillin	1.56	1.56	1.56	6.25	6.25	3.12
Kanamycin	3.12	1.56	0.39	3.12	3.12	1.56

Most compounds in series 1, which contained a 2-carbon spacer, displayed good activities against the test microorganisms. In this series, compound 3b showed pronounced activity against B. subtilis and S. faecalis with MIC values of 1.56 μg/mL and 0.78 μg/mL, respectively. In addition, compound 3e showed good activity against S. aureus.

A few compounds in series 2, which contain a 3-carbon spacer, exhibited great activities against the test microorganisms. In this series, compounds 4b, 4d, and 4e showed great activities against S. aureus, B. subtilis, and E. coli. Among them, 4b and 4e showed strong activities against E. coli with their MIC value (1.56 μg/mL) superior to the positive control kanamycin.

Similarly, several compounds in series 3, which contain a 4-carbon spacer, showed good activities against the test microorganisms. Among them, compounds 5a and 5e exhibited strong activities against S. aureus with their MIC values of 1.56 μg/mL and 0.78 μg/mL, respectively. Compounds 5b and 5c displayed great activities against B. subtilis. Also, compounds 5d and 5f showed strong activities against P. aeruginosa with the MICs (3.12 μg/mL), which were comparable to the positive control kanamycin.

Conclusion

To investigate the biological activities of formononetin derivatives, we had synthesized three series of formononetin derivatives. For all the compounds synthesized, antiproliferative and antibacterial activities against two cancer cell lines (Jurkat and HepG-2) and six bacterial strains (three Gram-positive bacterial strains: Bacillus subtilis, Staphylococcus aureus, and Streptococcus faecalis and three Gram-negative bacterial strains: Escherichia coli, Pseudomonas aeruginosa, and Enterobacter cloacae) were determined. From the bioassay results, it may be concluded that those containing long alkyl amino side chains exhibited better activities against gram-positive bacteria than those containing short ones, while it had the opposite rule for gram-negative. Specifically, the derivatives with dipropylamine moiety were more active than most of the other analog. In this study, we focused our attention on the structure-activity relationships. The work was of interest because this is a preliminary investigation of SAR, serving as the basis of further more detailed work.

Acknowledgement

Financial support by the National Natural Science Foundation of China (no. 30772627) is kindly acknowledged.

References

1. V. M. Malikov and M. P. Yuldashev, "Phenolic compounds of plants of the Scutellaria genus. Distribution, structure, and properties," Chemistry of Natural Compounds, vol. 38, no. 5, pp. 473–519, 2002.
2. M. J. del Baño, J. Lorente, J. Castillo, et al., "Flavonoid distribution during the development of leaves, flowers, stems, and roots of Rosmarinus officinalis. Postulation of a biosynthetic pathway," Journal of Agricultural and Food Chemistry, vol. 52, no. 16, pp. 4987–4992, 2004.
3. P. L. Whitten, S. Kudo, and K. K. Okubo, "Isoflavonoids," in Handbook of Plant and Fungal Toxicants, pp. 117–137, CRC Press, Boca Raton, Fla, USA, 1997.

4. G. M. Boland and D. M. X. Donnelly, "Isoflavonoids and related compounds," Natural Product Reports, vol. 15, no. 3, pp. 241–260, 1998.

5. K. T. Papazisis, D. Zambouli, O. T. Kimoundri, et al., "Protein tyrosine kinase inhibitor, genistein, enhances apoptosis and cell cycle arrest in K562 cells treated with γ-irradiation," Cancer Letters, vol. 160, no. 1, pp. 107–113, 2000.

6. A. Ullrich and J. Schlessinger, "Signal transduction by receptors with tyrosine kinase activity," Cell, vol. 61, no. 2, pp. 203–212, 1990.

7. S. R. Hubbard and J. H. Till, "Protein tyrosine kinase structure and function," Annual Review of Biochemistry, vol. 69, pp. 373–398, 2000.

8. F. M. Uckun and C. Mao, "Tyrosine kinases as new molecular targets in treatment of inflammatory disorders and leukemia," Current Pharmaceutical Design, vol. 10, no. 10, pp. 1083–1091, 2004.

9. P. Traxler, J. Green, H. Mett, U. Séquin, and P. Furet, "Use of a pharmacophore model for the design of EGFR tyrosine kinase inhibitors: isoflavones and 3-phenyl-4(1H)-quinolones," Journal of Medicinal Chemistry, vol. 42, no. 6, pp. 1018–1026, 1999.

10. X. Yu, W. Wang, and M. Yang, "Antioxidant activities of compounds isolated from Dalbergia odorifera T. Chen and their inhibition effects on the decrease of glutathione level of rat lens induced by UV irradiation," Food Chemistry, vol. 104, no. 2, pp. 715–720, 2007.

11. Y. Ungar, O. F. Osundahunsi, and E. Shimoni, "Thermal stability of genistein and daidzein and its effect on their antioxidant activity," Journal of Agricultural and Food Chemistry, vol. 51, no. 15, pp. 4394–4399, 2003.

12. S. Sato, J. Takeo, C. Aoyama, and H. Kawahara, "Na+-glucose cotransporter (SGLT) inhibitory flavonoids from the roots of Sophora flavescens," Bioorganic & Medicinal Chemistry, vol. 15, no. 10, pp. 3445–3449, 2007.

13. Z.-N. Ji, W. Y. Zhao, G. R. Liao, et al., "In vitro estrogenic activity of formononetin by two bioassay systems," Gynecological Endocrinology, vol. 22, no. 10, pp. 578–584, 2006.

14. S. Medjakovic and A. Jungbauer, "Red clover isoflavones biochanin A and formononetin are potent ligands of the human aryl hydrocarbon receptor," The Journal of Steroid Biochemistry and Molecular Biology, vol. 108, no. 1-2, pp. 171–177, 2008.

15. L.-N. Zhang, Z.-P. Xiao, H. Ding, et al., "Synthesis and cytotoxic evaluation of novel 7-O-modified genistein derivatives," Chemistry & Biodiversity, vol. 4, no. 2, pp. 248–255, 2007.

16. J. Meletiadis, J. F. G. M. Meis, J. W. Mouton, J. P. Donnelly, and P. E. Verweij, "Comparison of NCCLS and 3-(4,5-dimethyl-2-thiazyl)-2,5-diphenyl-2-H-tetrazolium bromide (MTT) methods of in vitro susceptibility testing of filamentous fungi and development of a new simplified method," Journal of Clinical Microbiology, vol. 38, no. 8, pp. 2949–2954, 2000.
17. M. Hajdúch, V. Mihál, J. Minařík, et al., "Decreased in vitro chemosensitivity of tumour cells in patients suffering from malignant diseases with a poor prognosis," Cytotechnology, vol. 19, no. 3, pp. 243–245, 1996.
18. National committee for clinical laboratory standards, "Reference methed for broth dilution antifungal susceptibility testing of yeasts: approved standard," M27-A, NCCLS, Wayne, Pa, USA, 1997.
19. National committee for clinical laboratory standards, "Development of in vitro susceptibility testing criteria and quality control parameters: tentative guideline," M23-T3, NCCLS, Villanova, Pa, USA, 1998.

Expedient Syntheses of the N-Heterocyclic Carbene Precursor Imidazolium Salts Ipr ·Hcl, Imes ·Hcl and Ixy ·Hcl

Lukas Hintermann

ABSTRACT

The 1,3-diaryl-imidazolium chlorides IPr·HCl (aryl = 2,6-diisopropylphenyl), IMes·HCl (aryl = 2,4,6-trimethylphenyl) and IXy·HCl (aryl = 2,6-dimethylphenyl), precursors to widely used N-heterocyclic carbene (NHC) ligands and catalysts, were prepared in high yields (81%, 69% and 89%, respectively) by the reaction of 1,4-diaryl-1, 4-diazabutadienes, paraformaldehyde and chlorotrimethylsilane in dilute ethyl acetate solution. A reaction mechanism involving a 1,5-dipolar electrocyclization is proposed.

Background

Imidazolylidene carbenes have been investigated as ligands in coordination chemistry, as powerful steering/controlling elements in transition-metal catalysis,[1,2] and more recently as metal-free catalysts for organic reactions[3,4]. Some prominent members of the family of N-heterocyclic carbenes (NHC) are the sterically encumbered imidazolylidenes IPr and IMes (Figure 1), which can also be considered as analogues of bulky and electron-rich tertiary phosphanes. In contrast to the latter, their synthesis does not involve air-sensitive or pyrophoric organometallic reagents, and they are conveniently stored and used in the form of their air-stable precursors, namely the imidazolium salts IPr·HCl (1,3-bis-{2,6-diisopropylphenyl}imidazolium chloride; 1) or IMes·HCl (1,3-bis-{2,4,6-trimethylphenyl} imidazolium chloride; 2) (Figure 1).

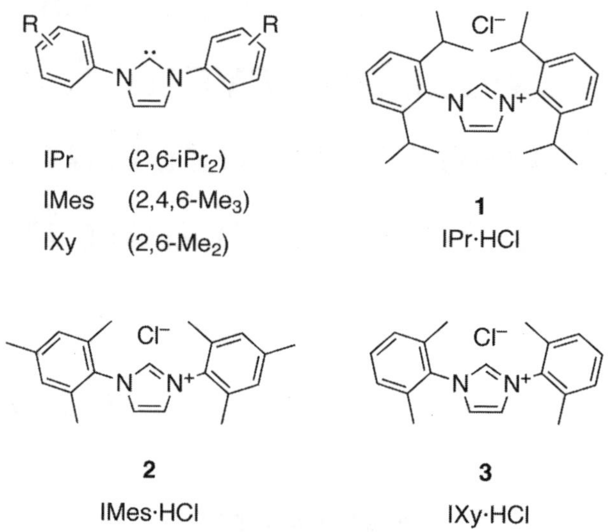

Figure 1. The carbenes IPr, IMes, IXy and their imidazolium salt precursors

At present, there is a slightly ironic situation: even though salts 1 and 2 are in high demand and used by many researchers, information about their synthesis is scattered over the literature, and standard references for satisfactory, high-yielding (e.g., >60%) preparations are not available. The present work introduces a simple, reliable and economic synthesis of IPr·HCl (1), IMes·HCl (2) and IXy·HCl (3). The procedure should be especially valuable to those who want to prepare these useful salts on large scale, rather than obtaining them from commercial suppliers.

A synthesis of N,N-disubstituted symmetric imidazolium salts by condensation of glyoxal, two equivalents of aliphatic or aromatic amine, and an equivalent of paraformaldehyde in the presence of hydrochloric acid was reported by Arduengo in a patent in 1991 (Scheme 1a). [5] It is successful for a range of alkyl- and arylamines, but the original document gave only a few examples and limited details on product purification. In fact, one characteristic of the protocol is the generation of dark-brown impurities which render product purification tedious. Thus, the synthesis of IMes·HCl (2) according to the Arduengo protocol is followed by extensive solvent washes, resulting in a low yield (40%). [6,7] Both the Arduengo and Nolan groups noted that this synthetic protocol cannot be extended to the sterically more hindered IPr·HCl (1). [8,9] Instead, Arduengo and coworkers found that the reaction of the 1,4-diaryl-1,4-diazadienes (DADs, or glyoxal imines) 4 and 5 with chloromethyl-ethylether (THF, 40°C/16 h or 23°C/5 d) gave pure IPr·HCl (1) or IMes·HCl (2) in 47% or 40% yield, respectively (Scheme 2b). [8]

Scheme 1. Synthetic routes to and diazadiene precursors for imidazolium salts.

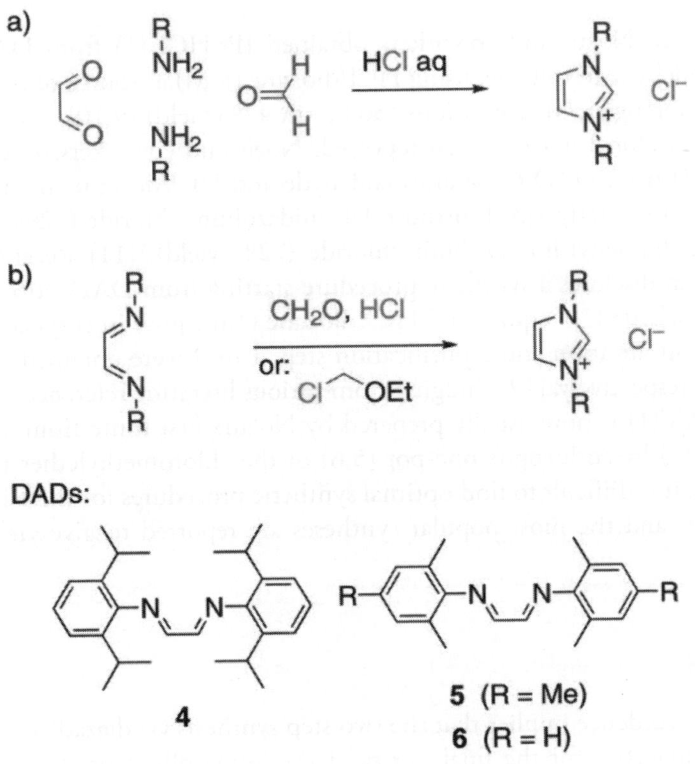

Scheme 2. The imidazolium salt synthesis as a 1, 5-dipolar electrocyclization.

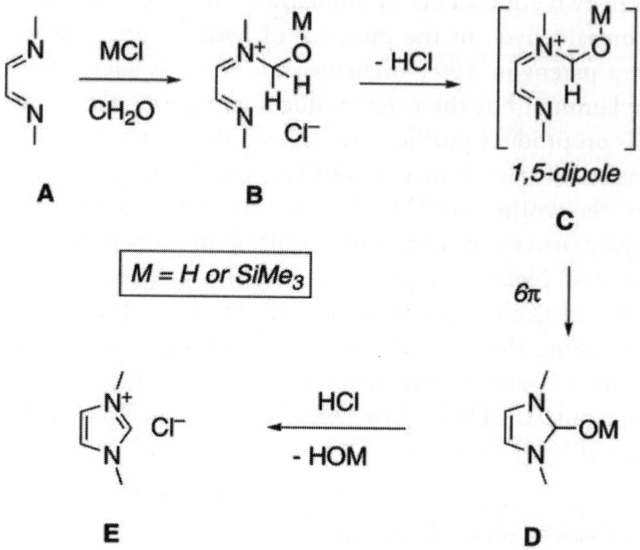

In parallel, Nolan and coworkers obtained IPr·HCl (1) from DAD 4 and paraformaldehyde in toluene, using HCl/dioxane (4 M) as source of the counter-ion under homogeneous conditions (36 h, r.t., 47% yield) [9,10]. A corresponding procedure for 2 has not been reported. Noels and coworkers have prepared IXy·HCl (3) from DAD 6, paraformaldehyde and HCl/dioxane in THF (r.t., 4 h, 55%), and similarly 1,3-di-ortho-tolyl-imidazolium chloride (32% yield) and 1,3-bis-para-biphenyl-imidazolium chloride (72% yield). [11] Recently, a patent by Nolan discloses a synthetic procedure starting from DADs 4/5 (1 equiv), paraformaldehyde (1.3 equiv) and HCl/dioxane (1.6 equiv) in ethyl acetate. Following a neutralization and a purification step, 1 or 2 were obtained in 70% or 66% yield, respectively. [12]. Judging from various literature references, it appears that IPr·HCl (1) is now usually prepared by Nolan's first route from 4,[10] and IMes·HCl (2) by Arduengo's one-pot [5,6] or the chloromethylether route. [8]. In any case, it is difficult to find optimal synthetic procedures for these important compounds, and the most popular syntheses are reported to give yields in the 40% range.

Results

Literature precedence implies that the two step synthesis via diazadienes (Scheme 2b) is advantageous for the final purity of the imidazolium products. We have obtained the glyoxal imines 4–6 following established procedures.[8,10,13-15] Our experiments on the synthesis of 1 started with the Nolan procedure (condensation of 4 and paraformaldehyde in toluene), but we whished to replace the reagent HCl/dioxane (4 M) by a more readily available and cheaper source of the chloride counter ion. The choice fell on chlorotrimethylsilane (TMSCl), the "silyl version" of HCl, which is easily measured and handled. The curr ent pricing of TMSCl is 8 €/mol, as opposed to 80–100 €/mol for HCl/dioxane (4 M). The reaction optimization went as follows: Initial experiments used two equivalents of TMSCl in hot toluene, assuming the validity of equation 1 for the reaction:

(RN)2C2H2 + CH2O + 2 TMSCl = R2N2C3H3 + /Cl- + TMS2O + HCl (1)

This protocol gave IPr·HCl (1) in satisfactory yield as almost colorless microcristalline powder (Table 1, entry 1), and also worked well for the synthesis of IMes·HCl (2; entry 7). However, it gave mixed results when applied to IXy·HCl (3), with reaction solutions turning deep brown, and the solid precipitate being accompanied by impurities. We suspected that HCl as a strongly acidic byproduct (see equation 1) was responsible for side reactions. The amount of TMSCl was therefore reduced to one equivalent, which assures neutral reaction conditions, but also means that water is liberated in the course of the reaction (equation 2):

(RN)2C2H2 + CH2O + TMSCl = R2N2C3H3+/Cl-
+ 0.5 TMS2O + 0.5 H2O (2)

DAD 4 was still cyclized to 1 (entry 4), but in reactions of 6, the water of condensation separated as drops which accumulated both acid (from TMSCl + H2O) and reaction intermediates, producing brown resins. We reasoned that the water of condensation had to be removed either chemically or simply by homogeneous dissolution in the reaction mixture. Indeed, by adding 30% of tetrahydrofurane (THF) to the reaction solvent (entry 13), IXy·HCl (3) was precipitated in very high yield and satisfactory purity. Since THF is relatively expensive, we hoped to find another one-component reaction solvent instead, whose polarity would be low enough to precipitate the imidazolium salts in pure form, but also sufficiently high in order to dissolve the liberated water of condensation. Ethyl acetate fulfilled these requirements and gave optimal results in all imidazolium salt syntheses (entries 6,8,14). In the course of scale-up, the reactant concentration was identified as another important parameter (entry 2 vs 3; entries 5,10). The final, recommended procedure for the synthesis of 1,3-diarylimidazolium chlorides consists in combining the reactants in stoichiometric quantities in ethyl acetate (7–10 mL/mmol) at 70°C. This simple protocol gave IPr·HCl (1), IMes·HCl (2) and IXy·HCl (3) in pure form and high yields (Table 1, entries 6,8,14).

Table 1. Optimization of the TMSCl-induced imidazolium salt synthesis

entry	starting material	stoichiometry[a]	scale [mmol]	solvent [mL/mmol]	T [°C]	time [h]	product	yield [%]
1	**4**	1:1:2[b]	6.6	PhMe (7.5)	80+r.t.	4+10	**1**	61
2	**4**	1:1:1.9	26.6	PhMe (5.6)	75–80	3	**1**	67
3	**4**	1:1:1.9	122	PhMe (2.9)	70	5	**1**	<40[c]
4	**4**	1:1:1.05	135.5	PhMe (7.4)	85	2.5	**1**	65
5	**4**	1:1:1.04	5.3	THF (2.8)	70	4	**1**	31[d]
6	**4**	1:1:1	134	EtOAc (9.5)	70	2.75	**1**	81
7	**5**	1:1:2	7.5	PhMe (4)	75–80	3	**2**	64
8	**5**	1:1:1	7.5	EtOAc (9)	70	1.5	**2**	69
9	**6**	1:1:2	25	PhMe (6)	80	3.5	**3**	43[e]
10	**6**	1:1.05:1.35	10	PhMe (2)	80	2.25	**3**	–[f]
11	**6**	1.04:1:1.09	59	PhMe (7)	85	1	**3**	–[f]
12	**6**	1:1:1.2	10	THF (2)	r.t.	24	**3**	53
13	**6**	1:1:1	19	PhMe/THF (8.4)	85	3.5	**3**	88
14	**6**	1:1:1	19	EtOAc (8.4)	80	1.5	**3**	89.5

a) Diazadiene/paraformaldehyde/TMSCl. b) Addition of TMSCl in two portions, in all other cases slow, dropwise addition. c) Impure product. d) Yield after recrystallization from acetone/tBuOMe; raw yield 60%. e) Losses due to washing away of colored impurities with acetone. f) Brown, impure material was obtained.

Discussion

The synthesis of N,N-diarylimidazolium salts according to the Arduengo three component condensation [5] produces strongly colored reaction mixtures, from which the products are obtained as impure raw materials. [5-8,11] The biphasic aqueous/organic reaction conditions of that protocol have been identified as problematic, and more recent syntheses start from preformed 1,4-diazabutadienes, and either paraformaldehyde/HCl/dioxane [9,10] or chloromethyl ethers [8] or other chloromethyl-derivatives (e.g., $ClCH_2OPiv$) in combination with silver salts of non-coordinating counter ions. [16,17] We have now found that the use of TMSCl as chloride donor instead of HCl, ethyl acetate as the solvent, and an appropriate dilution of the reactants lead to a reliable synthetic procedure for the imidazolium salts 1–3. However, in terms of generality, syntheses of imidazolium salts remain capricious in nature, as our conditions were not successful for other DADs such as those derived from ortho-toluidine, para-toluidine or para-chloroaniline. These gave strongly colored reaction mixtures, from which dark solids separated, which we have not attempted to purify. What is the mechanistic basis behind such differences of the reaction outcome? To our knowledge, the mechanism of the Arduengo imidazolium salt synthesis has not been discussed in the literature. We propose that alkylation of the DADs A by formaldehyde and HCl (or TMSCl) leads to an iminium salt B (Scheme 2); analogous alkylations can be induced by chloromethylethers. Loss of HCl from B (presumably to A, which is a monobasic species)[13] gives an imino-azomethin-ylide 1,5-dipole C (only one mesomeric structure shown), which will undergo 1,5-dipolar cyclization (6π electrocyclization) [18-20] to an oxy-imidazoline D. Elimination of the oxy group then generates the imidazolium salt E. The electrocyclic mechanism proposed here, though not backed up by experimental data, is plausible considering related precedence

[18-20] and because it proceeds via a favorable electroneutral intermediate (C). Alternative ionic mechanisms require a double alkylation at nitrogen, which leads to an energetically unfavorable dication. While this might be prevented through charge-quenching reversible 1,2-additions of chloride (e.g., to the iminium function in B), the resulting mechanistic schemes require additional steps and intermediates which are not backed up by sideproducts.

The rate-determining step might well be the alkylation of A to B, and competing processes will lead to side-products. In 1986, tom Dieck and coworkers described that the DAD from tert-butylamine reacted with HCl gas in toluene to give an imidazolium salt (Scheme 3a). [21] This is one potential side-reaction, but brown colored side-products can also derive from electrophilic aromatic alkylations of arylamine units by iminium salts from formaldehyde or protonated DADs (as in the bakelite reaction). Such side-reactions will be most prominent with DADs from electron-rich anilines, and with sterically less hindered DADs which are prone to hydrolysis (Scheme 3b). The clean reactions observed under our conditions with the DADs 4–6 are thus probably a consequence of their kinetic stability relative to hydrolysis, in combination with steric hindrance that prevents them from participating as electrophiles in aromatic substitution reactions.

Scheme 3. Potential side-reactions in the imidazolium salt synthesis.

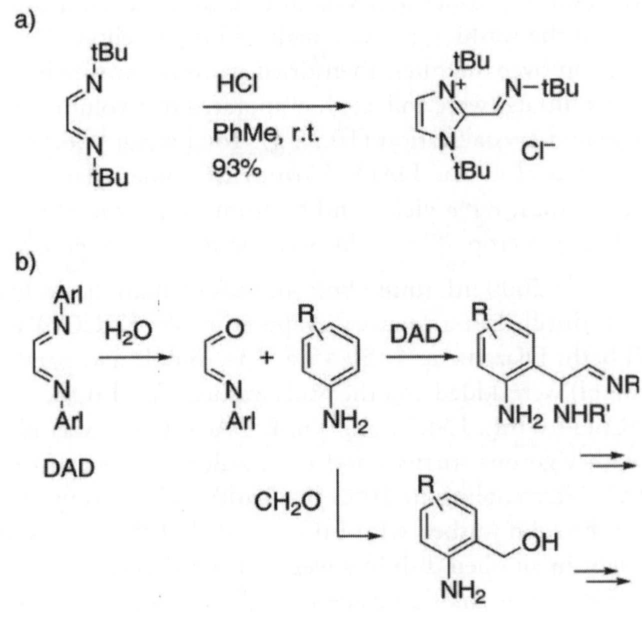

Conclusion

The useful imidazolium salts IPr·HCl (1), IMes·HCl (2) and IXy·HCl (3) have previously been synthesized by a range of methods giving the products in variable purity and yields. We now present a protocol for the condensation of 1,4-diaryl-diazabutadienes and paraformaldehyde with the following characteristics: a) chlorotrimethylsilane (TMSCl) serves as a cheap and easy to handle source of the chloride counter ion, b) ethyl acetate is the optimal reaction solvent with respect to product purity and precipitation, and c) the reaction requires an appropriate dilution to give high yields. The starting materials and solvents for this synthesis are readily available. The imidazolium salts precipitate from the reaction mixture as pure microcrystalline powders in high yields, considerably surpassing those previously reported in the literature. No product purification is necessary, and the procedure is amenable to large-scale.

Experimental

A) Synthesis of DAD 4:[8,10] A solution of glyoxal (72.55 g, 40% in water, 0.50 mol) in MeOH (250 mL) was added with vigorous stirring to a warmed (50°C) solution of 2,6-diisopropylaniline (197 g, purity 90%, 1 mol) and HOAc (1 mL) in MeOH (250 mL). A slightly exothermic reaction commenced and the product started to crystallize after 15 min. The mixture was stirred for 10 h at r.t., after which the resulting suspension was filtered and the solid product washed with MeOH, until the washing phase remained bright yellow. The product was pre-dried by suction over the filter, then dried to constant weight (158.29 g) in high vacuum. The filtrates were collected, evaporated to a volume of 100 mL and set aside for a second crystallization (10.28 g). Total yield: 168.57 g (89.5%) of bright yellow crystals of 4. The DADs 5 (from 2,6-dimethylaniline, in MeOH, 0.25 mol scale, 2 crops, 84% yield) and 6 (from 2,4,6-trimethylaniline, in iP-rOH, 50 mmol scale, 1 crop, 87% yield) were prepared analogously.

B) IPr·HCl (1): A 2000 mL round bottom flask containing EtOAc (1200 mL, technical quality, distilled in a rotatory evaporator over K2CO3) was heated to 70°C in an oil bath. Diazadiene 4 (50.45 g, 134 mmol) and paraformaldehyde (4.06 g, 135 mmol) were added and the walls washed with EtOAc (50 mL). A solution ofTMSCl (17.0 mL, 134.5 mmol) in EtOAc (20 mL) was added dropwise over 45 min with vigorous stirring, and the resulting yellow suspension stirred for 2 h at 70°C. After cooling to 10°C (ice-bath) with stirring, the suspension was filtered and the solid washed with EtOAc and tBuOMe. The solid was dried to constant weight in an open dish in a well-ventilated oven at 100°C (1 d), giving 46.04 g (81%) of 1 as colorless microcrystalline powder. The salts IMes·HCl

(2; 69%) and IXy·HCl (3: 89.5%) were prepared analogously, see also Table 1 for conditions.

The reaction products were identified by comparison of 1H and 13C NMR data to literature values.

Acknowledgements

Prof. Carsten Bolm, RWTH Aachen University, the Deutsche Forschungsgemeinschaft (Emmy Noether Programm) and the Fonds der Chemischen Industrie are thanked for generous support.

References

1. Glorius F, (Ed): N-Heterocyclic Carbenes in Transition Metal Catalysis. Topics Organomet Chem 2007, 21:1–218.
2. Herrmann WA: Angew Chem Int Ed. 2002, 41:1290–1309.
3. Marion N, Diéz-González S, Nolan SP: Angew Chem Int Ed. 2007, 46:2988–3000.
4. Enders D, Balensiefer T: Accounts Chem Res. 2004, 37:534–541.
5. Arduengo AJ: US Patent 5077414. 1991.
6. Voges MH, Rømming C, Tilset M: Organometallics. 1999, 18:529–533.
7. Cole ML, Junk PC: Cryst Eng Comm. 2004, 6:173–176.
8. Arduengo AJ, Krafczyk R, Schmutzler R, Craig HA, Goerlich JR, Marshall WJ, Unverzagt M: Tetrahedron. 1999, 55:14523–14534.
9. Huang J, Nolan SP: J Am Chem Soc. 1999, 121:9889–9890.
10. Jafarpour L, Stevens ED, Nolan SP: J Organomet Chem. 2000, 606:49–54.
11. Delaude L, Szypa M, Demonceau A, Noels AF: Adv Synth Catal. 2002, 344:749–756.
12. Nolan SP: US Patent 7109348. 2006.
13. Kliegman JM, Barnes RK: Tetrahedron. 1970, 26:2555–2560.
14. Kliegman JM, Barnes RK: J Org Chem. 1970, 35:3140–3143.
15. tom Dieck H, Renk IW: Chem Ber. 1971, 104:92–109.
16. Glorius F: EP 1521745. 2004.
17. Kison C, Opatz T: Synthesis. 2006, 3727–3738.

18. Huisgen R: Angew Chem Int Ed Eng. 1980, 19:947–1034.
19. Reimlinger H: Chem Ber. 1970, 103:1900–1907.
20. Taylor EC, Turchi IJ: Chem Rev. 1979, 79:181–231.
21. Zettlitzer H, tom Dieck H, Haupt ETK, Stamp L: Chem Ber. 1986, 119:1868–1875.

CS_2CO_3/[bmim]Br as an Efficient, Green, and Reusable Catalytic System for the Synthesis of N-Alkyl Derivatives of Phthalimide under Mild Conditions

Alireza Hasaninejad, Abdolkarim Zare, Ahmad Reza Moosavi-Zare, Fatemeh Khedri, Rahimeh Rahimi and Ali Khalafi-Nezhad

ABSTRACT

Aza-conjugate addition of phthalimide to α,β-unsaturated esters efficiently achieves in the presence of catalytic amount of Cs_2CO_3 and ionic liquid 1-butyl-3-methylimidazolium bromide ([bmim]Br) under mild reaction *conditions (70 °C) to afford N-alkyl phthalimides in high yields and relatively short reaction times.*

Introduction

Over the past decade, room temperature ionic liquids have been emerged as a new class of stable and inert solvents. They exhibit high thermal stability, high polarity due to their ionic nature and a great ability to solubilize polar and nonpolar organic compounds. To date, various organic reactions have been carried out and investigated in ionic liquids, such as conjugate addition of sulfonamides [1, 2], azide ion [3], amines and N-heterocycles [4, 5], indoles [6], thiocyanide ion [7], active methylenes [8, 9], and thiols [10] to electrophilic alkenes, and other carbon-carbon, carbon-nitrogen, carbon-oxygen as well as carbon-sulfur bonds formation [11, 12].

N-alkyl derivatives of phthalimide have attracted much interest due to their potential use as antipsychotic [13], anti-inflammatory [14], hypolipidemic [15], agents and receptors [16], and so on. Moreover, these compounds are very useful intermediates in organic synthesis as they can be easily converted to primary amines (Gabriel synthesis) [17]. Therefore, there is a great deal of interest in the synthesis of this class of compounds. The aza-conjugate addition of phthalimide to electrophilic alkenes can be used as a useful synthetic route toward N-alkylated phthalimides [18–21]. Several catalysts have been applied to achieve this transformation such as Na/EtOH [18], AlMe2Cl [19], ZnO [20], and 1,4-diazabicyclo[2,2,2]octane [21]. However, these reported methods are associated with one or more of the following drawbacks: (i) moderate yield, (ii) relatively long reaction time, (iii) harsh conditions, (iv) the use of more reactive phthalimide salts instead of phthalimide, (v) difficult experimental procedure, and (vi) the necessity of stoichiometric amount of catalyst. Moreover, the aza-conjugate addition reaction of phthalimide has been scarcely studied. Therefore, it seems highly desirable to find an efficient new protocol for this reaction.

Cesium carbonate is a commercially available, heterogeneous, and environmentally benign basic catalyst that has been used in various organic transformations [22–28].

Considering the above subjects and also in continuation of our previous studies on green organic synthesis [20, 21, 29–32], we report here an efficient, green, and simple method for the preparation of N-alkyl phthalimides via aza-conjugate addition of phthalimide to α,β-unsaturated esters in the presence of catalytic amount of Cs2CO3 in [bmim]Br at 70°C (Scheme 1). This present method has none of the above disadvantages at all.

Scheme 1

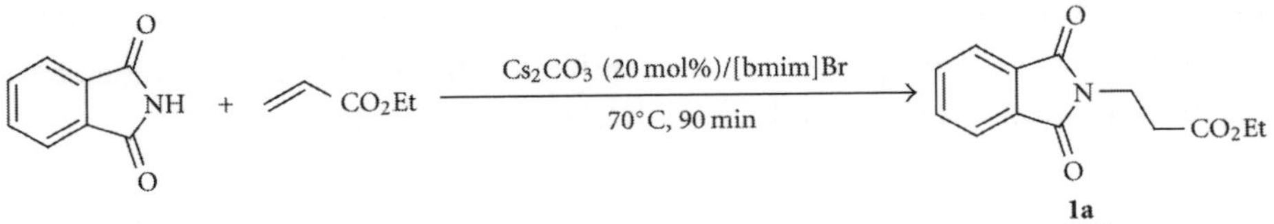

1a
98%

Results and Discussion

We have found previously Cs_2CO_3 acts as an efficient basic reagent for N3-alkylation of N1-substituted pyrimidines [22], and N-arylation of nucleobases [23]. Moreover, this base has been frequently applied for alkylation and arylation reactions [24–28]. These subjects encouraged us to use Cs_2CO_3 as catalyst for N-alkylation of phthalimide via aza-conjugate addition reaction. Therefore, firstly we used different amounts of Cs_2CO_3 to accomplish aza-conjugate addition of phthalimide (2 mmol) to ethyl acrylate (2.4 mmol) in [bmim]Br (1 g) at range of 25–100°C to provide compound 1a (Scheme 1). The results showed that the reaction proceeded efficiently in the presence of 20 mol% of Cs_2CO_3 at 70°C and the product was obtained in 98% yield after 90 minutes. We also examined the reaction in the presence of sodium and potassium carbonate in which the product was produced in 59 and 82% within 240 and 180 minutes, respectively. Thus, we selected Cs_2CO_3 as catalyst for our reaction.

To compare the efficiency of ionic liquid versus the conventional solvents, we examined the reaction between phthalimide (2 mmol) and ethyl acrylate (2.4 mmol) using Cs2CO3 (20 mol%) in some conventional solvents (10 mL) at 70°C (Table 1). As it can be seen from Table 1, higher yield and shorter reaction time were obtained in [bmim]Br. The reaction was also tested in solvent-free conditions; however, these conditions were not efficient (Table 1). Therefore, ionic liquid is an essential factor to promote our reaction.

Table 1. Comparing the reaction of phthalimide with ethyl acrylate in conventional solvents versus [bmim]Br.

Entry	Solvent	Time (min)	Yield (%)[a]
1	[bmim]Br	90	98
2	CH_3CN	180	49
3	EtOH	120	61
4	H_2O	120	37
5	DMSO	120	89
6	DMF	120	86
7	HMPTA	120	75
8	Solvent-free	240	13

[a] Isolated yield.

After optimization of the reaction conditions, we reacted phthalimide with different α,β-unsaturated esters. The results are summarized in Table 2. As Table 2 indicates, all reactions proceeded efficiently and the N-alkyl phthalimides were produced in excellent yields and relatively short reaction times.

Table 2. Synthesis of N-alkyl phthalimides using Cs_2CO_3/[bmim]Br catalytic system.

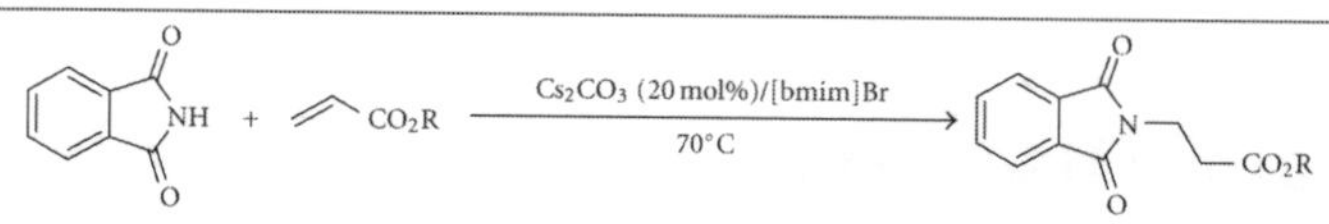

1a-i

R	Product[a]	Time (min)	Yield (%)[b]
CH_3CH_2	**1a**	90	98
$CH_3(CH_2)_2CH_2$	**1b**	90	98
$CH_3(CH_2)_4CH_2$	**1c**	90	97
$C_6H_5CH_2$	**1d**	100	96
$C_6H_5CH_2CH_2$	**1e**	100	96
$C_6H_5CH{=}CHCH_2$	**1f**	100	95
o-CH_3O-$C_6H_4OCH_2CH(OH)CH_2$	**1g**	130	93
C_6H_5	**1h**	70	96
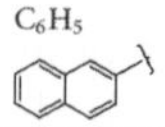	**1i**	70	94

[a]All compounds are known and their structures were identified by comparison of their melting points and spectral data with those in the authentic samples.
[b]Isolated yield.

The interesting behavior of [bmim]Br/Cs2CO3 system lies in the fact that it can be reused after simple washing with Et2O, rendering this process more economical. The yields of compound 1a (model compound) in the 2nd and 3rd uses of the [bmim]Br/Cs2CO3 were almost as high as in the first use (see Table 3).

Table 3. Aza-conjugate addition of phthalimide to ethyl acrylate in the presence of recycled Cs2CO3/[bmim]Br.

Entry	Cycle	Time (min)	Yield (%)[a]
1	1st use	90	98
2	2nd use	90	96
3	3rd use	100	95

[a]Isolated yield.

Experimental

All chemicals were purchased from Merck, (Germany) or Fluka, (Switzerland) Chemical Companies. The H1 NMR (250 MHz) and ^{13}C NMR (62.5 MHz) were run on a Bruker Avance DPX-250, FT-NMR spectrometer. Mass spectra were recorded on a Shimadzu GC MS-QP 1000 EX apparatus. Melting points were recorded on a Büchi B-545 apparatus in open capillary tubes.

General Procedure for the Synthesis of N-Alkyl Phthalimides

To a well-ground mixture of phthalimide (0.294 g, 2 mmol) and Cs_2CO_3 (0.130 g, 0.4 mmol) in a 10 mL round-bottomed flask connected to a reflux condenser [bmim]Br (1 g) and α,β-unsaturated ester (2.4 mmol) were added. The resulting mixture was stirred in an oil bath (70°C) for the times reported in Table 1. Afterward, the reaction mixture was cooled to room temperature and was extracted with Et_2O (3 × 50 mL). The organic extracts were then combined. After removal of the solvent, the crude product was purified by short column chromatography on silica gel eluted with EtOAc/n-hexane (1/3). After isolation of the product and evaporation of the remaining Et_2O in ionic liquid, the ionic liquid containing the catalyst (Cs_2CO_3/[bmim]Br) was used for the next run under identical reaction conditions.

Selected Physical and Spectral Data

Ethyl 3-Phthalimido Propanoate (1a)

Colorless solid; mp 59-60°C (Lit. [20] mp 60-61°C); IR (KBr): 3051, 2968, 1774, 1716 cm^{-1}; ^1H NMR ($CDCl_3$): δ 1.11 (t, 3H, J = 7.1 Hz, CH_3), 2.57 (t, 2H, J = 7.2 Hz, $O{=}CCH_2$), 3.86 (t, 2H, J = 7.2 Hz, NCH_2), 4.06 (q, 2H, J = 7.1 Hz, OCH_2), 7.57–7.70 (m, 4H); ^{13}C NMR ($CDCl_3$): δ 13.8, 32.7, 33.5, 64.3, 122.9, 131.7, 133.8, 167.5, 170.6; MS (m/z): 247 (M^+). 2-Hydroxy-3-(2-Methoxyphenoxy)Propyl 3-Phthalimido Propanoate (1g) Pale yellow oil (Lit. [21] oil); IR (neat): 3480, 3049, 2948, 1770, 1715 cm^{-1}; ^1H NMR (CDCl3): δ 2.63 (t, 2H, J = 7.0 Hz, O=CCH2), 3.67 (s, 3H, CH3), 3.80–390 (m, 5H), 4.11–4.19 (m, 3H), 6.73–6.80 (m, 4H), 7.54 (m, 2H), 7.68 (m, 2H); ^{13}C NMR ($CDCl_3$): δ 32.8, 33.6, 55.7, 65.7, 67.9, 70.5, 111.9, 114.2, 120.9, 121.8, 123.2, 131.7, 134.0, 147.8, 149.3, 168.0, 170.8; MS (m/z): 399 (M^+).

Conclusions

In summary, we have developed a new method for the synthesis of N-alkyl phthalimides as biologically interesting compounds via aza-conjugate addition reaction. This new strategy has the advantage of high yield, short reaction time, mild conditions, ease of product isolation, potential for recycling of the catalytic system, and compliance with the green chemistry protocols.

Acknowledgements

The authors appreciate Persian Gulf University and Payame Noor University Research Councils for the financial support of this work.

References

1. A. Zare, A. Hasaninejad, A. Khalafi-Nezhad, A. R. Moosavi Zare, and A. Parhami, "Organic reactions in ionic liquids: MgO as efficient and reusable catalyst for the Michael addition of sulfonamides to α,β-unsaturated esters under microwave irradiation," Arkivoc, vol. 2007, no. 13, pp. 105–115, 2007.
2. A. Zare, A. Hasaninejad, A. R. Moosavi Zare, A. Parhami, H. Sharghi, and A. Khalafi-Nezhad, "Zinc oxide as a new, highly efficient, green, and reusable catalyst for microwave-assisted Michael addition of sulfonamides to α,β-unsaturated esters in ionic liquids," Canadian Journal of Chemistry, vol. 85, no. 6, pp. 438–444, 2007.
3. L.-W. Xu, L. Li, C.-G. Xia, S.-L. Zhou, and J.-W. Li, "The first ionic liquids promoted conjugate addition of azide ion to α,β-unsaturated carbonyl compounds," Tetrahedron Letters, vol. 45, no. 6, pp. 1219–1221, 2004.
4. L. Yang, L.-W. Xu, W. Zhou, L. Li, and C.-G. Xia, "Highly efficient aza-Michael reactions of aromatic amines and N-heterocycles catalyzed by a basic ionic liquid under solvent-free conditions," Tetrahedron Letters, vol. 47, no. 44, pp. 7723–7726, 2006.
5. J.-M. Xu, C. Qian, B.-K. Liu, Q. Wu, and X.-F. Lin, "A fast and highly efficient protocol for Michael addition of N-heterocycles to α,β-unsaturated compound using basic ionic liquid [bmIm]OH as catalyst and green solvent," Tetrahedron, vol. 63, no. 4, pp. 986–990, 2007.
6. J. S. Yadav, B. V. S. Reddy, G. Baishya, K. V. Reddy, and A. V. Narsaiah, "Conjugate addition of indoles to α,β-unsaturated ketones using Cu(OTf)2 immobilized in ionic liquids," Tetrahedron, vol. 61, no. 40, pp. 9541–9544, 2005.
7. L. D. S. Yadav, R. Patel, V. K. Rai, and V. P. Srivastava, "An efficient conjugate hydrothiocyanation of chalcones with a task-specific ionic liquid," Tetrahedron Letters, vol. 48, no. 44, pp. 7793–7795, 2007.
8. H. Hagiwara, T. Okabe, T. Hoshi, and T. Suzuki, "Catalytic asymmetric 1,4-conjugate addition of unmodified aldehyde in ionic liquid," Journal of Molecular Catalysis A, vol. 214, no. 1, pp. 167–174, 2004.
9. B. Ni, Q. Zhang, and A. D. Headley, "Pyrrolidine-based chiral pyridinium ionic liquids (ILs) as recyclable and highly efficient organocatalysts for the asymmetric

Michael addition reactions," Tetrahedron Letters, vol. 49, no. 7, pp. 1249–1252, 2008.

10. B. C. Ranu and S. S. Dey, "Catalysis by ionic liquid: a simple, green and efficient procedure for the Michael addition of thiols and thiophosphate to conjugated alkenes in ionic liquid, [pmIm]Br," Tetrahedron, vol. 60, no. 19, pp. 4183–4188, 2004.
11. R. D. Rogers and K. R. Seddon, Eds., Ionic Liquids As Green Solvents: Progress and Prospects, R. D. Rogers and K. R. Seddon, Eds., An American Chemical Society Publication, Washington, DC, USA, 2005.
12. P. Wasserscheid and T. Welton, Ionic Liquids in Synthesis, Wiley-VCH, Weinheim, Germany, 2003.
13. M. H. Norman, D. J. Minick, and G. C. Rigdon, "Effect of linking bridge modifications on the antipsychotic profile of some phthalimide and isoindolinone derivatives," Journal of Medicinal Chemistry, vol. 39, no. 1, pp. 149–157, 1996.
14. L. M. Lima, P. Castro, A. L. Machado, et al., "Synthesis and anti-inflammatory activity of phthalimide derivatives, designed as new thalidomide analogues," Bioorganic & Medicinal Chemistry, vol. 10, no. 9, pp. 3067–3073, 2002.
15. J. M. Chapman, Jr., G. H. Cocolas, and I. H. Hall, "Hypolipidemic activity of phthalimide derivatives. 3. A comparison of phthalimide and 1,2-benzisothiazolin-3-one 1,1-dioxide derivatives to phthalimidine and 1,2-benzisothiazoline 1,1-dioxide congeners," Journal of Medicinal Chemistry, vol. 26, no. 2, pp. 243–246, 1983.
16. A. Raasch, O. Scharfenstein, C. Tränkle, U. Holzgrabe, and K. Mohr, "Elevation of ligand binding to muscarinic M2 acetylcholine receptors by bis(ammonio) alkane-type allosteric modulators," Journal of Medicinal Chemistry, vol. 45, no. 17, pp. 3809–3812, 2002.
17. M. B. Smith and J. March, Advanced Organic Chemistry: Reactions, Mechanisms and Structures, John Wiley & sons, New York, NY, USA, 4th edition, 2001.
18. O. A. Moe and D. T. Warner, "1,4 Addition reactions. III. The addition of cyclic imides to α,β-unsaturated aldehydes. A synthesis of β-alanine hydrochloride," Journal of the American Chemical Society, vol. 71, no. 4, pp. 1251–1253, 1949.
19. G. Cardillo, A. De Simone, L. Gentilucci, P. Sabatino, and C. Tomasini, "Michael-type addition of phthalimide salts to chiral α,β-unsaturated imides," Tetrahedron Letters, vol. 35, no. 28, pp. 5051–5054, 1994.

20. A. Zare, A. Hasaninejad, A. Khalafi-Nezhad, A. R. Moosavi Zare, A. Parhami, and G. R. Nejabat, "A green solventless protocol for Michael addition of phthalimide and saccharin to acrylic acid esters in the presence of zinc oxide as a heterogeneous and reusable catalyst," Arkivoc, vol. 2007, no. 1, pp. 58–69, 2007.
21. G. H. Imanzadeh, A. Khalafi-Nezhad, A. Zare, A. Hasaninejad, A. R. Moosavi Zare, and A. Parhami, "Michael addition of phthalimide and saccharin to α,β-unsaturated esters under solvent-free conditions," Journal of the Iranian Chemical Society, vol. 4, no. 2, pp. 229–237, 2007.
22. A. Khalafi-Nezhad, A. Zare, A. Parhami, and M. N. Soltani Rad, "Practical synthesis of some novel unsymmetrical 1,3-dialkyl pyrimidine derivatives at room temperature," Arkivoc, vol. 2006, no. 12, pp. 161–172, 2006.
23. A. Khalafi-Nezhad, A. Zare, A. Parhami, M. N. Soltani Rad, and G. R. Nejabat, "Microwave-assisted N-nitroarylation of some pyrimidine and purine nucleobases," Canadian Journal of Chemistry, vol. 84, no. 7, pp. 979–985, 2006.
24. S. Kotha and K. Singh, "N-Alkylation of diethyl acetamidomalonate: synthesis of constrained amino acid derivatives by ring-closing metathesis," Tetrahedron Letters, vol. 45, no. 52, pp. 9607–9610, 2004.
25. F. Chu, E. E. Dueno, and K. W. Jung, "Cs2CO3 promoted O-alkylation of alcohols for the preparation of mixed alkyl carbonates," Tetrahedron Letters, vol. 40, no. 10, pp. 1847–1850, 1999.
26. F. Churruca, R. SanMartin, I. Tellitu, and E. Domínguez, "Regioselective diarylation of ketone enolates by homogeneous and heterogeneous catalysis: synthesis of triarylethanones," Tetrahedron Letters, vol. 44, no. 31, pp. 5925–5929, 2003.
27. R. N. Salvatore, R. A. Smith, A. K. Nischwitz, and T. Gavin, "A mild and highly convenient chemoselective alkylation of thiols using Cs2CO3-TBAI," Tetrahedron Letters, vol. 46, no. 51, pp. 8931–8935, 2005.
28. C. Geraci, M. Piattelli, and P. Neri, "Regioselective synthesis of calix[8]crowns by direct alkylation of p-tert-butylcalix[8]arene," Tetrahedron Letters, vol. 37, no. 22, pp. 3899–3902, 1996.
29. A. Khalafi-Nezhad, A. Parhami, A. Zare, A. R. Moosavi Zare, A. Hasaninejad, and F. Panahi, "Trityl chloride as a novel and efficient organic catalyst for room temperature preparation of bis(indolyl)methanes under solvent-free conditions in neutral media," Synthesis, vol. 617, pp. 617–621, 2008.
30. A. Zare, A. Hasaninejad, M. H. Beyzavi, et al., "Zinc oxide-tetrabutylammonium bromide tandem as a highly efficient, green, and reusable catalyst for the Michael addition of pyrimidine and purine nucleobases to α,β-unsaturated

esters under solvent-free conditions," Canadian Journal of Chemistry, vol. 86, no. 4, pp. 317–324, 2008.

31. A. Zare, A. Hasaninejad, A. Khalafi-Nezhad, A. Parhami, and A. R. Moosavi Zare, "A solventless protocol for the Michael addition of aromatic amides to α,β-unsaturated esters promoted by microwave irradiation," Journal of the Iranian Chemical Society, vol. 5, no. 1, pp. 100–105, 2008.
32. A. Hasaninejad, A. Zare, H. Sharghi, and M. Shekouhy, "P2O5/SiO2 an efficient, green and heterogeneous catalytic system for the solvent-free synthesis of N-sulfonyl imines," Arkivoc, vol. 2008, no. 11, pp. 64–74, 2008.

Conformational Rigidity of Silicon-Stereogenic Silanes in Asymmetric Catalysis: A Comparative Study

Sebastian Rendler and Martin Oestreich

ABSTRACT

In recent years, cyclic silicon-stereogenic silanes were successfully employed as stereoinducers in transition metal-catalyzed asymmetric transformations as exemplified by (1) the hydrosilylation of alkenes constituting a chirality transfer from silicon to carbon and (2) the kinetic resolution of racemic mixtures of alcohols by dehydrogenative silicon-oxygen coupling. In this investigation, a cyclic and a structurally related acyclic silane with silicon-centered chirality were compared using the above-mentioned model reactions. The stereochemical outcome of these pairs of reactions was correlated with and rationalized by the current mechanistic pictures. An acyclic silicon-stereogenic silane is also capable of inducing excellent chirality transfer (ct) in a palladium-catalyzed intermolecular carbon-silicon bond formation yet silicon incorporated into a cyclic framework is required in the copper-catalyzed silicon-oxygen bond forming reaction.

Findings

Within the last decade, several asymmetric transformations based on silicon-stereogenic reagents or substrates were revisited or invented. [1-4] Aside from the use of silicon-stereogenic chiral auxiliaries in substrate-controlled reactions, [5] a still limited number of remarkable stereoselective processes with a stereogenic silicon as the reactive site were reported, [6] namely the inter- [7] as well as intramolecular [8] chirality transfers from silicon to carbon. Moreover, we had demonstrated that chiral silanes resolve racemic mixtures of alcohols in a non-enzymatic, transition metal-catalyzed kinetic resolution. [9]

During our ongoing investigations directed towards the mechanistic elucidation of the origin of the chirality transfer in a palladium-catalyzed hydrosilylation, [10] we had to perform an extensive screening of silicon-stereogenic tertiary silanes. On that occasion, we became aware that a similar level of stereoselection was obtained when priveleged cyclic system 1a [11] was exchanged for the important acyclic congener 1b [12-15] (Figure 1). We had erroneously missed this known tertiary silane. This was particularly unfortunate in the light of the fact that these silanes are both decorated with three substituents of different steric demand and, therefore, display marked stereochemical differentiation around silicon.

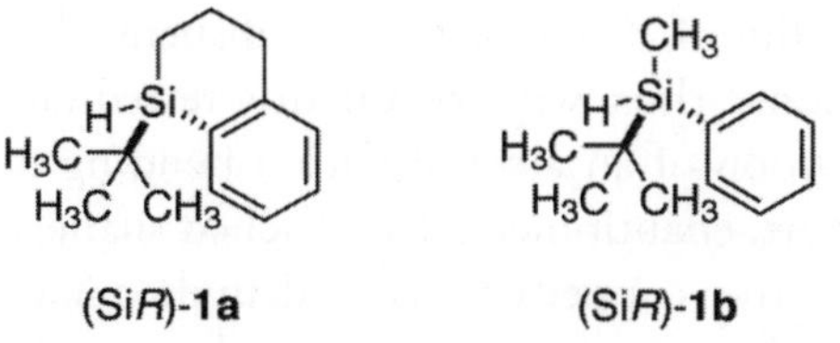

Figure 1. Cyclic and acyclic sterically encumbered silanes.

In this preliminary communication, we wish to report a comparison of cyclic 1a and acyclic 1b as stereoinducers in the palladium-catalyzed chirality transfer from silicon to carbon and in the copper-catalyzed kinetic resolution of donor-functionalized alcohols capable of two-point binding.

The reagent-controlled hydrosilylation of norbornene derivative 2 with silane 1a proceeds with a perfect chirality transfer (rac-1a → rac-3a, Scheme 1). [8] Mechanistic investigation of the nature of the stereochemistry-determining step in this catalysis required a silane, which would produce slightly diminished diastereoselectivity and, hence, attenuated chirality transfer from silicon to carbon. [10] It was that situation that prompted us to investigate a considerable range of silicon-stereogenic silanes initially varied in ring size and exocyclic substituent; this was not met with satisfactory success. Based on the assumption that less rigid acyclic silanes would induce lower levels of diastereoselection, previously reported silane rac-1b – readily prepared in its racemic form [13] – was then supposed to serve such purpose. To our surprise, the palladium-catalyzed hydrosilylation of 2 with rac-1b gave almost perfect diastereoselectivity and good yield (rac-1b → rac-3b, Scheme 1).

Scheme 1. Cyclic and acyclic chiral silanes as potent reagents for the silicon-to-carbon chirality transfer.

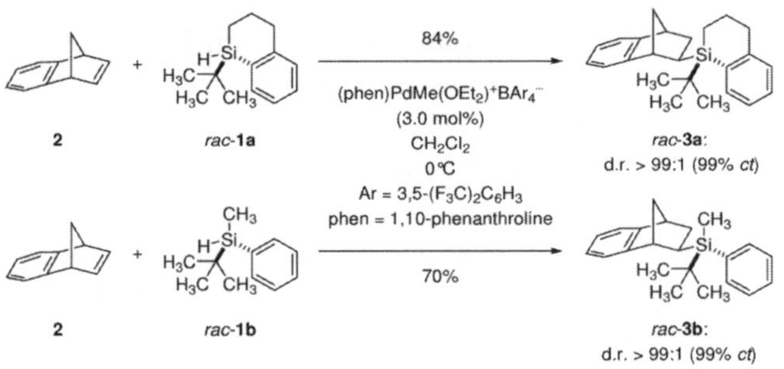

This unexpected result inevitably introduced the pivotal question whether conformational rigidity of chiral silanes is a dispensible characteristic for asymmetric transformations. Thus, we subsequently tested rac-1b as resolving reagent in the kinetic resolution of an alcohol with a pending nitrogen donor (Scheme 2). In an earlier report, enantiomerically enriched silane 1a (96% ee) was applied in this diastereoselective copper-catalyzed dehydrogenative silicon-oxygen coupling affording promising optical purities for the unreacted alcohol ent-4 (84% ee) along with 5 (d.r. = 84:16) at 56% conversion. [9] For the present study, the diastereoselectivity of the formed ethers 5 is conclusive, which, in turn, allows for working with racemic silanes rac-1 (rac-1a → rac-5a versus rac-1b → rac-5b, Scheme 2). This is sufficient since the d.r. of 5 will be identical to the e.r. of the remaining alcohol 4 at exactly 50% conversion when using enantiopure silane 1. It must be noted that that diastereoselectivity is not dependent on conversion when using racemic silanes rac-1; conversely, using enantioenriched 1 it is.

Scheme 2. Kinetic resolution of secondary alcohols using a dehydrogenative coupling reaction.

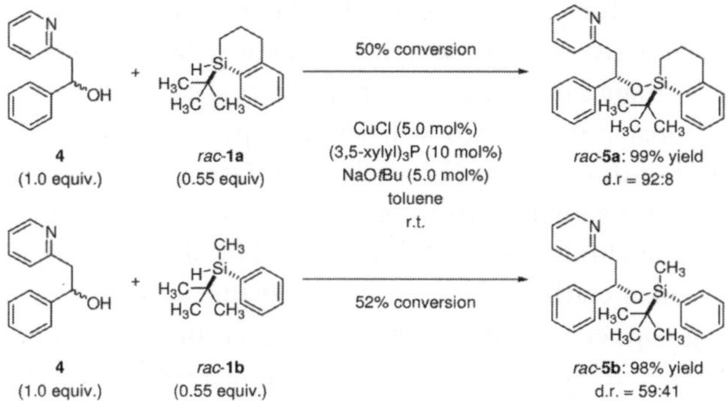

Whereas rac-5a was formed highly diastereoselectively (d.r. = 92:8) at 50% conversion, [9] the analogous reaction of rac-1b yielded rac-5b in a poor diastereomeric ratio (d.r. = 59:41) at comparable conversion. In sharp contrast to the results obtained in the hydrosilylation, embedding the asymmetrically substituted silicon into a cyclic framework appears to be an essential feature.

A comparison of the mechanisms of each reaction might serve as an explanation for this unexpected divergence. As outlined in Scheme 3, the hydrosilylation proceeds via a three-step catalytic cycle: (i) Reversible coordination of cationic silyl palladium species 6 by the alkene 2 ($6 \rightarrow 7$), followed by (ii) fast and reversible migratory insertion forming β-silyl alkyl palladium intermediate 8 ($7 \rightarrow 8$), and (iii) the involvement of a second silane moiety in the irreversible σ-bond metathesis. [10,16] Recent results clearly indicate step (ii) as diastereoselectivity-determining, revealing a thermodynamically controlled, reversible but highly diastereoselective migratory insertion step. [10]

Scheme 3. Catalytic cycle for hydrosilylation.

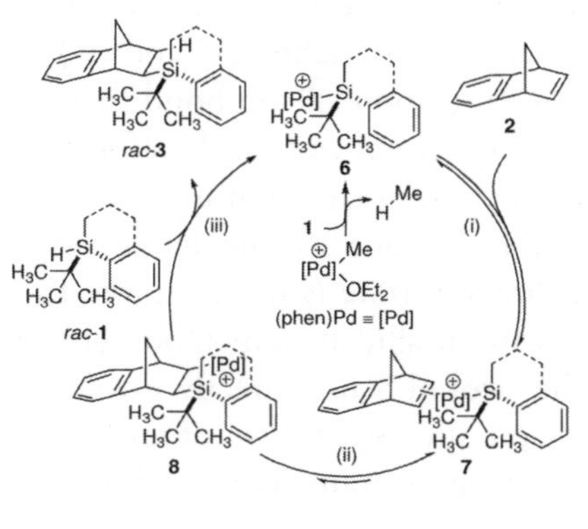

A different scenario might apply to the copper-catalyzed kinetic resolution of alcohols (Scheme 4). The phosphine-stabilized copper hydride 12 [17] is likely to be the catalytically active species, which is generated by alkoxide exchange (9 → 10) followed by a single catalytic turnover. The actual catalytic cycle then proceeds in a four-step propagation: (i) Coordination of pyridyl alcohol rac-4 accompanied by liberation of dihydrogen (12 → 10), (ii) rate-limiting dissociation of one phosphine ligand to generate a free coordination site, [18] (iii) coordination of the weakly donating chiral silane (10 → 11), followed by (iv) an exothermic and irreversible σ-bond metathesis [19] establishing the silicon-oxygen linkage in 5 and regenerating copper hydride 12 after coordination of another phosphine ligand (11 → 12). With steps (ii) and (iii) being reversible and chelate 10 being capable of alkoxide exchange, that is exchange of the optical antipodes of 4, one enantiomer of 4 is preferentially funnelled out via diastereomeric transition states (11 → 12).

Scheme 4. Postulated catalytic cycle for dehydrogenative coupling.

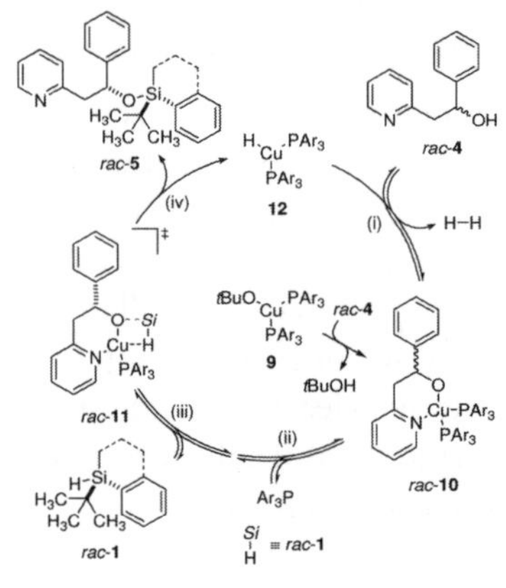

There is one major difference between the diastereoselectivity-determining steps in these catalytic cycles: (ii) in Scheme 3 and (iv) in Scheme 4. In the migratory insertion (ii, 7 → 8), carbon-silicon bond formation occurs between the stereogenic silicon and the prochiral carbon therefore entailing their close proximity. The newly formed stereogenic carbon is directly connected to the former source of chiral information. In contrast, the decisive asymmetrically substituted carbon atom in the alcohol substrate is more remote from the stereoselectivity-controlling silicon moiety in the silicon-oxygen bond formation (iv, 11 → 5). The stereogenic carbon in the alcohol is not directly involved in the actual bond formation. This mechanistic picture might account for the more demanding requirements to chiral silane 1: A cyclic framework leading to a locked conformation

[11] improving the degree of organization in the stereochemistry-determining transition state 11.

In summary, we have shown for the first time that an excellent chirality transfer from silicon to carbon is also realized with suitably substituted acyclic silanes such as 1b. Our survey, however, underscores once more that cyclic silane 1a is a priveleged structure and certainly generally more applicable to catalytic asymmetric processes than 1b. The current mechanistic pictures provide a rather simple explanation for the observed stereochemical outcome of both diastereoselective carbon-silicon and silicon-oxygen bond formation. Based on this insight, further research will be devoted to the extension chiral silicon-based asymmetric catalysis.

Acknowledgements

The research was supported by the Deutsche Forschungsgemeinschaft (Emmy Noether program, Oe 249/2-3 and Oe 249/2-4), the Fonds der Chemischen Industrie (pre-doctoral fellowship to S. R.), and the Aventis Foundation (Karl Winnacker fellowship to M. O.). The authors thank Oliver Plefka for an orientating experiment. Generous donations of chemicals from Wacker AG (Burghausen/Germany) and Lanxess AG (Leverkusen/Germany) are gratefully acknowledged.

References

1. Maryanoff CA, Maryanoff BE: Synthesis and Utilization of Compounds with Chiral Silicon Centers. Asymmetric Synthesis. Volume 4. Edited by: Morrison JD, Scott JW. Orlando: Academic Press; 1984:355–374.
2. Sommer LH: Stereochemistry, Mechanism and Silicon. New York: McGraw-Hill; 1965.
3. Sommer LH: Intra-Sci Chem Rep. 1973, 7:1–44.
4. Corriu RJP, Guérin C, Moreau JJE: Stereochemistry at Silicon. Topics in Stereochemistry. Volume 15. Edited by: Eliel EL. New York: Wiley; 1984:43–198.
5. Bienz S: Chimia. 1997, 51:133–139.
6. Oestreich M: Chem Eur J. 2006, 12:30–36.
7. Schmidt DR, O'Malley SJ, Leighton JL: J Am Chem Soc. 2003, 12:1190–1191.
8. Oestreich M, Rendler S: Angew Chem Int Ed. 2005, 44:1661–1664.
9. Rendler S, Auer G, Oestreich M: Angew Chem Int Ed. 2005, 44:7620–7624.

10. Rendler S, Oestreich M, Butts CP, Lloyd-Jones GC: J Am Chem Soc. 2007, 129:502–503.
11. Rendler S, Auer G, Keller M, Oestreich M: Adv Synth Catal. 2006, 348:1171–1182.
12. Bertrand G, Dubac J, Mazerolles P, Ancelle J: Nouv J Chim. 1982, 6:381–386.
13. Larson GL, Torres E: J Organomet Chem. 1985, 293:19–27.
14. Jankowski P, Schaumann E, Wicha J, Zarecki A, Adiwidjaja G: Tetrahedron: Asymmetry. 1999, 10:519–526.
15. Jankowski P, Schaumann E, Wicha J, Zarecki A, Adiwidjaja G, Asztemborska M: Chem Commun. 2000, 1029–1030.
16. LaPointe AM, Rix FC, Brookhart M: J Am Chem Soc. 1997, 119:906–917.
17. Rendler S, Oestreich M: Angew Chem Int Ed. 2007, 46:498–504.
18. Rendler S, Oestreich M 2006 unpublished results.
19. Grimme S, Oestreich M 2006 unpublished results.

Photochromism of Dihydroindolizines Part XI: Synthesis of Novel Carbon-Rich Photochromic Dihydroindolizines-Based Potential Electronic Devices

Saleh Abdel-Mgeed Ahmed

ABSTRACT

Novel carbon-rich photochromic dihydroindolizine (DHI) derivatives substituted in the fluorene part (region A) in addition to the new spi-rocyclopropene 6 have been synthesized. The synthesis of dimethyl 2′,7′-diethynylspiro[cycloprop[2]ene-1,9′-fluorene]-2,3-dicarboxylate precursor 6 was accomplished in five steps, starting with the literature known conversion of fluorene to 2,7-dibromo-9H-fluoren-9-one in 56% yield over three steps. The chemical structures of the new synthesized materials have been elucidated by both analytical and spectroscopic tools. Three alterative synthetic pathways for the synthesis of DHI 9 have been established.

Introduction

Molecules that respond to the application of external stimuli by undergoing reversible transformations between two distinct structures have the potential to significantly influence the development of numerous important materials science- and structural biology technologies [1, 2]. This potential is based on the fact that, because the molecules typically undergo dramatic changes in their electronic and topological characteristics, they can act as switching elements and other dynamic components in various optoelectronic devices and functional materials. Photons are particularly appealing stimuli because modern lasers can be used to achieve fast response times, to focus a fine-tuned stimulus on small localized domains without significant diffusion and to trigger photochemical events under conditions mild enough to pose minimal danger to sensitive biomaterials. Compounds that interconvert between different isomers having unique absorption spectra when stimulated with light are referred to as photochromic, and the process is called photochromism. In these systems, the changes in the electronic patterns responsible for the changes in color also result in variations in other practical physical properties such as luminescence [3], electronic conductance [4, 5], refractive index [6], optical rotation [7], and viscosity [8, 9]. These materials, based on the 1,5-electrocyclization between two distinct isomeric states: ring-opening form (betaine-form) and ring-closed form (DHI-form), are promising candidates for optical storage media and electronic devices [10–29].

As a continuation of our research on photochromic dihydroindolizines (DHIs), this article is devoted to the synthesis of carbon-rich fluorenyl-dihydroindolizines derivatives and represents the first step toward the application of photochromic dihydroindolizines in electronic devices. Different synthetic approaches will be described.

Results and Discussion

Synthesis of Fluorenylacetylene Spirocyclopropene Precursor 6

The synthesis of spirocyclopropene 6 was accomplished in five steps, starting with the previously known conversion of fluorene to 2,7-dibromo-9H-fluoren-9-one in 56% yield in over three steps [10–13] (see Scheme 1). The Sonogashira-coupling of 2,7-dibromo-9H-fluoren-9-one with trimethylsilylacetylene (TMSA) in the presence of $Pd(PPh)_3Cl_2$ (5 wt%)/CuI/Et_2NH in THF at room temperature for 24 hours afforded the coupling product 2,7-bis((trimethylsilyl)ethynyl)-9H-fluoren-9-one 2 in good yield (79%). Interestingly, condensation of compound 2 with hydrazine hydrate in boiling ethanol for 6 hours leads not only to the formation of the condensation product (2,7-bis((trimethylsilyl)ethynyl)-9H-fluoren-9--hydrazone but also to the occurrence of dimethyl silylation, and to the obtainment of (2,7-di(prop-1-ynyl)-9H-fluoren-9-ylidene)hydrazone 3 in 47% yield. The evidence for the formation of compound 3 was established by 1HNMR which showed the complete disappearance of the trimethylsilyl protons which appear as singlet at 0.4 ppm in compound 2 and the appearance of a singlet at 4.05 ppm in compound 3 which is related to the acetylenic proton in addition to the presence of a broad singlet at 5.32 ppm corresponding to the NH_2 protons which disappeared upon treatment with deuterium oxide. Oxidation of the hydrazone 3 with manganese dioxide in dry ether at room temperature in the absence of light afforded the 9-diazo-2,7-di(prop-1-ynyl)-9H-fluorene 4 in moderate yield (56%). Addition of methyl acetylenedicarboxylate (MADC) to the 9-diazofluorene derivative 4 in dry ether in dark condition for 24 hours led to the formation of dimethyl 2,7-diethynylspiro[fluorene-9,3′-pyrazole]-4′,5′-dicarboxylate 5 pyrazole derivative 6 in 37% yield.

Scheme 1. The synthesis of dimethyl 2′,7′-diethynylspiro[cycloprop[2]ene-1,9′-fluorene]-2,3-dicarboxylate precursor 6.

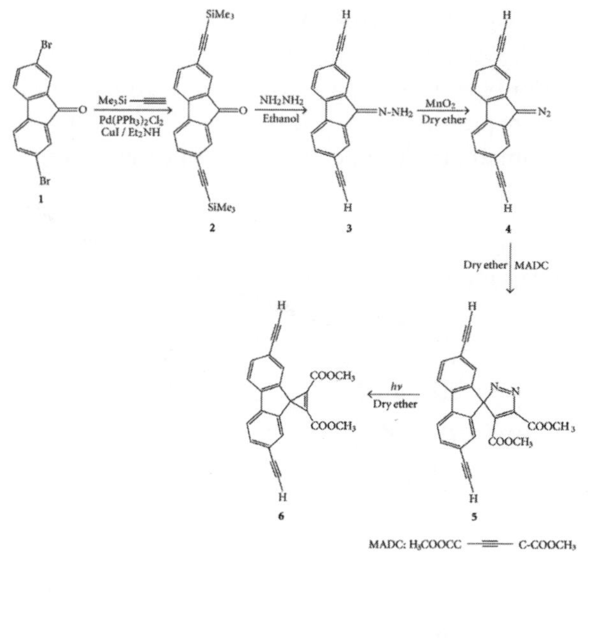

Photolysis of the pyrazole derivative 5 with high pressure mercury lamp (125 W) in dry ether solution for two hours under nitrogen atmosphere gave the target dimethyl 2′,7′-diethynylspiro[cycloprop[2]ene-1,9′-fluorene]-2,3-dicarboxylate 6 in low yield (22%). The reaction was accompanied with the formation of some unidentified product which is expected to be related to some addition polymerization on the two acetylenic groups in the fluorene part (region A) which leads to the decreasing of the yield of the desired spirocyclopropene derivative 6. The chemical structure of the newly synthesized compounds 2–6 (see Scheme 1) was confirmed and established by both spectroscopic (NMR, IR, and mass spectrometry) and analytical tools (give satisfactory elemental analysis data). For example, the 1HNMR (400 MHz, CDCl3) of the spirocyclopropene precursor 6 showed the following signals: δ 7.82–7.86 (d, J = 1.76 Hz, 1H, CH-arom.), 7.63–7.67 (d, J = 1.32 Hz, 1H, CH-arom.), 7.52–7.57 (m, 4H, CH-arom.), 4.11 (s, 2H, acetylenic protons in 2,7 position), 3.82 (s, 6H, 2′,3′-CH3) ppm.

Different Attempts for Synthesis of the Photochromic Dihydroindolizine 9

Nucleophilic addition of pyridazine 7 to spirocyclopropene 6 using the cyclopropene route [10–29] (see Scheme 2) in dry ethereal solution at room temperature under dry nitrogen in the absence of light (TLC-controlled using CH2Cl2 as eluent) led to the formation of the photochromic dihydroindolizine (DHI) 9 in low yield (26%). The reaction occurs through the electrophilic addition of the electron-deficient spirocyclopropenes 6 tothe nitrogen of the N-heterocyclic pyridazine 7 which led to ring opening via a cyclopropyl-allyl conversion 8′ to the colored betaines 8. A subsequent ring closure to DHI 9 results in a slow thermal 1,5-electrocyclization back reaction (see Scheme 2) which can be reversed upon exposure to light. Pure photochromic DHI 9 was obtained in all cases by two successive column chromatography operations on silica gel using dichloromethane as the eluent.

Scheme 2. Preparation outline of photochromic DHI 9 from spirocyclopropene 6.

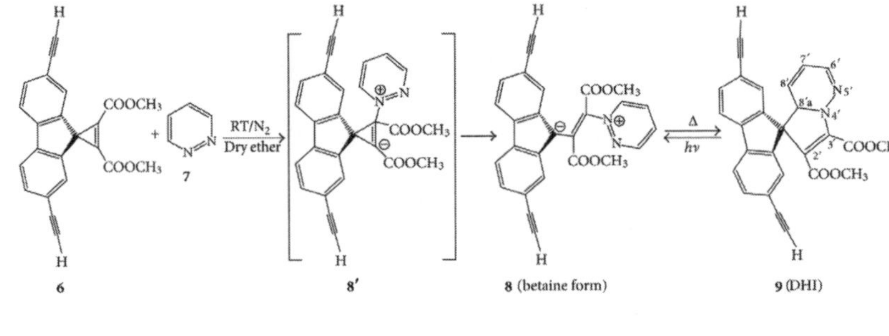

6 8′ 8 (betaine form) 9 (DHI)

Another successful alternative method for the synthesis of the target photochromic DHI 9 was achieved through the following multistep synthesis (see Scheme 3). The Sonogashira coupling of dimethyl 2,7-dibromo-4a′H-spiro[fluorene-9,5′-pyrrolo[1,2-b]pyridazine]-6′,7′-dicarboxylate which was previously prepared by us [10–13] with trimethylsilylacetylene in presence of palladium-catalyzed reaction (5%) and CuI/Et2N in T = dry THI for 12 hours [29, 30] yielded the desired product dimethyl 2,7-bis((trimethylsilyl)ethynyl)-4a′H-spiro[fluorene-9,5′-pyrrolo[1,2-b]pyridazine]-6′,7′-dicarboxylate 11 in 34% yield after purification by flash chromatography on silica gel and CH2Cl2 as eluent. Treatment of DHI 11 with tetrabutyl ammonium fluoride (TBAF) in dry THF for 17 hours afforded the trimethylsilylated product 9 in 68% yield. A good proof for the trimethylsilylation that occurs during the condensation of compound 2 with hydrazine hydrate is that detrimethylsilylation occurs when DHI 11 was treated with hydrazine hydrate in ethanol under mild conditions and at low temperature (0°C) for 2 hours in 43% yield. Thus, acetylenic DHI 9 could successfully prepared through three reactions' pathways as shown in Schemes 2 and 3. The three products obtained from the different pathways showed the same analytical and spectroscopic data as well as m.p and m.m.p.

Scheme 3. Another reaction pathway for the synthesis of the target photochromic DHI 9.

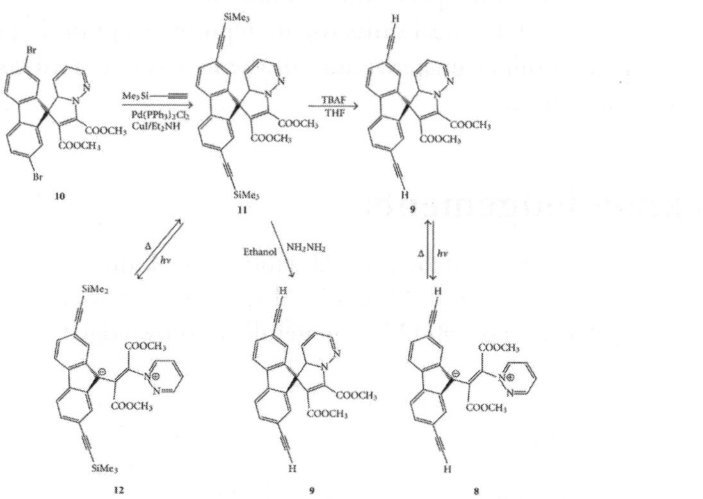

Conclusion

We have successfully extended the photochromism of photochromic DHI through the coupling reactions in the fluorene part. New spirocyclopropene and photochromic dihydroindolizines (DHIs) substituted in the fluorene part (region A) with acetylenic bridge for future using in electronic devices have been furnished. Further modification of the chemical structure of DHI system and their photochromic properties as well as supporting onto the surface of metals such as gold, silicon, and titanium will be discussed in details in the forthcoming article.

Experimental

Spirocyclopropene derivatives were obtained via photolysis of the corresponding pyrazoles prepared according to reported procedures [10–23]. Photolysis was carried out in the photochemical reactor of Schenck [31] made from Pyrex (λ > 290 nm). The source of irradiation was a high-pressure mercury lamp Philips HPK 125 W. Solutions to be photolyzed were flushed with dry nitrogen for 30 minutes before switching on the UV lamp. The progress of the reaction and the purity of the products isolated were monitored using TLC. Separation and purification of all synthesized photochromic materials were carried out using column chromatography (80 cm length × 2 cm diameter) on silica gel and CH_2Cl_2 as eluent. Melting points were determined on (Electrothermal Eng. Ltd., Essex, UK) melting point apparatus and are uncorrected. All NMR spectra were collected on a Brüker DRX-400 spectrometer (400 MHz) in $CDCl_3$ using TMS as the internal standard. Chemical shifts (δ) are reported in ppm. Experimental details, procedures, and full characterizations of the new synthesized compounds will be described elsewhere.

Acknowledgements

The author is highly indebted to Alexander von Humboldt Foundation (AvH) for the financial support of this work. Also, the financial support from the Taibah University (Project no. 48/427) is gratefully acknowledged.

References

1. M. Irie, in Molecular Switches, B. L. Feringa, Ed., pp. 37–60, Wiley-VCH, Weinheim, Germany, 2001.
2. M. Irie, in Organic Photochromic and Thermochromic Compounds, J. C. Crano and R. J. Gugielmetti, Eds., vol. 1, pp. 207–221, Plenum Press, New York, NY, USA, 1999.

3. M. Irie, T. Fukaminato, T. Sasaki, N. Tamai, and T. Kawai, "A digital fluorescent molecular photoswitch," Nature, vol. 420, no. 6917, pp. 759–760, 2002.

4. T. Tsujioka, Y. Hamada, K. Shibata, A. Taniguchi, and T. Fuyuki, "Nondestructive readout of photochromic optical memory using photocurrent detection," Applied Physics Letters, vol. 78, no. 16, pp. 2282–2284, 2001.

5. T. Kawai, T. Kunitake, and M. Irie, "Novel photochromic conducting polymer having diarylethene derivative in the main chain," Chemistry Letters, vol. 28, no. 9, pp. 905–906, 1999.

6. C. Bertarelli, A. Bianco, F. D'Amore, M. C. Gallazzi, and G. Zerbi, "Effect of substitution on the change of refractive index in dithienylethenes: an ellipsometric study," Advanced Functional Materials, vol. 14, no. 4, pp. 357–363, 2004.

7. R. A. van Delden, M. K. J. ter Wiel, and B. L. Feringa, "A chiroptical molecular switch with perfect stereocontrol," Chemical Communications, vol. 10, no. 2, pp. 200–201, 2004.

8. M. Moniruzzaman, C. J. Sabey, and G. F. Fernando, "Synthesis of azobenzene-based polymers and the in-situ characterization of their photoviscosity effects," Macromolecules, vol. 37, no. 7, pp. 2572–2577, 2004.

9. L. N. Lucas, J. van Esch, R. M. Kellogg, and B. L. Feringa, "Photocontrolled self-assembly of molecular switches," Chemical Communications, no. 8, pp. 759–760, 2001.

10. S. A.-M. Ahmed, T. Hartmann, V. Huch, H. Dürr, and A. A. Abdel-Wahab, "Synthesis of IR-sensitive photoswitchable molecules: photochromic 9′-styryl quinolinedihydroindolizines," Journal of Physical Organic Chemistry, vol. 13, no. 9, pp. 539–548, 2000.

11. Y. Tan, S. A.-M. Ahmed, H. Dürr, V. Huch, and A. Abdel-Wahab, "First intramolecular trapping and structural proof of the key intermediate in the formation of indolizine photochromics," Chemical Communications, no. 14, pp. 1246–1247, 2001.

12. S. A.-M. Ahmed, "Photochromic spirotetrahydroazafluorenes: part IV. First trapping of novel type of cis-fixed photochromes based on pyridazinopyrrolo[1,2-b] isoquinoline," Molecular Crystals and Liquid Crystals, vol. 430, pp. 295–300, 2005.

13. S. A.-M. Ahmed and H. Dürr, "Photochromic spirotetrahydroazafluorenes: part V. Why photochromic molecules with rigid region B exhibiting extremely fast bleaching process?," Molecular Crystals and Liquid Crystals, vol. 431, pp. 275–280, 2005.

14. S. A.-M. Ahmed, "Photochromism of dihydroindolizines. Part III [1]. Synthesis and photochromic behavior of novel photochromic dihydroindolizines incorporating a cholesteryl moiety," Monatshefte für Chemie, vol. 135, no. 9, pp. 1173–1188, 2004.
15. S. A.-M. Ahmed, A.-M. Abdel-Wahab, and H. Dürr, "Steric substituent effects of new photochromic tetrahydroindolizines leading to tunable photophysical behavior of the colored betaines," Journal of Photochemistry and Photobiology A, vol. 154, no. 2-3, pp. 131–144, 2003.
16. S. A.-M. Ahmed, "Photochromism of dihydroindolizines. Part II. Synthesis and photophysical properties of new photochromic IR-sensitive photoswitchable substituted fluorene-9′-styrylquinolinedihydroindolizines," Journal of Physical Organic Chemistry, vol. 15, no. 7, pp. 392–402, 2002.
17. S. A.-M. Ahmed, "Photochromism of dihydroindolizines part VI: synthesis and photochromic behavior of a novel type of IR-absorbing photochromic compounds based on highly conjugated dihydroindolizines," Journal of Physical Organic Chemistry, vol. 19, no. 7, pp. 402–414, 2006.
18. S. A.-M. Ahmed, "Photochromism of dihydroindolizines part VII: multiaddressable photophysical properties of new photochromic dihydroindolizines bearing substituted benzo[i]phenanthridine as a fluorescing moiety," Journal of Physical Organic Chemistry, vol. 20, no. 8, pp. 574–588, 2007.
19. S. A.-M. Ahmed, "Photochromism of dihydroindolizines part VI: synthesis and photochromic behavior of a novel type of IR-absorbing photochromic compounds based on highly conjugated dihydroindolizines," Journal of Physical Organic Chemistry, vol. 19, no. 7, pp. 402–414, 2006.
20. S. A.-M. Ahmed, T. Hartmann, and H. Dürr, "Photochromism of dihydroindolizines. Part VIII. First holographic image recording based on di- and tetrahydroindolizines photochromes," Journal of Photochemistry and Photobiology A, vol. 200, no. 1, pp. 50–56, 2008.
21. S. A.-M. Ahmed and J.-L. Pozzo, "Photochromism of dihydroindolizines Part IX. First attempts towards efficient self-assembling organogelators based on photochromic dihydroindolizines and N -acyl-1,ω-amino acid units," Journal of Photochemistry and Photobiology A, vol. 200, no. 1, pp. 57–67, 2008.
22. H. Dürr, C. Schommer, and T. Münzmay, "Dihydropyrazolopyridine und Bis(dihydroindolizine) - neuartige mono- und difunktionelle photochrome Systeme," Angewandte Chemie, vol. 98, no. 6, pp. 565–567, 1986.
23. H. Dürr, C. Schommer, and T. Münzmay, "Dihydropyrazolopyridine and bis(dihydroindolizine)-novel mono- and bifunctional photochromic systems,"

Angewandte Chemie International Edition in English, vol. 25, no. 6, pp. 572–574, 1986.

24. R. Fromm, S. A.-M. Ahmed, T. Hartmann, V. Huch, A. A. Abdel-Wahab, and H. Dürr, "A new photochromic system based on a pyridazinopyrrolo[1,2-b] pyridazine with ultrafast thermal decoloration," European Journal of Organic Chemistry, vol. 2001, no. 21, pp. 4077–4080, 2001.

25. S. A.-M. Ahmed, A. A. Abdel-Wahab, and H. Dürr, "Photochromic nitrogen-containing compounds," in CRC Handbook of Organic Photochemistry and Photobiology, W. M. Horspool and F. Lenci, Eds., pp. 1–25, CRC press, New York, NY, USA, 2nd edition, 2003.

26. H. Dürr, "Perspectives in photochromism: a novel system based on the 1,5-electrocyclization of heteroanalogous pentadienyl anions," Angewandte Chemie International Edition in English, vol. 28, no. 4, pp. 413–431, 1989.

27. S. A.-M. Ahmed, C. Weber, Z. A. Hozien, Kh. M. Hassan, A. A. Abdel-Wahab, and H. Dürr, unpublished results.

28. S. A.-M. Ahmed, Ph. D. thesis, Saarland-Assiut Universities, 2000.

29. S. Zarwell and K. Rück-Braun, "Synthesis of an azobenzene-linker-conjugate with tetrahedrical shape," Tetrahedron Letters, vol. 49, no. 25, pp. 4020–4025, 2008.

30. H. Jian and J. M. Tour, "En route to surface-bound electric field-driven molecular motors," Journal of Organic Chemistry, vol. 68, no. 13, pp. 5091–5103, 2003.

31. A. Schönberg, Präparative Organische Photochemie, Springer, Berlin, Germany, 1958.

8-Epi-Salvinorin B: Crystal Structure and Affinity at the K Opioid Receptor

Thomas A. Munro, Katharine K. Duncan, Richard J. Staples, Wei Xu, Lee-Yuan Liu-Chen, Cécile Béguin, William A. Carlezon Jr. and Bruce M. Cohen

ABSTRACT

There have been many reports of epimerization of salvinorins at C-8 under basic conditions, but little evidence has been presented to establish the structure of these compounds. We report here the first crystal structure of an 8-epi-salvinorin or derivative: the title compound, 2b. The lactone adopts a boat conformation with the furan equatorial. Several lines of evidence suggest that epimerization proceeds via enolization of the lactone rather than a previously proposed indirect mechanism. Consistent with the general trend in related compounds, the title compound showed lower affinity at the kappa opioid receptor than the natural epimer salvinorin B (2a). The related 8-epi-acid 4b showed no affinity.

Introduction

Salvinorin A (1a), isolated from the hallucinogenic sage Salvia divinorum,[1] is a potent and selective κ opioid receptor (KOR) agonist.[2] Because it is the first known non-nitrogenous compound to have biologically significant actions at mammalian opioid receptors, 1a enables new approaches to studies of endogenous opioid receptor systems. KOR ligands, in particular, have attracted considerable interest because of their effects on mood states.[3-6] Recently, numerous synthetic derivatives of 1a have been prepared and evaluated for activity at opioid receptors. Some potent agonists have been identified which are expected to show increased stability or solubility.[7] Others have increased affinity and potency, [8] or altered subtype selectivity.[9] As yet, however, no derivatives of 1a appear to be KOR partial agonists or antagonists, classes of agents that may have utility in the treatment of psychiatric conditions such as depression or mania.[4,5,10]

Salvinorins tend to isomerize under basic conditions. Valdés reported that borohydride reduction of 1a gave an unidentified stereoisomeric byproduct, which could be converted to an undetermined stereoisomer of 1a.[11] The latter compound was subsequently identified by Brown as 8-epi-salvinorin A (1b).[12] Brown also reported that deacetylation of 1a under basic conditions gave 8-epi-salvinorin B (2b), but did not characterize either compound. Several further reports of epimerization at C-8 appeared over the following decade, [13,14] but no characterization data was presented. Valdés later identified the byproduct mentioned above as 8-epi-diol 3.[15] Characterization data was given, but the basis of the structure assignment was not stated.

The first structure elucidation of one of these compounds was of 8-epi-salvinorin A (1b).[16] The trans-diaxial H-8 coupling constant found in 1a was absent in 1b, establishing an equatorial configuration. Also, irradiation of H-12 in 1b gave a strong nOe enhancement of H-8. The corresponding experiment on 1a gave instead an enhancement of H-20. These findings can be extrapolated to 2b, since acetylation gives 1b quantitatively.[9,17] Conflicting 1H NMR data for 2b itself were later reported by two groups.[8,9] The 1H NMR spectrum of 2b is reproduced in 1; the corresponding amended data have been reported previously.[17] Interestingly, epimerization has also recently been reported under acidic conditions.[18]

The epimers can be readily identified by TLC: the unnatural compounds almost invariably spot above the natural compounds in EtOAc/hexanes, and give a blue rather than pink/purple colour when visualized with vanillin.[19] The unnatural epimers are also recognizable by their distinctive H-12 multiplet in 1H NMR, which resembles a broad doublet shifted upfield to ~δ 5.30 ppm. Many 8-epi-salvinorin derivatives have now been reported, although many have not been fully characterized.[7-9,17,18,20-24] Thus, the many reports of 8-epi-salvinorins and derivatives have been based on limited data.

Results and Discussion

The crystal structure presented here (Figure 1) is the first reported for an 8-epi-salvinorin or derivative. It firmly establishes the structure of 2b, and therefore of 1b. The lactone carbonyl C-17 is axial with respect to the B ring (C6-7-8-17 torsion angle 77° versus 173° in 1a).[1] The lactone itself adopts a boat conformation with the furan equatorial (C9-11-12-13 torsion angle 179°). This is as predicted in solution, on the basis of a trans-diaxial coupling constant for H-12.[17] This is also consistent with the crystal structures of furanolactones with all other possible C8/9/12 stereochemistries (trans/anti, trans/syn and cis/syn) – the furan is equatorial in all cases.[17] The rest of the structure is very similar to the crystal structure of 1a.[1] The hydroxyl group participates in an intramolecular hydrogen bond with the ketone (O2-H2···O1, 2.12 Å). There are no intermolecular hydrogen bonds. The asymmetric unit consists of two molecules; the only substantial difference between them is in the rotation of the furan ring (C11-12-13-14 torsion angle -87° (A) versus 53° (B)). The crystals are monoclinic, space group P21 (see Figure 2). The crystallographic data can be found in 2; the structure factors are in 3. The crystallographic data have also been deposited with the Cambridge Crystallographic Data Centre (CCDC 626179).[25] 8-epi-Salvinorins and derivatives have a much weaker tendency to crystallize than their natural counterparts. Unsurprisingly, therefore, 2b has a lower melting point (192–196°C) than 2a (239–240°C).[17]

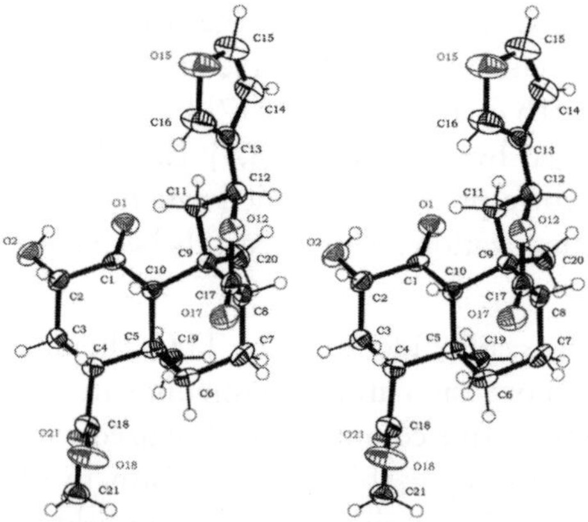

Figure 1. Stereoview of the molecular structure of 2b, showing 50% probability displacement ellipsoids and the atom-numbering scheme. Only one of the two molecules in the asymmetric unit is shown.

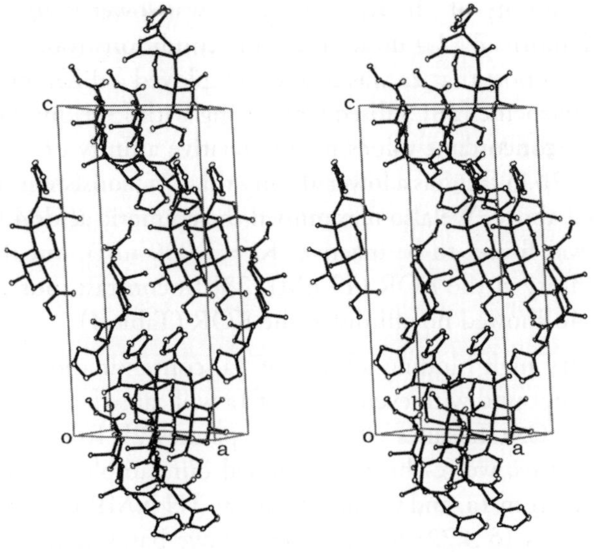

Figure 2. Stereoview of the packing of 2b. H atoms are not shown.

Configuration at C-8 is biologically significant. The affinity and potency of 8-epi-salvinorin A (1b) at the KOR are dramatically lower than those of 1a.[16] This finding has been replicated several times.[8,9,20] The same trend is evident with many salvinorin derivatives: epimerization of active compounds at C-8 reduces affinity and potency.[8,9,20,23,24] Very few exceptions to this trend have been reported to date.[8,23] These include 8-epi-salvinorin B (2b) itself, whose binding affinity (Ki = 43 nM) was reportedly greater than that of the natural epimer 2a (111 nM).[8] To explore this anomaly, we submitted a new sample of 2b for in vitro testing at the KOR. Binding affinity, potency and efficacy were determined as previously described (Table 1).[26]

Table 1. Affinities (Ki), potencies (EC50), and efficacies at the KOR.

Compound	K_i ± SEM [a,b] nM	EC_{50} ± SEM[b,c] nM	E_{max} ± SEM[d] %
1a	2.4 ± 0.4	1.8 ± 0.5	98 ± 3
2b	304 ± 46	214 ± 33	90 ± 2
4a	>10,000	-	-
4b	>10,000	-	-
U50,488H	2.2 ± 0.3	1.4 ± 0.3	100

[a]Inhibition of [^{3}H]diprenorphine binding to membranes of Chinese hamster ovary cells stably transfected with the human KOR (CHO-hKOR). [b]Mean ± SEM of three independent experiments performed in duplicate. [c]Enhancement of [^{35}S]GTPγS binding to CHO-hKOR membranes. [d]Relative to that of U50,488H control.

The binding affinity of 2b (Ki = 304 nM) was lower than those previously reported for salvinorin B (2a) under the same conditions (66, 111 or 155 nM).[7,8,27] An early report that 2a was inactive employed a different radiolabeled ligand, [3H] bremazocine.[28] Subsequent testing with [3H]diprenorphine by the same group gave concordant values for the relative affinity of 2a.[17] Thus, our data suggest that 2b in fact has a lower affinity than 2a, consistent with the general trend mentioned above. We also reexamined the epimeric acids 4.[16] In a previous report, 4a was found to be inactive (Ki > 1,000 nM), but the 8-epimer 4b showed high affinity at the KOR (49 nM).[23] In contrast, our current samples of both 4a and 4b showed no affinity at the KOR (Table 1).

Given the very high binding affinity of 1a, contamination of an inactive or weakly active compound with even traces of 1a will cause large errors. Flash chromatography in EtOAc/hexanes effectively separates 2b from 2a, but not from 1a. To overcome this, we re-chromatographed our sample in acetone/CH2Cl2, which resolves 2b from 1a, and verified purity by 1H NMR [1]. No methoxy peak corresponding to 1a (δ 3.72) was apparent above baseline noise. We separated 4a and 4b with difficulty by repeated chromatography in EtOAc/hexanes. The sample of 4a contained traces of an inseparable impurity, which if active might artificially elevate the apparent binding affinity. Since the sample showed no affinity, however, this problem does not arise.

There is no consensus on the mechanism of base-catalyzed epimerization at C-8. Koreeda and coworkers proposed a complex mechanism, initiated by ketone enolate formation. The configuration of H-8 is inverted indirectly, without exchange, by cleavage of the C-8/9 bond (see Scheme 1). [11-13] The simpler mechanism of enolization of the lactone itself has also been proposed.[16] A detailed case for this mechanism has been presented, giving evidence that H-8 exchanges under mildly basic conditions, and that similar furanolactones lacking the ketone also undergo epimerization.[17] Other workers remain undecided.[8,18]

Scheme 1. Koreeda et al.'s proposed mechanism for the epimerization.

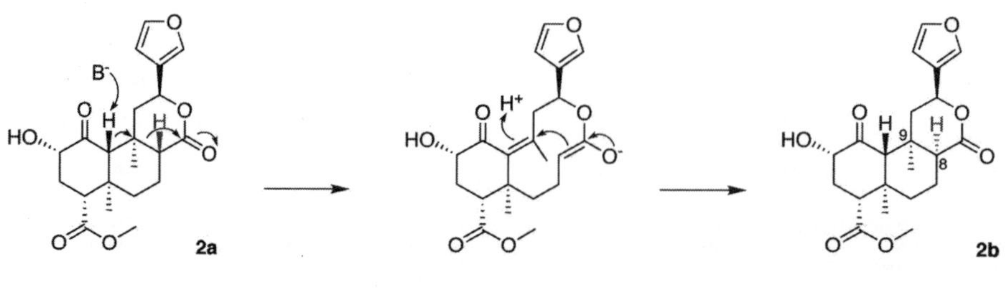

Acknowledgements

This work was supported by grants from the Stanley Medical Research Institute, the National Institute of Mental Health (MH63266), NARSAD and the Engelhard Foundation.

References

1. Ortega A, Blount JF, Manchand PS: J Chem Soc, Perkin Trans 1. 1982, 2505–2508.
2. Roth BL, Baner K, Westkaemper R, Siebert D, Rice KC, Steinberg S, Ernsberger P, Rothman RB: Proc Natl Acad Sci USA. 2002, 99:11934–11939.
3. Beardsley PM, Howard JL, Shelton KL, Carroll FI: Psychopharmacology (Berl). 2005, 183:118–26.
4. Carlezon WA Jr, Béguin C, DiNieri JA, Baumann MH, Richards MR, Todtenkopf MS, Rothman RB, Ma Z, Lee DY, Cohen BM: J Pharmacol Exp Ther. 2006, 316:440–7.
5. Mague SD, Pliakas AM, Todtenkopf MS, Tomasiewicz HC, Zhang Y, Stevens WC Jr, Jones RM, Portoghese PS, Carlezon WA Jr: J Pharmacol Exp Ther. 2003, 305:323–30.
6. Todtenkopf MS, Marcus JF, Portoghese PS, Carlezon WA Jr: Psychopharmacology (Berl). 2004, 172:463–70.
7. Béguin C, Richards MR, Wang Y, Chen Y, Liu-Chen L-Y, Ma Z, Lee DYW, Carlezon J, William A, Cohen BM: Bioorg Med Chem Lett. 2005, 15:2761–2765.
8. Lee DYW, Karnati VVR, He M, Liu-Chen L-Y, Kondareti L, Ma Z, Wang Y, Chen Y, Béguin C, Carlezon WA, Cohen B: Bioorg Med Chem Lett. 2005, 15:3744–3747.
9. Harding WW, Tidgewell K, Byrd N, Cobb H, Dersch CM, Butelman ER, Rothman RB, Prisinzano TE: J Med Chem. 2005, 48:4765–4771.
10. Ma J, Ye N, Lange N, Cohen BM: Neuroscience. 2003, 121:991–8.
11. Valdés LJJ III, Butler WM, Hatfield GM, Paul AG, Koreeda M: J Org Chem. 1984, 49:4716–4720.
12. Brown L: [http://wwwlib.umi.com/dissertations/fullcit/8422201] PhD Thesis. University of Michigan; 1984.
13. Koreeda M, Brown L, Valdés LJJ III: Chem Lett. 1990, 2015–2018.

14. Valdés LJJ III: J Psychoact Drugs. 1994, 26:277–283.

15. Valdés LJJ III, Chang HM, Visger DC, Koreeda M: Org Lett. 2001, 3:3935–3937.

16. Munro TA, Rizzacasa MA, Roth BL, Toth BA, Yan F: J Med Chem. 2005, 48:345–348.

17. Munro TA: [http://eprints.infodiv.unimelb.edu.au/archive/00002327] PhD Thesis. University of Melbourne; 2006.

18. Harding WW, Schmidt M, Tidgewell K, Kannan P, Holden KG, Dersch CM, Rothman RB, Prisinzano TE: Bioorg Med Chem Lett. 2006, 16:3170–3174.

19. Munro TA, Goetchius GW, Roth BL, Vortherms TA, Rizzacasa MA: J Org Chem. 2005, 70:10057–10061.

20. Béguin C, Richards MR, Li J-G, Wang Y, Xu W, Liu-Chen L-Y, Carlezon WA, Cohen BM: Bioorg Med Chem Lett. 2006, 16:4679–4685.

21. Béguin C, Carlezon W, Cohen BM, He M, Lee DY-W, Richards MR, Liu-Chen L-Y: Salvinorin derivatives and uses thereof. US Patent Application 20060052439. 2006.

22. Harding WW, Schmidt M, Tidgewell K, Kannan P, Holden KG, Gilmour B, Navarro H, Rothman RB, Prisinzano TE: J Nat Prod. 2006, 69:107–112.

23. Lee DY, He M, Kondaveti L, Liu-Chen LY, Ma Z, Wang Y, Chen Y, Li JG, Béguin C, Carlezon WA Jr, Cohen B: Bioorg Med Chem Lett. 2005, 15:4169–4173.

24. Lee DY, He M, Liu-Chen LY, Wang Y, Li JG, Xu W, Ma Z, Carlezon WA Jr, Cohen B: Bioorg Med Chem Lett. 2006, 16:5498–5502.

25. CCDC CIF Depository Request Form [http://www.ccdc.cam.ac.uk/data_request/cif]

26. Wang Y, Tang K, Inan S, Siebert DJ, Holzgrabe U, Lee DYW, Huang P, Li J-G, Cowan A, Liu-Chen L-Y: J Pharmacol Exp Ther. 2005, 312:220–230.

27. Lee DYW, Ma Z, Liu-Chen L-Y, Wang Y, Chen Y, Carlezon J, William A, Cohen B: Bioorg Med Chem. 2005, 13:5635–5639.

28. Chavkin C, Sud S, Jin W, Stewart J, Zjawiony JK, Siebert DJ, Toth BA, Hufeisen SJ, Roth BL: J Pharmacol Exp Ther. 2004, 308:1197–1203.

Effect of Transannular Interaction on the Redox-Potentials in a Series of Bicyclic Quinones

Grigoriy Sereda, Jesse Van Heukelom, Miles Koppang, Sudha Ramreddy and Nicole Collins

ABSTRACT

Background

Better understanding of the transannular influence of a substituent on the redox-potentials of bicyclo[2.2.2]octane-derived quinones will help in the design of new compounds with controlled biological activity. However, attempts to directly relate the reduction potentials of substituted triptycene-quinones to the electronic effects of substituents are often unsuccessful.

Results

First and second redox-potentials of a series of bicyclic quinones are compared to computed energies of their LUMO, LUMO+1, and energies of reduction. Transannular influence of substituent on the redox-potentials is rationalized in terms of MO theory. Acetoxy-substituents in the 5,8-positions of the triptycene-quinone system selectively destabilize the product of the two-electron reduction.

Conclusion

We have shown that first redox-potentials of substituted bicyclic quinones correlate with their calculated LUMO energies and the energies of reduction. The second redox-potentials correlate with calculated LUMO+1 energies. As opposed to the LUMO orbitals, the LUMO+1 orbital coefficients are weighted significantly on the non-quinone part of the bicyclic system. This accounts for: (1) significantly larger substituent effect on the second redox-potentials, than on the first redox-potentials; (2) lack of stability of the product of two electron reduction of 5,8-diacetoxy-9,10-dihydro-9,10-[1,2]benzenoanthracene-1,4-dione 5.

Background

It has been shown that 9,10-dihydro-9,10-[1,2]benzenoanthracene-1,4-dione (triptycene-quinone, 1) exhibits anti-leukemia activity, comparable with activity of substituted triptodiquinones [1]. One of the reasons for such activity is believed to be caused by the oxidizing properties of the quinone ring [1]. A recent study has revealed significant anti-inflammatory activity of the substituted triptycene-quinones 2 and 3 (Figure 1), which is also believed to be linked to the free radical redox-processes, involving triptycene-quinones and reactive oxygen species [2]. Better understanding of the transannular influence of a substituent on the redox-potentials of bicyclo[2.2.2]octane-derived quinones will help in the design of new compounds with controlled biological activity. However, attempts to directly relate the reduction potentials of substituted triptycene-quinones to the electronic effects of substituents are often unsuccessful. Thus, the negative shift of the reduction potential, caused by two methoxy-groups at the 5,8-positions (compound 2), was surprisingly only half the decrease caused by the 6,7-methoxy-groups, which are more distant from the quinone fragment [3].

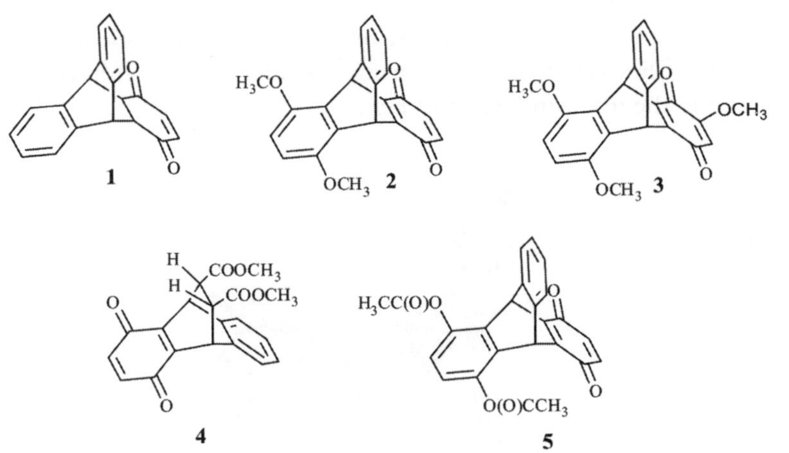

Figure 1. Bicyclic quinones explored for the transannular interaction.

Here we report cyclic voltammetric data and DFT (Density Functional Theory) calculations of five bicyclic quinones 1–5 (Figure 1) with the purpose to relate their redox-potentials to the calculated parameters and to the nature and positions of substituents in the bicyclic system.

Results and Discussion

Accurate computational prediction of redox-potentials requires comparison of energies for both the starting quinone and its reduced forms. The open-shell nature of the reduced species and often the necessity to take into account solvation makes the prediction of the redox-potentials a challenging and time consuming computational problem. However, Koopmans' theorem [4] enabled us to relate redox-potentials of bicyclic quinones with their LUMO energies, which characterize solely the starting compound. Despite the neglected orbital relaxation that immediately follows the reduction, such correlations have proved to be an efficient tool for prediction of redox-potentials of anthracyclines [5], substituted anthracenes [6], and oligothiophenes [7].

For all chemical species, the computations were performed for the global minimum conformation. These conformations for the methoxy-derivatives 2, 3, 2.-, and 3.- correspond to the α-methyl groups, symmetric with respect to the plane of the benzene ring (conformation A, Figure 2). In addition, we present computational results for the alternative conformation of 2 with two methyl groups oriented toward the quinone ring (conformation B, Figure 3). To minimize steric repulsion in the alternative conformation of the trimethoxy-derivative 3, only the methyl group remote from the quinone methoxy-group was oriented toward the quinone ring. These alternative conformations are marked with asterisk in the Table 1 and in the following text and Figures.

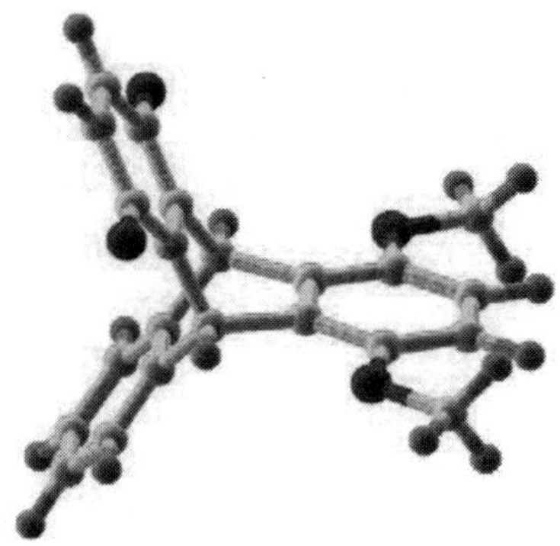

Figure 2. Conformation A of compound 2.

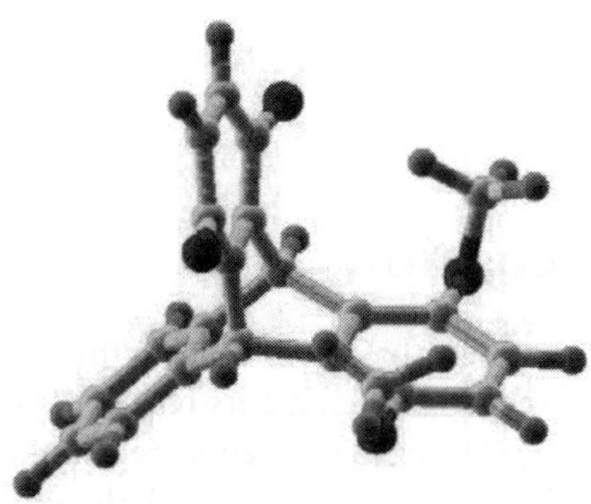

Figure 3. Conformation B of compound 2.

Table 1. Cyclic Voltammetry data for compounds 1–5.

Compound	E_{pr1}, V	E_{pox1}, V	$E^{o'}_{1}$, V	E_{pr2}, V	E_{pox2}, V	$E^{o'}_{2}$, V
1	-0.453	-0.369	-0.411	-1.065	-0.976	-1.0205
2	-0.435	-0.360	-0.398	-0.980	-0.894	-0.937
3	-0.550	-0.470	-0.510	-1.090	-0.994	-1.042
4	-0.441	-0.360	-0.401	-1.010	-0.923	-0.9665
5	-0.376	-0.296	-0.336	-0.804	-0.610	-0.707

The Figure 4 shows correlation between the first redox-potentials and calculated LUMO energies for compounds 1–5.

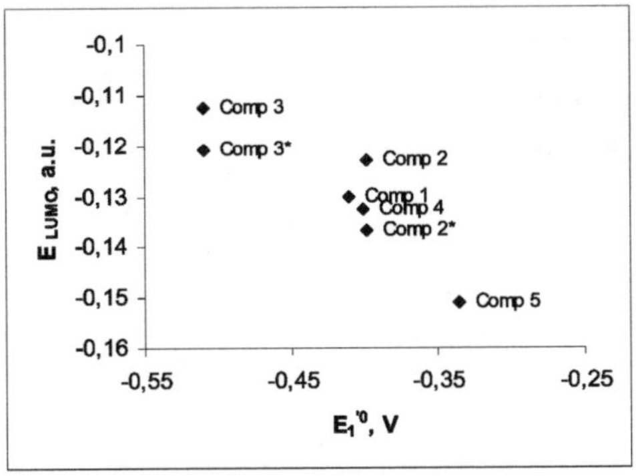

Figure 4. First formal redox-potentials of compounds 1–5 vs. their calculated LUMO energies.

The experimental redox-potential of the quinone 2 is 50 mV higher, than expected, based on this correlation and the calculated LUMO energy for its conformation A. This unexpectedly high redox-potential of 2 is consistent with the LUMO energy, calculated for the conformation B with two methyl groups, turned toward the quinone ring. This conformation is stabilized by weak H-bonds with the quinone carbonyl oxygens (2.5 A). Due to the lack of conjugation between the lone electron pairs of the methoxy groups and the benzene ring, conformation B is 5.5 kcal/mol higher in energy than conformation A. This value calculated for vacuum, can be greatly affected by solvation. Therefore, the energy difference may fall below the threshold that would warrant sufficient concentrations of the conformation B to account for the experimental redox-potential of 2.

Yamamura and co-authors [3] also noticed that the reduction potential of the quinone 2 was higher than expected from the electronic effects of the methoxy-substituents. They explained this difference by the parallel alignment of the C-O-bond with the π-system of the benzene ring, which amplifies the inductive effect of the methoxy-group. In other words, quinone 2 assumes the conformation B. As we move from the conformation A to conformation B, changing of the C-C-O-C dihedral angle from 0° to 90° enhances the inductive effect of the methoxy-group and weakens its counterbalancing resonance effect. Therefore, our computations provide additional support for the assumption of Yamamura.

For quinone 3, both the global minimum and the alternative conformations fit well into the correlation (Figure 4). A more precise approach to prediction of redox-potential should involve comparison of energies of both the original quinone and its reduced form. Figure 5 shows correlation between the first redox-potentials and calculated energies of reduction for compounds 1–5. The energies of reduction were calculated as a difference between the energy of the reduced form and the original quinone.

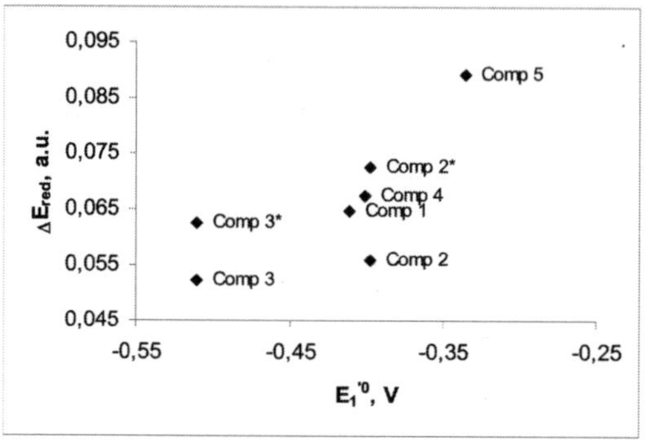

Figure 5. First formal redox-potentials of compounds 1–5 vs. their calculated energies of reduction.

Due to the computational challenges of optimization open-shell structures 1.—5.-, correlation between first redox-potentials of quinones and their reduction energies takes significantly more computational time, but does not substantially improve the quality of prediction.

As opposed to the first redox-potentials, we did not find any correlation between the second redox-potentials and the LUMO energies for the reduced species 1.—5.-, computed at the time permissible level of theory. This computational challenge may be partly due to the degenerate nature of the LUMO and LUMO+1 orbitals of 1.—5.-. However, the second redox-potentials can be easily predicted due to their correlation with the calculated LUMO+1 energies of the starting quinones, shown in Figure 6.

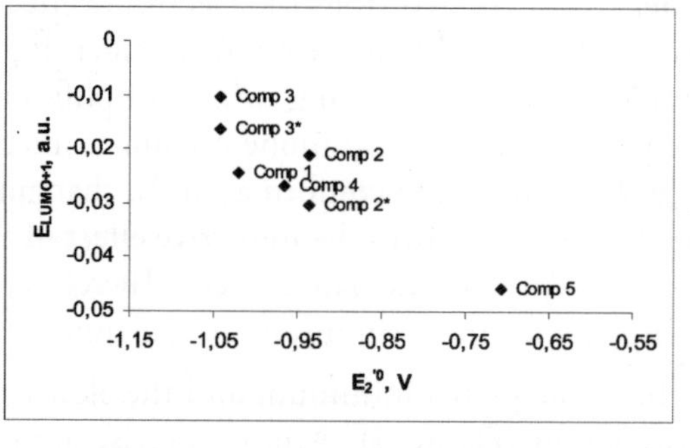

Figure 6. Second formal redox-potentials of compounds 1–5 vs. their calculated LUMO^{+1} energies.

This correlation illustrates that similarly to the first redox-potentials, prediction of second redox-potentials should be performed with consideration of the conformation with the highest oxidation potential, which is conformation B for quinones 2 and 3. The correlation on Figures 5 and 6 demonstrate that the Koopmans' theorem provides us with the useful tool to evaluate both the first and the second redox-potentials for the series of bicyclic quinones.

It is worthwhile to note that substituents in the non-quinone ring exert significantly stronger influence (by the factor of 4 to 6, see Table 1 in the Experimental section) on the second redox-potential, than on the first potential. Contrary, the methoxy-group, attached to the quinone ring in the compound 3 has similar effect (about 0.1 V) on both the first and second redox-potentials. To understand the reason of such behavior, we need to consider the transannular orbital interaction in the quinones 1–5.

Due to the Mobius-type transannular orbital overlap in the triptycene-quinone system, each π-orbital of the quinone ring (which is always anti-symmetric with respect to the plane of the ring) may interact only with out-of-phase combinations of the group orbitals of the other two benzene rings. Conversely, interactions of the π-quinone orbitals with the in-phase combinations (they have slightly lower energies, than the out-of-phase combinations and are likely to contribute to the next lower energy molecular orbital) are not allowed by symmetry. This situation is clearly illustrated by the comparison of the calculated LUMO of the triptycene-quinone 1, lacking noticeable contribution from the non-quinone π-system of the molecule (Figure 7), and the LUMO+1, heavily weighed on the non-quinone benzene rings due to the involvement of the out-of-phase combination of their π-orbitals (Figure 8). Additionally, higher LUMO+1 energy matches better with the antibonding orbital energies of the rest of the bicyclic system.

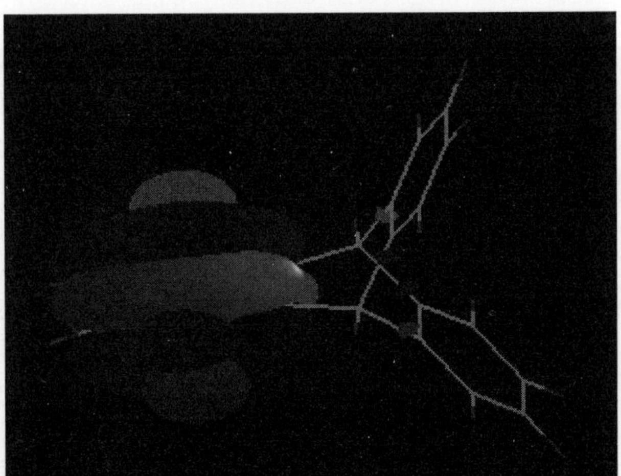

Figure 7. LUMO of Compound 1.

Figure 8. LUMO+1 of Compound 1.

In the dimethoxy-derivative 2, the LUMO has some contribution from the bridgehead σ-bonds of the non-quinone part of the molecule, whereas the LUMO+1 orbital is heavily weighted on the substituted benzene ring. This phenomenon is general for the whole series of substituted triptycene-quinones 1–3 and 5 and explains why the substituents in the none-quinone ring influence the second redox-potentials significantly more, than the first redox-potentials. The generality of this orbital overlap pattern is illustrated by Figures 9 and 10.

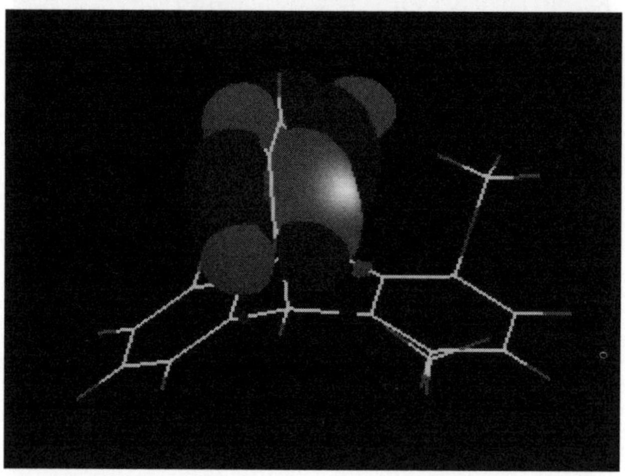

Figure 9. LUMO of Compound 2.

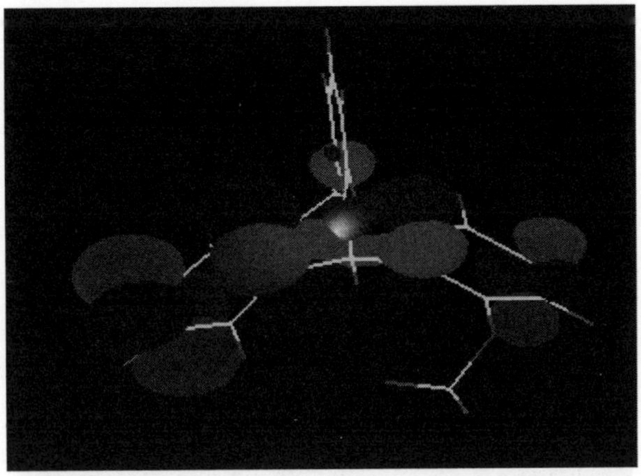

Figure 10. LUMO+1 of Compound 2.

In the compound 3, the lone electron pair of the methoxy-group, attached to the quinone ring, makes a major contribution to the LUMO which explains the highest LUMO energy and the lowest redox-potential of 3 in the whole series.

The non-aromatic fragment with two electron-withdrawing carbomethoxy-groups, attached to the bridgehead σ-bonds and contributing to the LUMO of the quinone 4 (Figure 11), slightly lowers the LUMO energy and hence increases the first redox-potential by 10 mV, compared with the triptycene-quinone 1.

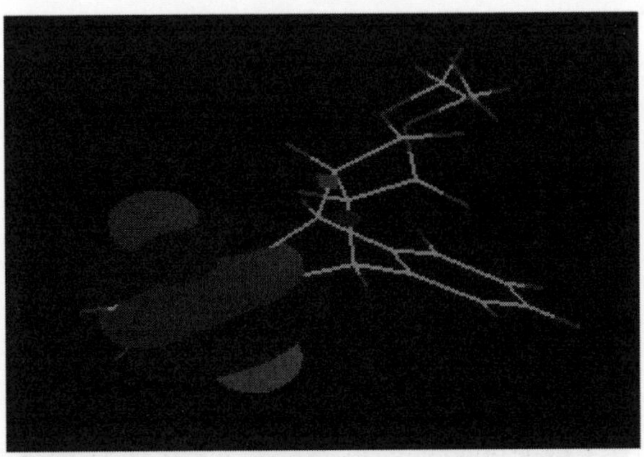

Figure 11. LUMO of Compound 4.

The significant contribution of the non-aromatic moiety to the LUMO+1 of compound 4 (Figure 12) accounts for the much larger increase (by 50 mV, see Table 1) of the second redox-potential.

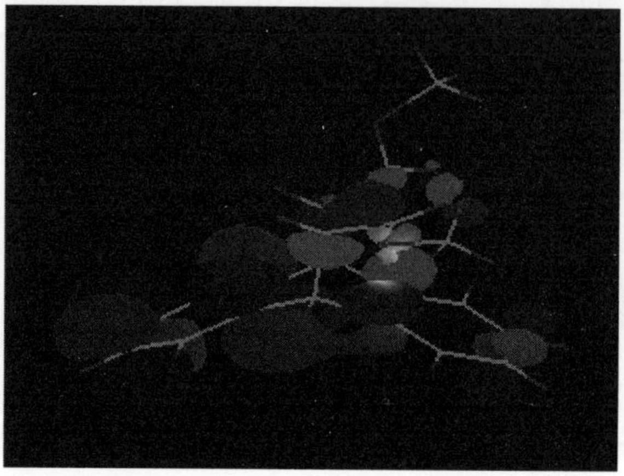

Figure 12. LUMO+1 of Compound 4.

The different modes of the transannular interaction of orbitals are best illustrated by the different stabilities of the products of one- and two-electron reduction of the 5,8-diacetoxy-derivative 5. Because of the low contribution of the substituted benzene ring to the LUMO, placing an electron on this orbital does not activate the leaving acetoxy-anions, keeping the product of one-electron reduction stable (Figure 13). The second electron placed on the LUMO+1 orbital of 5, activates the acetoxy-groups, which causes decomposition of the product of the two electron reduction (Figure 14) and makes the second reduction chemically irreversible.

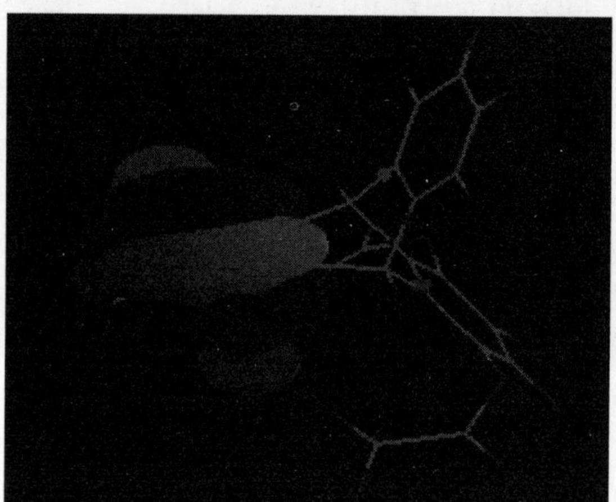

Figure 13. LUMO of Compound 5.

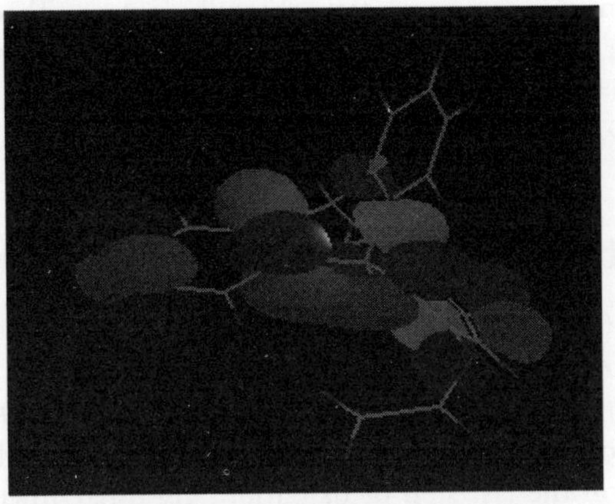

Figure 14. LUMO+1 of Compound 5.

The LUMO of the reduced species 5.- is also mostly located at the substituted benzene ring (Figure 15), additionally illustrating the reason of the low stability of the dianion 52-.

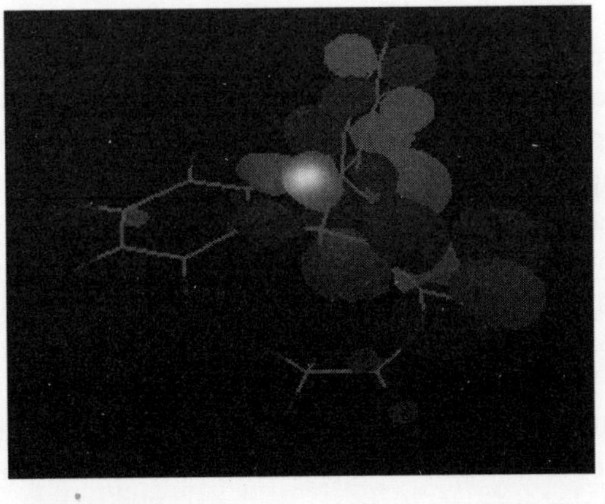

Figure 15. LUMO of the reduced species $5^{\cdot-}$.

Conclusion

We have shown that first redox-potentials of substituted bicyclic quinones correlate with their calculated LUMO energies and the energies of reduction. The second redox-potentials correlate with calculated LUMO+1 energies. As opposed to the LUMO orbitals, the LUMO+1 orbital coefficients are weighted significantly on the non-quinone part of the bicyclic system. This accounts for: (1) significantly larger substituent effect on the second redox-potentials, than on the first redox-potentials; (2) lack of stability of the product of two electron reduction of 5,8-diacetoxy-9,10-dihydro-9,10-[1,2]benzenoanthracene-1,4-dione 5.

Experimental

9,10-Dihydro-9,10-[1,2]benzenoanthracene-1,4-dione 1 and 2,5,8-trimethoxy-9,10-dihydro-9,10-[1,2]benzenoanthracene-1,4-dione 3 were synthesized as described in the literature [8,9]. 5,8-Dimethoxy-9,10-dihydro-9,10-[1,2]benzenoanthracene-1,4-dione 2 was synthesized as described [2], but with the use of silver oxide in acetone on the last step of oxidation. cis-anti-Dimethyl 1,2,3,4-Tetrahydro-1,4-[1,2]benzenonaphthalene-5,8-dione-2,3-dicarboxylate 4 and 5,8-diacetoxy-9,10-dihydro-9,10-[1,2]benzenoanthracene-1,4-dione 5 were synthesized for the first time in our laboratory [10]. A set of redox-potentials was obtained for each of the bicyclic quinones 1–5 by the following procedure. A 2 mmol portion of the compound was dissolved in 25 mL of 0.1 M $(n\text{-}C_4H_9)_4N$+BF4- (electrochemical grade from Southwestern Analytical) in acetonitrile (HPLC grade) and placed in a three electrode electrochemical cell. The working electrode was a BAS platinum electrode (Bioanalytical Systems, West Lafayette, IN, area ca. 0.02 cm2), the auxiliary electrode was a carbon rod and the reference electrode was a BAS Ag/AgCl. To eliminate the influence of oxygen, the solution was degassed with argon gas prior to the experiment and a blanket of argon was maintained over the solution during the experiment. From an initial applied voltage of 0 V, the working electrode's potential was scanned to -1.5 V and then back to 0 V at a rate of 0.1 V/s. For each of the compounds 1–4, we observed two reduction and two oxidation waves. The formal redox-potentials ($E^{o'}$) were calculated as the average of the complementary peak reduction (Ep_{red}) and peak oxidation potentials (Ep_{ox}) where ($E^{o'} = 1/2(Ep_{red} + Ep_{ox})$ [11]. In order to check our process, we measured the first reduction potential of p-benzoquinone to be -0.507 V, which is exactly the same as the value reported in the literature [3]. The cyclic voltammograms (CV) for p-benzoquinone and the quinone 3 are presented in Figure 16.

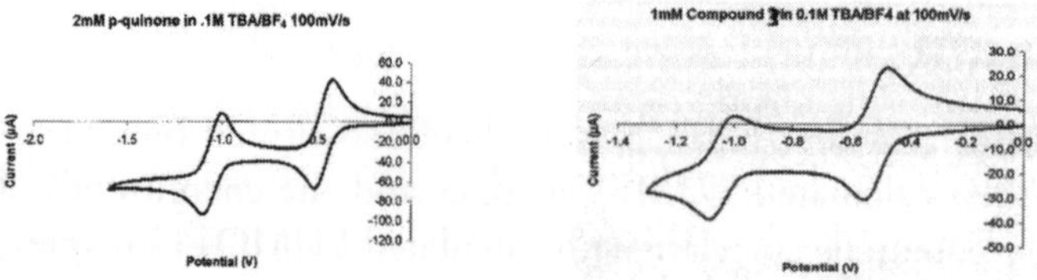

Figure 16. CV for p-benzoquinone and quinone 3.

For the quinone 5, lack of a well-defined second oxidation wave indicated that its two electron reduction was chemically irreversible. The potential measurements were not corrected for IR drop. Electrochemical control of the experiment was achieved with a PAR (Princeton Applied Research) model 273 potentiostat equipped with PAR model 270 computer controlled software.

All computational methods were used as implemented to the GAUSSIAN 98W package [12], running on a PC Pentium 4 computer. The molecular structures 1–5 were pre-optimized in vacuum at the AM1 semi-empirical level and then geometry optimized by the B3LYP density functional method (3-21G basis set). The molecular orbitals were calculated at the B3LYP/6-31G** level for the optimized geometries. Consistency of the computational results was checked with a double-split basis set with added diffuse functions (6-311+G**). The product of one-electron reduction of the quinone 5 was treated at the restricted open shell density functional level (ROB3LYP). The MO images were visualized with the Orb Draw 3.00.1 program [13].

The cyclic voltammetry data (first reduction peak potential Epr1, first oxidation peak potential Epox1, first formal redox-potential Eo'1, second reduction potential Epr2, second oxidation potential Epox2 and second redox-potential Eo'2) for compounds 1–5 are presented in Table 1.

The computed parameters for compounds 1–5 and for their reduced species 1.-–5.- are presented in Table 2.

Table 2. Computed parameters for compounds 1–5 and for their reduced species $1^{\cdot-}$–$5^{\cdot-}$.

Species	E_{LUMO}, a.u.	E_{LUMO+1}, a.u.	E_{SOMO}, a.u.	E, a.u.	ΔE_{red}, a.u.
1	-0.130	-0.024	-	-919.794	0.065
$1^{\cdot-}$	0.099	0.099	0.017	-919.860	
2	-0.123	-0.021	-	-1148.839	0.056
$2^{\cdot-}$	0.099	0.099	0.030	-1148.894	
2*	-0.137	-0.030	-	-1148.830	0.073
$2^{\cdot-}$*	0.098	0.098	0.007	-1148.902	
3	-0.112	-0.010	-	-1263.367	0.052
$3^{\cdot-}$	0.100	0.100	0.021	-1263.419	
3*	-0.121	-0.016	-	-1263.359	0.063
$3^{\cdot-}$*	0.100	0.107	0.015	-1263.422	
4	-0.132	-0.027	-	-1223.109	0.068
$4^{\cdot-}$	0.094	0.094	0.015	-1223.176	
5	-0.150	-0.046	-	-1375.532	0.089
$5^{\cdot-}$	0.073	0.073	-0.007	-1375.622	

* – Conformation B (Figure 2)

Acknowledgements

We thank the NSF (DUE-0311303 Grant), the State of South Dakota (2010 Research Initiative) and the University of South Dakota (Nelson Research Grant) for financial support of this work.

References

1. Perchellet EM, Magill MJ, Huang X, Brantis CE, Hua DH, Perchellet JP: Anti-Cancer Drugs. 1999, 10:749–766.
2. Xanthopoulou NJ, Kourounakis AP, Spyrouis S, Kourounakis PN: European J Med Chem. 2003, 38:621–626.
3. Yamamura K, Miyake H, Himeno S, Inagaki S, Nakasuji K, Murata I: Chem Lett. 1988, 1819–1822.
4. Pilar FL: J Chem Ed. 1978, 55:2–6.
5. Kawakami Y, Hopfinger AJ: Chem Res Toxicol. 1990, 3:244–247.
6. Okazaki S, Oyama M, Nomura S: Electroanalysis. 1997, 9, 16:1242–1246.
7. Casado J, Delgado M, Merchan M, Hernandez V, Navarrete J, Pappenfus T, Williams N, Stegner W, Johnson J, Edlund B, Janzen D, Mann K, Orduna J, Viullacampa B: Chem Eur J. 2006, 12:5458–5470.
8. Bartlett PD, Ryan MJ, Cohen SG: J Am Chem Soc. 1942, 64:2649–2653.
9. Hua DH, Tamura M, Huang X, Stephany HA, Helfrich BA, Perchellet EM, Sperfslage BJ, Perchellet JP, Jiang S, Kyle DE, Chiang PK: J Org Chem. 2002, 67:2907–2912.
10. Sereda GA, Van Heukelom J, Ramreddy S: Tetrahedron Lett. 2006, 47:8901–8903.
11. Heineman WR, Kissinger PT: Laboratory Techniques in Eletroanalytical Chemistry. Volume 3. Marcel Dekker: New York; 1984:90.
12. Frisch MJ, Trucks GW, Schlegel HB, Scuseria GE, Robb MA, Cheeseman JR, Zakrzewski VG, Montgomery JA Jr, Stratmann RE, Burant JC, Millam JM, Daniels AD, Kudin KN, Strain MC, Farkas O, Tomasi J, Barone V, Cossi M, Cammi R, Mennucci B, Pomelli C, Adamo C, Clifford C, Ochterski J, Petersson GA, Ayala PY, Cui QK, Morokuma K, Malick DK, Rabuck AD, Raghavachari K, Foresman JB, Cioslowski J, Ortiz JV, Stefanov BB, Liu G, Liashenko A, Piskorz P, Komaromi I, Gomperts R, Martin RL, Fox DJ, Keith T, Al-Laham MA, Peng CY, Nanayakkara A, Gonzalez C, Challacombe M, Gill PM, Johnson B, Chen W, Wong MW, Andres JL, Gonzalez C, Head-Gordon M, Replogle ES, Pople JA: Gaussian 98, Revision A.6. Gaussian, Inc., Pittsburgh, PA; 1998.
13. Orb Draw 3.00.1 June 24, 2003. Serena Software.

Three-Component One-Pot Synthesis of Novel Benzo[b]1,8-naphthyridines Catalyzed by Bismuth(III) Chloride

Tangali R. Ravikumar Naik, Halehatty S. Bhojya Naik, Halehatty R. Prakasha Naik and P. J. Bindu

ABSTRACT

A novel and efficient three-component one-pot synthesis of benzo[b]1,8-naphthyridines by 2-amino-4-methylquinoline, aromatic aldehydes, and malononitrile was done. The reaction was catalyzed by an acidic Bismuth(III) chloride, functionalized Bismuth(III) chloride, at room temperature to give various benzo[b]1,8-naphthyridines in high yields. The Bismuth(III) chloride is an environmentally friendly catalyst.

Introduction

One of the most important reactions in organic chemistry for the formation of carbon—carbon bonds—is the multicomponent reactions (MCRs) and much effort has been devoted to the development of this reactions in recent years [1, 2].

Among the nitrogen heterocycles, naphthyridines and their derivatives represent an important class of organic molecules that attract the interest of both synthetic and medicinal chemists. Functionalized naphthyridines have found applications as pharmaceuticals, fungicides, bactericides, herbicides, and insecticides as well as useful synthetic blocks in the preparation of several alkaloids [3–6]. Many syntheses of naphthyridines are known, but due to their importance, the development of new synthetic approaches remains an active research area [7, 8].

Various procedures are available for the synthesis of benzo[b]1,8-naphthyridines. However, there are several disadvantages associated with these methodologies including unsatisfactory yields, long conversion times and inflammable organic solvents. In recent years, we have initiated our efforts on the development of simple methods for the synthesis of naphthyridines derivatives [9–11]. Hence, now we report herein a useful, general approach towards the formation of benzo[b]1,8-naphthyridines in one-pot operation (Scheme 1). This method permits the condensation of aromatic 2-amino-4-methylquinoline 1, aldehyde 2, and malononitrile 3using Lewis acid catalysts under mild conditions to afford diverse benzo[b]1,8-naphthyridines (4a–h) and is also amenable to small library production using solid-phase techniques.

Scheme 1. The synthesis of benzo[b]1,8-naphthyridines.

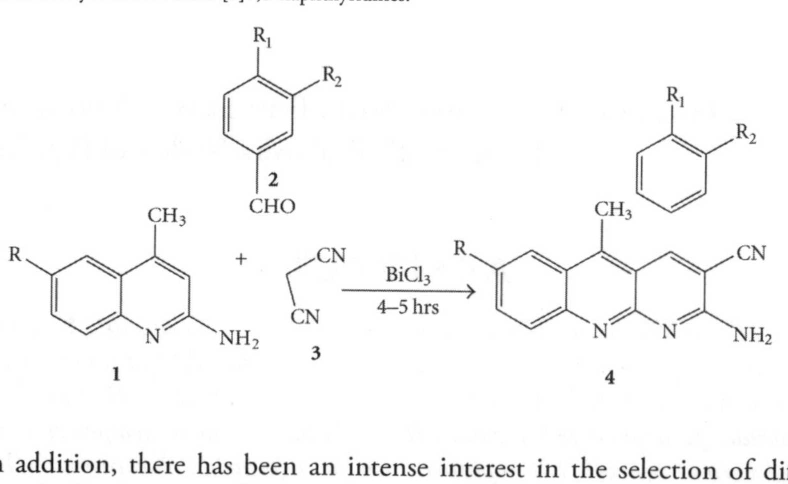

In addition, there has been an intense interest in the selection of different catalysts in development of new methods for MCRs. The utility of different metal salts such as mercuric salts, lead salts, zinc chloride, cadmium chloride, and indium chloride, lanthanide compounds, InBr3, ZrCl4 [12, 13], as potential catalyst in variety of synthetic reactions was recognized in recent years. However, the expensiveness and toxicity of some metal salts hampers their wide applications in organic synthesis. On the other hand, in recent years, economically affordable ecofriendly catalysts received some interest in various organic reactions. The application of some of these catalysts such as Cu(II) salts [14], Fe(II)/Fe(III) salts [15], nickel salts, and Bismuth(III) salts [16, 17] as catalysts in organic synthesis, has been investigated extensively. Bismuth trichloride has received considerable attention as an inexpensive, nontoxic, and readily available catalyst for various organic transformations, affording the corresponding products in excellent yields with high selectivity (Scheme 1).

Equimolar of various substituted aromatic aldehyde 2 reacted well with 2-amino-4-methylquinoline 1 and malononitrile 3 in presence of 10 mol% of BiCl3 catalyst in CH3CN solvent to give the corresponding substituted naphthyridine 4a-h in 93–97% yields. In each experiment, molar ratio 1:1:1.5 of the three components 1, 2, and 3 were used as reactants. The method is very simple and it can be used as derivatives of substituted 2-amino-4-methylquinoline 1, different substituted aromatic aldehyde 2, and malononitrile 3 to prepare different substituted naphthyridine derivatives (Table 1).

Table 1. One-pot synthesis of benzo[b]1,8-naphthyridines.

Entry	R	R_1	R_2	Product[a]	Yield (%)[b]
1	H	H	H	4a	93
2	CH_3	H	H	4b	95
3	OCH_3	H	H	4c	95
4	Cl	H	H	4d	97
5	CH_3	CH_3	H	4e	93
6	H	H	CH_3	4f	95
7	H	H	OCH_3	4g	96
8	H	H	Cl	4h	96

[a] All the products were characterised by IR, ^1H NMR, and mass spectra.
[b] Yields of isolated products.

Experimental

Melting points were recorded on an open capillary tube with a Buchi melting point apparatus and are uncorrected. Elemental analyses were carried out using Perkin-Elmer 24°C CHN-analyzer. IR spectra were recorded on an FT-IR infrared spectrophotometer. H1-NMR spectra were obtained using a 300 MHz on a Bruker spectrometer (chemical shifts in δ ppm). Mass spectra were recorded on an LC MS Mass spectrometer.

General Procedure

A mixture of 2-amino-4-methylquinoline (1 mmol), aromatic aldehyde (1.5 mmol), and malononitriles (1 mmol) in anhydrous CH3CN (15 mL) was stirred at room temp for 30 minutes. $BiCl_3$ (20 mol%) was added over a period of 20 minutes. The resulting mixture was stirred for 4-5 hours. After completion of the reaction as indicated by TLC, the reaction mixture was diluted with water (30 mL) and extracted with ethyl acetate (3×15 mL). The organic layer was separated, dried (Na_2SO_4), and concentrated, and the resulting product was recrystalized to afford pure benzo[b]1,8-naphthyridines 4a. The same procedure was used for the synthesis of 4b–h. The physicochemical data for the synthesized compounds are as shown below.

4a: M.p. 160–163°C. IR (KBr): 3251; 1723; cm−1. 1H NMR (DMSO d6): 1.20 (s, 3H, CH3), 6.33 (s, 2H, NH2), 6.80–7.40 (m, 5Harom), 7.45 (d, 1H, J=8.5), 7.51 (d, 1H, J=8.5), 7.70–7.72 (m, 2Harom), 21.5, 89.2, 118.5 (CN), 121.4, 124.5, 126.7, 127.5, 127.6, 127.7, 129.4, 129.5, 129.6, 129.5, 129.4, 138.3, 138.8, 146.8, 152.6, 160.6, 162.5 (NH2), mass m/z:310.

4b: M.p. 162–165°C. IR (KBr): 3251; 1723; cm−1. 1H NMR (DMSO d6): 1.26 (s, 3H, CH3), 1.80 (s, 3H, CH3), 6.30 (s, 2H, NH2), 6. 68–7.25 (m, 5Harom), 7.36 (d, 1H, J=8.5), 7.42–7.45 (m, 3Harom), 21.3, 24. 9, 89.0, 118.4 (CN), 121.3, 124.2, 126.8, 127.3, 127.6, 127.4, 129.2, 129.5, 129.3, 129.4, 129.7, 138.0, 138.7, 146.7, 152.4, 160.3, 162.4 (NH2), mass m/z:324.

4c: M.p. 171–173°C. IR (KBr): 3251; 1723; cm−1. 1H NMR (DMSO d6): 1.22 (s, 3H, CH3), 3.10 (s, 3H, OCH3), 6.25 (s, 2H, NH2), 6.70–7.13 (m, 5Harom), 7.26 (d, 1H, J=8.5), 7.33–7.40 (m, 3Harom), 20.8, 25.4, 88.9, 118.3 (CN), 121.5, 124.5, 126.7, 127.6, 127.6, 127.8, 129.4, 129.7, 129.8, 129.8, 129.9, 138.3, 138.5, 146.8, 152.5, 160.4, 162.7 (NH2), mass m/z:340.

4d: M.p. 178–180°C. IR (KBr): 3251; 1723; cm−1. 1H NMR (DMSO d6): 1.15 (s, 3H, CH3), 6.20 (s, 2H, NH2), 6.80–7.36 (m, 5Harom), 7.40 (d, 1H, J=8.5), 7.58–7.60 (m, 3Harom), 21.7, 89.5, 118.2 (CN), 121.3, 124.4, 126.5, 127.3, 127.6, 127.5, 129.5, 129.6, 129.5, 129.5,

129.7, 132.5(Cl), 138.5, 138.8, 146.8, 152.6, 160.5, 162.7 (NH2), mass m/z:344.

4e: M.p. 170–173°C. IR (KBr): 3251; 1723; cm−1. 1H NMR (DMSO d6): 1.23 (s, 3H, CH3), 1.65 (s, 3H, CH3), 1.93 (s, 3H, CH3), 6.20 (s, 2H, NH2), 6.70–6.92 (m, 5Harom), 7.08 (d, 1H, J=8.5), 7.20–7.30 (m, 3Harom), 20.5, 23. 5, 24. 7, 89.4, 118.4 (CN), 121.5, 124.4, 126.4, 127.5, 127.6, 127.7, 129.4, 129.5, 129.6, 129.8, 129.9, 138.5, 138.6, 146.8, 152.6, 160.5, 162.9 (NH2), mass m/z:338.

4f: M.p. 162–164°C. IR (KBr): 3251; 1723; cm−1. 1H NMR (DMSO d6): 1.16 (s, 3H, CH3), 1.95 (s, 3H, CH3), 6.20 (s, 2H, NH2), 6.66–6.90 (m, 5Harom), 6.95 (d, 1H, J=8.5), 6.98–7.20 (m, 3Harom), 21.6, 25.2, 88.7, 118.0 (CN), 121.5, 124.6, 126.7, 127.5, 127.6, 127.6, 129.5, 129.6, 129.6, 129.7, 129.8, 138.5, 138.7, 146.8, 152.6, 160.5, 162.6 (NH2), mass m/z:324.

4g: M.p. 171–173°C. IR (KBr): 3251; 1723; cm−1. 1H NMR (DMSO d6): 1.21 (s, 3H, CH3), 3.00 (s, 3H, OCH3), 6.20 (s, 2H, NH2), 6.71–7.04 (m, 5Harom), 7.10 (d, 1H, J=8.5), 7.29–7.35 (m, 3Harom), 21.2, 26.3, 89.3, 117.9 (CN), 121.5, 124.2, 126.6, 127.5, 127.5, 127.6, 129.4, 129.4, 129.5, 129.6, 129.6, 138.7, 138.9, 146.7, 152.6, 160.5, 162.7 (NH2), mass m/z:340.

4h: M.p. 176–179°C. IR (KBr): 3251; 1723; cm−1. 1H NMR (DMSO d6): 1.15 (s, 3H, CH3), 6.20 (s, 2H, NH2), 6.75–7.20 (m, 5Harom), 7.32 (d, 1H, J=8.5), 7.35–7.38 (m, 3Harom), 21.7, 89.2, 118.3 (CN), 121.3, 124.2, 126.5, 127.3, 127.4, 127.4, 129.2, 129.3, 129.3, 129.3, 129.3, 131.5(Cl), 138.0, 138.7, 146.7, 152.4, 160.3, 162.2 (NH2), mass m/z:344.

Elemental Analysis

4a. $C_{20}H_{14}N_4$ Calc: C, 77.40; H, 4.55; N, 18.05; Found: C, 77.38; H, 4.53; N, 18.03.

4b. C21H16N4 Calc: C, 77.76; H, 4.97; N, 17.27; Found: C, 77.74; H, 4.96; N, 17.24.

4c. C21H16N4O Calc: C, 74.10; H, 4.74; N, 16.46; Found: C, 74.11; H, 4.72; N, 16.43.

4d. C20H13ClN4 Calc: C, 69.67; H, 3.80; N, 16.25; Found: C, 69.65; H, 3.78; N, 16.23.

4e. C22H18N4 Calc: C, 78.08; H, 5.36; N, 16.56; Found: C, 78.06; H, 5.34; N, 16.53.

4f. $C_{21}H_{16}N_4$ Calc: C, 77.76; H, 4.97; N, 17.27; Found: C, 77.75; H, 4.95; N, 17.25.

4g. $C_{21}H_{16}N_4O$ Calc: C, 74.10; H, 4.74; N, 16.46; Found: C, 74.07; H, 4.73; N, 16.47.

4h. $C_{20}H_{13}ClN_4$ Calc: C, 69.67; H, 3.80; N, 16.25; Found: C, 69.68; H, 3.79; N, 16.24.

Conclusion

In summary, we have established a new methodology, based on a three-component reaction to obtain new substituted benzo[b]1,8-naphthyridines derivatives catalyzed by $BiCl_3$. The simplicity of this elegant protocol and accessibility of the starting materials allowed us to prepare these new benzo[b]1,8-naphthyridines derivatives that should have wide applicability in heterocyclic and medicinal chemistry.

Acknowledgements

The authors would like to thank UGC, New Delhi, for awarding Rajiv Gandhi Research Fellowship and SIFC, IISc, Bangalore for 1H NMR and Mass Spectral Studies.

References

1. Y. K. Chen and P. J. Walsh, "A one-pot multicomponent coupling reaction for the stereocontrolled synthesis of (Z)-trisubstituted allylic alcohols," Journal of the American Chemical Society, vol. 126, no. 12, pp. 3702–3703, 2004.
2. I. Bae, H. Han, and S. Chang, "Highly efficient one-pot synthesis of N-sulfonylamidines by Cu-catalyzed three-component coupling of sulfonyl azide, alkyne, and amine," Journal of the American Chemical Society, vol. 127, no. 7, pp. 2038–2039, 2005.
3. V. P. Litvinov, S. V. Roman, and V. D. Dyachenko, "Naphthyridines. Structure, physicochemical properties and general methods of synthesis," Russian Chemical Reviews, vol. 69, no. 3, pp. 201–220, 2000.
4. G. B. Barlin and W.-L. Tan, "Potential antimalarials. I. 1,8-naphthyridines," Australian Journal of Chemistry, vol. 37, no. 5, pp. 1065–1073, 1984.

5. D. Ramesh and B. Sreenivasulu, "Synthesis and antimicrobial activity of 2-aryl-3-(2-methyl-1,8-h naphthyridin-3-YL)-thiazolidin-4-ones," Indian Journal of Heterocyclic Chemistry, vol. 15, no. 4, pp. 363–366, 2006.
6. B. Bachowska and T. Zujewska, "Chemistry and applications of benzonaphthyridines," Arkivoc, vol. 2001, no. 6, pp. 77–84, 2001.
7. K. Mogilaiah and J. U. Rani, "Microwave-assisted solvent-free Friedlander synthesis of 1,8-naphthyridines using ammonium acetate as catalyst," Indian Journal of Chemistry Section B, vol. 45, no. 4, pp. 1051–1053, 2006.
8. K. Mogilaiah, M. Prashanthi, and S. Kavitha, "Lithium chloride as an efficient catalyst for Friedlander synthesis of 1,8-naphthyridines via the use of microwave irradiation and pestle/mortar," Indian Journal of Chemistry Section B, vol. 45, no. 1, pp. 302–304, 2006.
9. T. R. R. Naik, H. S. B. Naik, M. Raghavedra, and S. G. K. Naik, "Synthesis of thieno[2,3-b]benzo[1,8]naphthyridine-2-carboxylic acids under microwave irradiation and interaction with DNA studies," Arkivoc, vol. 2006, no. 15, pp. 84–94, 2006.
10. T. R. R. Naik, H. S. B. Naik, and S. R. G. K. Naik, "One pot solvent-free synthesis of 2H-pyrano, 2H-thiopyrano, 2H-selenopyrano[2,3-b]-1,8-naphthyridin-2-ones on solid phase catalyst under microwave irradiation," Journal of Sulfur Chemistry, vol. 28, no. 4, pp. 393–400, 2007.
11. R. N. Butler, "Comparative reactions of nitrogen compounds with the isoelectronic series mercury(II), thallium(III), and lead(IV) acetates. Principles of oxidation reactions," Chemical Reviews, vol. 84, no. 3, pp. 249–276, 1984.
12. Q. Sun, Y. Yi, Z. Ge, T. Cheng, and R. Li, "A highly efficient solvent-free synthesis of dihydropyrimidinones catalyzed by zinc chloride," Synthesis, no. 7, pp. 1047–1051, 2004.
13. A. Manjula, B. Vittal Rao, and P. Neelakantan, "An inexpensive protocol for biginelli reaction," Synthetic Communications, vol. 34, no. 14, pp. 2665–2671, 2004.
14. J. S. Yadav, B. V. S. Reddy, P. Sridhar, et al., "Green protocol for the Biginelli three-component reaction: Ag3PW12O40 as a novel, water-tolerant heteropolyacid for the synthesis of 3,4-dihydropyrimidinones," European Journal of Organic Chemistry, vol. 2004, no. 3, pp. 552–557, 2004.
15. I. Mohammadpoor-Baltork, A. R. Khosropour, and H. Aliyan, "Efficient conversion of epoxides to 1,3-dioxolanes catalyzed by bismuth(III) salts," Synthetic Communications, vol. 31, no. 22, pp. 3411–3416, 2001.

16. H. Firouzabadi, I. Mohammadpoor-Baltork, and S. Kolagar, "A rapid, selective, and efficient method for deprotection of silyl ethers catalyzed by bismuth(III) salts," Synthetic Communications, vol. 31, no. 6, pp. 905–909, 2001.
17. S. H. Mashraqui and M. A. Karnik, "Bismuth nitrate pentahydrate: a convenient reagent for the oxidation of Hantzsch 1,4-dihydropyridines," Synthesis, no. 5, pp. 713–714, 1998.

The Vicinal Difluoro Motif: The Synthesis and Conformation of Erythro- and Threo-Diastereoisomers of 1,2-Difluorodiphenylethanes, 2,3-Difluorosuccinic Acids and their Derivatives

David O'Hagan, Henry S. Rzepa, Martin Schüler and Alexandra M. Z. Slawin

ABSTRACT

Background

It is well established that vicinal fluorines (RCHF-CHFR) prefer to adopt a gauche rather than an anti conformation when placed along aliphatic chains.

This has been particularly recognised for 1,2-difluoroethane and extends to 2,3-difluorobutane and longer alkyl chains. It follows in these latter cases that if erythro and threo vicinal difluorinated stereoisomers are compared, they will adopt different overall conformations if the fluorines prefer to be gauche in each case. This concept is explored in this article with erythro- and threo-diastereoisomers of 2,3-difluorosuccinates.

Results

A synthetic route to 2,3-difluorosuccinates has been developed through erythro- and threo- 1,2-difluoro-1,2-diphenylethane which involved the oxidation of the aryl rings to generate the corresponding 2,3-difluorosuccinic acids. Ester and amide derivatives of the erythro- and threo- 2,3-difluorosuccinic acids were then prepared. The solid and solution state conformation of these compounds was assessed by X-ray crystallography and NMR. Ab initio calculations were also carried out to model the conformation of erythro- and threo- 1,2-difluoro-1,2-diphenylethane as these differed from the 2,3-difluorosuccinates.

Conclusion

In general the overall chain conformations of the 2,3-difluorosuccinates diastereoisomers were found to be influenced by the fluorine gauche effect. The study highlights the prospects of utilising the vicinal difluorine motif (RCHF-CHFR) as a tool for influencing the conformation of performance organic molecules and particularly tuning conformation by selecting specific diastereoisomers (erythro or threo).

Background

Of the 298,876 registered fluorinated structures in the Beilstein Chemical Database (for 2005) only 279 compounds contain a genuine vicinal difluoro motif-CHF-CHF- and only 12 crystal structures of this motif are deposited in the Cambridge Structure Data Base. The relatively rare presence of this motif may partly be attributed to the difficulty of their selective synthesis. It remains a synthetic challenge to prepare vicinal difluorocompounds efficiently and particularly in a stereoselective manner. There are attractive reasons to explore this motif. It is well known that the conformation of 1,2-difluoroethane is influenced by the fluorine gauche effect, where the fluorines prefer to be gauche rather than anti to each other [1]. This preference extends to 2,3-difluorobutane [2], and we have shown that erythro- and threo-9,10 difluorostearic acids have very different physical properties [3], the origin of which appears to lie in the different conformational preferences associated with the vicinal difluoro- motif for each diastereoisomer. Early synthetic methods to vicinal difluoro compounds have involved direct fluorination of alkenes with for eg. elemental fluorine (F2) [4] or XeF2 [5]. Such methods however are either difficult to carry out in a standard laboratory environment or they suffer from very poor stereoselectivity. The direct conversion of vicinal diols to vicinal difluorides has been explored with some success. For example both erythro and threo stereoisomers of dimethyl 2,3-difluorosuccinic acid were obtained either from methyl esters of the L-tartrate 1 or the meso-tartrate 1 by treatment with SF4/HF (Scheme 2) [6,7].

Conversion to the product erythro-2 proved efficient (97%) but that to threo-2 was poor (23%) due to competing elimination. The preparation of the erythro isomer of 2 is attractive on a large scale although SF4 has to be used with care and it is not amenable to reactions on a small scale. Our attempts to replace SF4 with DAST failed in trying to develop an analogous small scale laboratory process. Deoxofluor is finding use in the stereoselective conversion of vicinal diols to vicinal difluorocompounds and seems less prone to elimination than DAST [8]. In addition, Deoxofluor has been reported to be thermally more stable than related aminosulfur trifluoride reagents which allows the conversions to be carried out safely at elevated temperatures [9]. The stereoselective conversion of vicinal ditriflates to vicinal difluorides by treatment with TBAF has also been reported, particularly for the synthesis of 3,4-difluoropyrrolidine ring systems, and these reactions are finding currency in pharmaceutical products [10,11]. Schlosser et al. [12] have developed the most practical and straightforward method to access a variety of erythro- or threo- vicinal difluoro compounds in a diastereoselective manner, using either cis- or trans- epoxides 3 obtained directly from either the Z- or the E- alkenes. (Scheme 1).

Ring opening of the epoxides 3 with HF-amine reagents generate the corresponding threo- and erythro- fluorohydrins 4 in largely a stereoselective manner. The resulting fluorohydrins 4 can then be converted to the erythro- or the threo-vicinal difluoro compounds 5 with reagents such as DAST [9] or Deoxofluor [8,13], although elimination products often compete with fluoride substitution depending on the nature of the substrate.

Scheme 1. Synthesis of vicinal dimethyl difluorosuccinates. The conversion of the tartrates 1 with SF4 and HF [6,7].

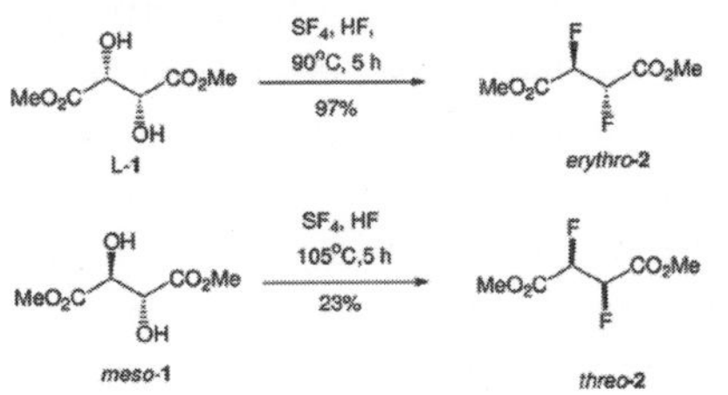

Scheme 2. Schlosser's route to vicinal erythro- or threo- difluoro alkanes 5 [12].

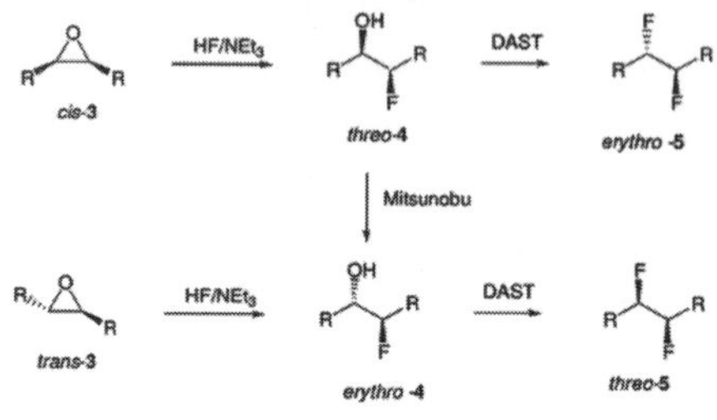

Vicinal difluoro compounds have been prepared by halo(bromo/iodo)fluorination of alkenes followed by halide substitution with silver fluoride [14]. The reaction has been applied to a variety of alkenes some of which (eg 6-9) are illustrated in Scheme 3.

Scheme 3. Halofluorination of electron-rich alkenes with in situ fluoride displacement generates vicinal difluoro products. PPHF is Olah's reagent, pyridinium poly(hydrogen fluoride) [14].

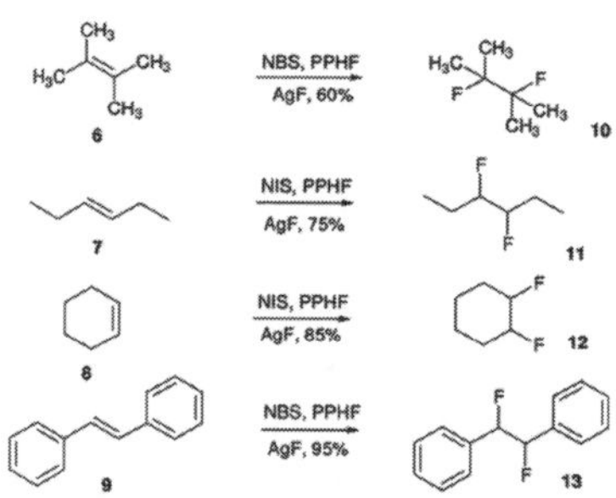

We were interested in accessing diastereomerically pure samples of erythro- and threo- 2,3-difluorosuccinic acids 19. The preparation of stereoisomers of 2,3-difluorosuccinic acids, has involved conversions of tartaric acids (esters) [6,7], as described above in Scheme 1. Other approaches have involved the direct fluorination of fumaric acid [15] and the catalytic hydrogenation of 2,3-difluoromaleic acid [16], but these processes result in significant by-product formation and gave only poor yields of the desired products. Our alternative approach chose to explore the oxidation of the aromatic rings of erythro- and threo- diastereoisomers of 1,2-diphenyl-1,2-difluoroethane 13, exploiting the ability of the phenyl group to act as a latent carboxylic acid [17]. This article describes these studies and we report the solid and solution state conformation of the erythro- and threo- diastereoisomers of 13 and the resultant 2,3-difluorosuccinic acid stereoisomers and some of their derivatives. Some of these results have recently been communicated

[18]. The study suggests that the vicinal fluorine gauche effect can have a significant influence on the conformation of the 1,2-difluorosuccinates.

Results and Discussion

Synthesis of Erythro- and Threo- 1,2-Diphenyl-1,2-Difluoroethanes 13

Stilbene 9 is readily converted to its bromofluoro adduct by treatment with NBS and pyridine:HF following Olah's method [19] (Scheme 4).

Scheme 4. Bromofluorination of stilbene [19].

In our hands product 14 was generated with a diastereoselectivity of 94%. The predominant anti stereochemistry of 14 was established from the coupling constants of the olefin products obtained after a dehydrobromination reaction. The elimination of hydrogen bromide from such β-fluorobromides had been explored previously, and the reaction proceeds in a stereospecific manner to generate either E or Z fluoroalkene products [20]. Accordingly treatment of 14 with potassium tert-butoxide in a refluxing solution of hexane or pentane lead to the exclusive formation of the E-alkene 15 as judged by the 3JHF coupling constant of 21.1 Hz obtained from 19F-NMR. This is indicative of a stereospecific anti-elimination of hydrogen bromide from 14 to generate 15 with a cisoid relationship between H and F, rather than compound 16 which would have a trans relationship and a much larger 3JHF coupling constant (~30 Hz), and reinforces the stereochemical assignment made to 14 as illustrated in Scheme 5[21].

Scheme 5. Treatment of anti-14 with base generated the E-fluorostilbene 15 by an anti elimination mechanism.

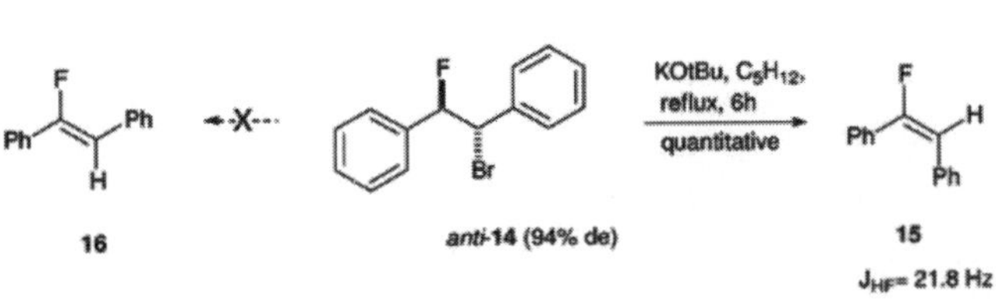

Substitution of the bromine in anti-14 with fluorine was accomplished by treatment with Ag(I)F in acetonitrile in the dark. Under these conditions, the substitution proceeds smoothly to erythro-13 but only in 56% de indicating a significant loss of stereochemical control during the reaction. The predominant stereochemical outcome of the fluorine substitution reaction suggests a double inversion mechanism as the major erythro-13 isomer must arise by replacement of the bromine of anti-14 by fluorine with an overall retention of configuration. Various examples of anchimeric assistance by phenyl groups have been reported [22] and in this case a carbocation is most reasonably generated which finds benzylic as well as anchimeric stabilisation via phenonium ring formation 18 with the β-phenyl group as illustrated in Scheme 6.

Scheme 6. Hypothesis for the predominent retention of configuration during fluoride substitution via phenonium intermediate 18.

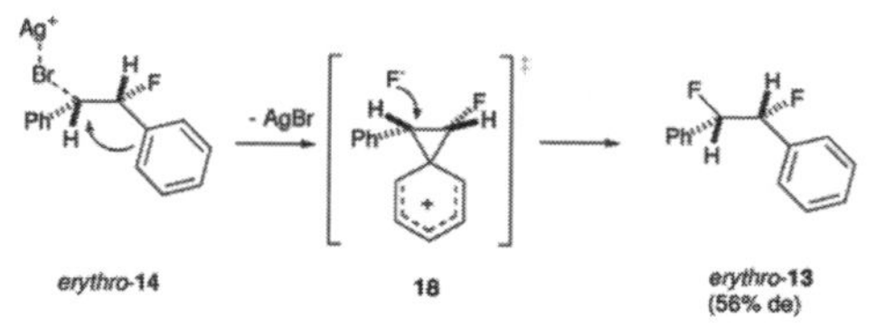

Isolation of the minor threo-13 isomer required careful chromatography. In order to improve the synthesis of threo-13 a reaction with cis-stilbene 17 was investigated. The one pot process with NBS, PPHF and Ag(I)F again proceeded smoothly however it also gave erythro-13 as the major product of the reaction, although with a reduced diastereoisomeric ratio (47% de) more suitable for threo-13 isolation. The bias towards erythro- 13 in this case is clearly a result of internal rotation about the central carbon-carbon bond, to relieve a steric clash between the vicinal phenyl groups, after initial formation of an intermediate bromonium ion 18 as illustrated in Scheme 7.

Scheme 7. Proposed C-C bond rotation during the preparation of 14 from cis-stilbene.

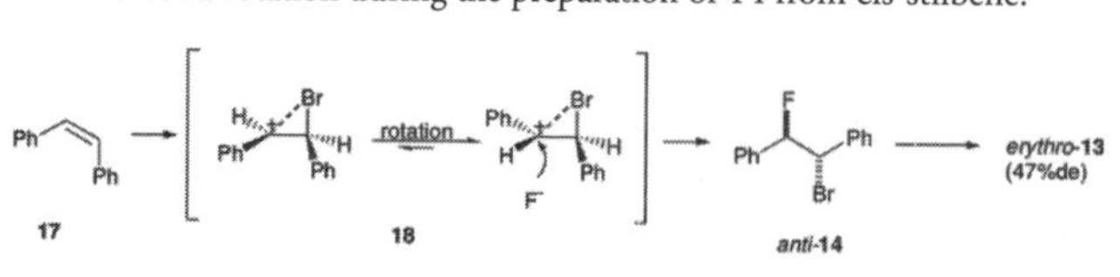

Erythro 13 was readily purified after several crystallisations whereas isolation of the threo isomer of 13 was more challenging. Partial separation of threo-13 was achieved by means of preparative thin layer chromatography. The enriched diasteroisomeric mixture could be crystallised to purity and crystals suitable for X-ray structure analysis were obtained (Figure 1). In the solid state erythro-13 adopts a conformation in which the phenyl substituents are anti to each other, with a Ph-C-C-Ph torsion angle of 180°. As a result the C-F bonds also align anti with respect to each other with a F-C-C-F torsion angle also close to 180°.

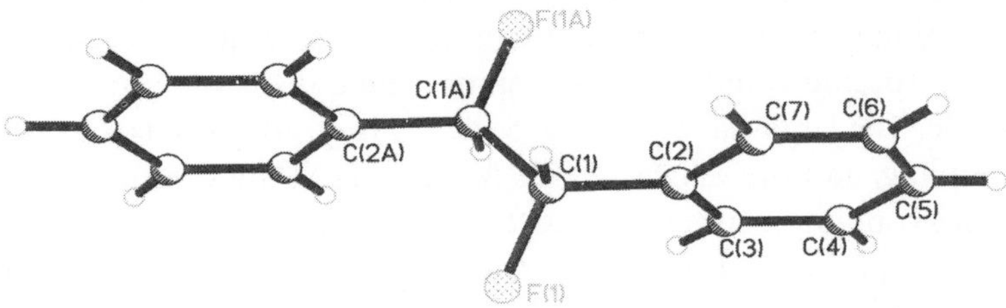

Figure 1. Crystal structure of erythro-13.

A stereochemical mixture enriched in favour of threo-13 was crystallised to purity and a suitable crystal was used for X-ray structure analysis. The resultant structure is shown in Figure 2.

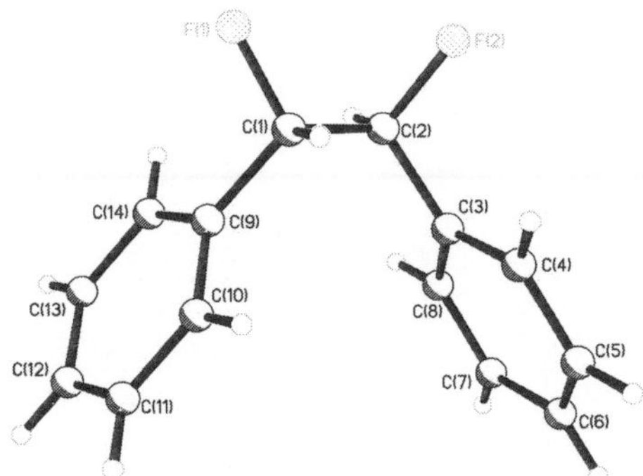

Figure 2. X-ray structure of threo-13.

The most obvious feature of this structure is the perhaps unexpected gauche relationship between the phenyl rings which places the fluorine atoms also gauche to each other. This superficially suggests that the fluorine gauche effect is over-riding any steric repulsion between the phenyl rings. To explore the significance of these solid state conformations further, NMR solution studies and ab initio analysis, exploring the preferred conformations for each of the diastereoisomers was carried out.

NMR Studies on Erythro- and Threo-13

The most obvious feature in the 1H- and 19F- NMR spectra of the diastereoisomers of 13 is the coupling pattern from the AA'XX' spin system (Fig 3). Due to the chemical equivalence but magnetic non-equivalence of the F and H atoms a second-order spectrum is generated in each case.

Measuring of coupling constants from such second-order spectra has been described by Abrahams et. al. [23] although the analysis requires an intuitive fitting of values to specific coupling relationships. These deduced values are tabulated in Figure 3c. The large values of 45.2 & 47.2 Hz clearly correlate to the geminal 2JHF coupling, and the values of 15.2 & 14.1 Hz to the vicinal 3JHF coupling. The smaller coupling constant of 2.6 & 6.0 Hz most appropriately correlate to the 3JHH couplings, and thus, the value of -16.5 & -17.3 Hz is assigned to the vicinal 3JFF coupling. The 19F NMR spectrum can similarly be assigned in each case and reinforced these values. The magnitude of the different vicinal NMR coupling constants can be rationalised in terms of rotational isomerism of the individual diastereoisomers. Only the three staggered conformations for erythro- and threo- 13 are considered (Figure 4).

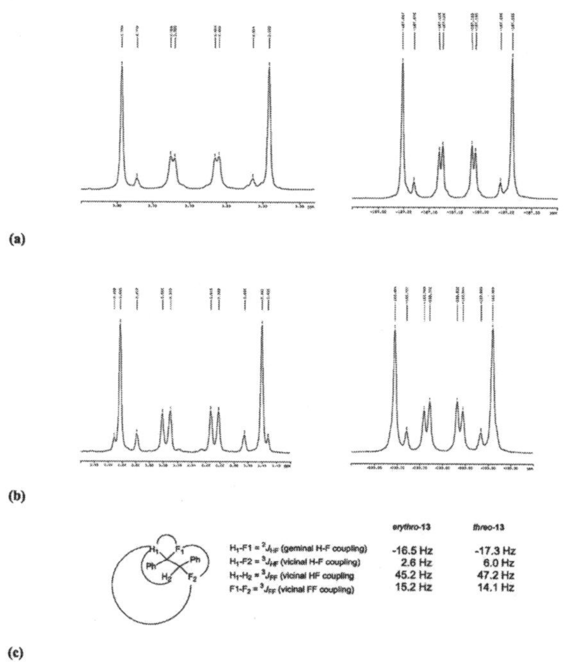

Figure 3. Expanded regions of the second order AA'XX' spin systems in the 1H-NMR (left) and 19F-NMR spectra (right) of erythro-13 (a), threo-13 (b) and the four individual coupling constants for the central 1H and 19F nuclei are given in (c).

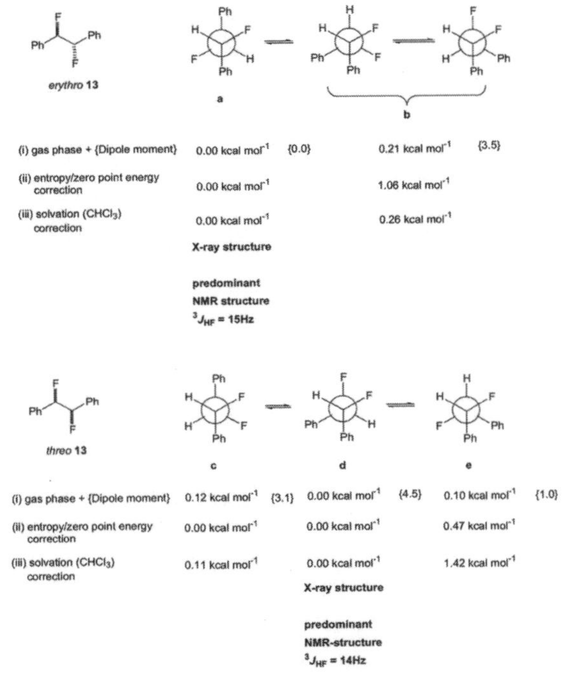

Figure 4. NMR coupling constants and calculated relative energies (kcalmol-1) of the staggered conformers of erythro- and threo- 13 calculated at the B3LYP//cc-pVTZ level. Relative energies (i) in the gas phase (ii) corrected for entropy and zero point energy differences and (iii) using a solvation model are reported. Calculated dipole moments {μ} are also given.

It is not obvious from the NMR data which of a or b is the favoured solution conformation for the erythro isomer. We infer a significant contribution from rotamer b where the C-H bonds are gauche on the basis of the small 3JHH value (2.6 Hz), however the relatively small 3JHF value (15 Hz) suggests two C-H and C-F gauche relationships implying a contribution from rotamer a. Rotamer a most closely resembles the X-ray structure for erythro-13 shown in Figure 1. The situation is much clearer for threo-13. The relatively large 3JHH coupling constant (6.0 Hz) and the small 3JHF coupling constant (14 Hz) suggests a significant population of rotatmer d. This isomer has the vicinal C-H bonds anti to each other and both of the C-F/C-H and C-F/C-F bonds gauche. This is also the preferred conformation for this compound in the solid state (X-ray structure in Figure 2).

Conformational Energy Calculations on Erythro and Threo-13

Due to the ambiguous solution state study particularly for erythro-13, ab initio calculations were carried out at the B3LYP//cc-pVTZ level exploring absolute energies of the three staggered conformers of both erythro- and threo- 13 [24,25]. The geometries were optimized at this level for a gas phase model, and corrected for entropy and zero-point energy differences at this level. A separate solvation correction (chloroform) was applied using a continuum model (PCPM) and the larger cc-pV5Z basis set (using pVTZ geometries). Chloroform was studied in an attempt to relate the calculated values to the NMR solution conformations (vide infra). The relative energy data and dipole moments for each diastereoisomer are presented in Figure 4.

Of the three staggered conformers of the erythro-13 isomer two are enantiomeric and have identical energies thus analysis of erythro-13 is reduced to a comparison of the energies of conformers a and b. Conformer a emerges as the more stable in the gas phase, with this stability originating predominantly from entropy and zero-point energy corrections (1.06 kcal/mol). This is also the conformer that most closely represents the X-ray structure (Fig 1). The solvent correction (which takes into account free energy differences associated with the solvent cavity, but does not allow for free energy differences arising from vibrational terms) does not alter the relative energies of a and b, despite a having a zero dipole moment and b having a relatively large value (3.5D) [26]. Although the more polar b should perhaps gain more from electrostatic solvation, it has a smaller solvent accessible surface area (239A2 vs 246 A2 for a) and these two appear to cancel in their overall effect on the relative energies. Our best estimate of the relative stability of a and b is about 1.0 kcal/mol in favour of the former as noted above. Thus structure a does not conform to a fluorine gauche effect and appears to be dominated by solvation of the trans relationship of the aryl rings and the zero dipole moment, although the smaller 3JHH coupling of 2.6 Hz and the slightly larger 3JHF coupling of 15 Hz in the NMR, measurement does suggest some contribution of conformer b in solution.

The threo-13 isomer has three distinct staggered conformations; c, d and e. Computationally, this requires modelling the subtle balance between the correlation effects due to gauche fluorine atoms and those due to gauche phenyl rings. In the gas phase (entropy and zero energy corrected) conformers c and d are isoenergetic. The dipole moments for these conformers vary significantly, with d > c > e. As with the erythro isomer, the greater solvation for d is partially offset by a smaller solvent-accessible surface (238 vs 247 Å 2 for c). Although d is slightly favoured in this model (by 0.11 kcal/mol), this is significantly smaller than the NMR estimate and may reflect a limitation of the solvation model. Taking all of the data together (theory, X-ray and NMR) conformer d appears to be the most favoured conformer for threo-13 with both the fluorine and the phenyl rings gauche, despite its larger dipole moment.

2,3-Difluorosuccinic Acids 19

The synthesis of the 2,3-difluorosuccinic acid diastereoisomers 19 was explored by the oxidation of the aryl rings of 13 to carboxylic acids. Oxidative degradation of aromatic rings has been achieved by $RuCl_3/NaIO_4$ oxidation [27] however this method proved unsatisfactory in our hands and lead to poor conversions and a complex product mixture. As an alternative strategy ozonolysis in acetic acid, with a hydrogen peroxide work-up was explored [28,29], and this proved successful as illustrated in Scheme 8. For example, reaction of a 4:1 mixture of erythro- and threo- 13 led to the formation of 19 also in a 4:1 ratio of diastereoisomers. Erythro 2,3-difluorosuccinic acid 19 was obtained in a modest yield as a single stereoisomer from a stereochemically pure sample of the erythro 13. A crystal of erythro-19 suitable for X-ray analysis was obtained after sublimation, and the resultant structure is shown in Figure 5.

Scheme 8. Synthesis of erythro-19 via ozonolysis of erythro-13.

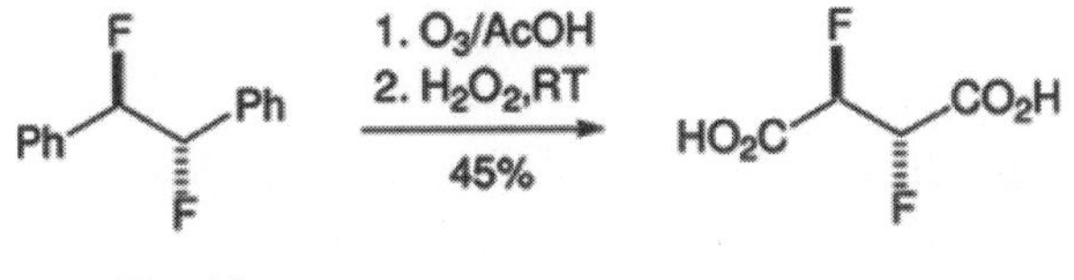

erythro-**13** *erythro*-**19**

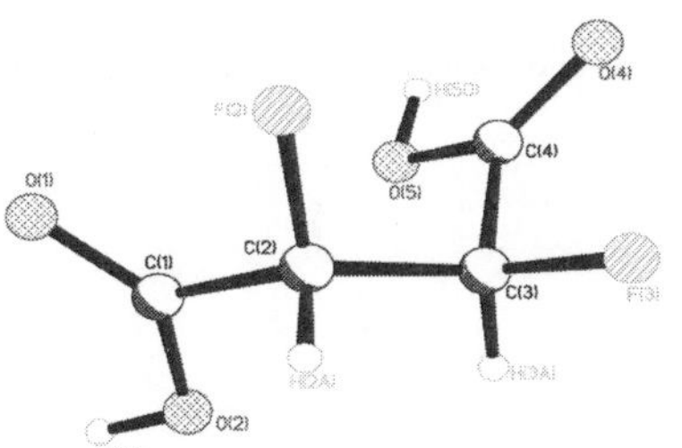

Figure 5. X-ray structure of erythro-19.

In the X-ray crystal structure of erythro-19 both of the carboxylic acid carbonyl oxygens adopt a syn periplanar conformation with respect to the C-F bonds. In the crystal packing, the carboxylate groups of two neighbouring molecules are hydrogen bonded and this clearly determines the three dimensional structure of the unit cell. The threo-19 diastereoisomer was prepared by a similar aryl oxidation reaction on a diastereomerically pure sample of threo-13 and this allowed crystallisation of a sample of racemic threo-19. The X-ray structure in Figure 6 shows the molecule in an extended chain conformation with both of the C-F bonds gauche to each other. One molecule of water is bound for every succinic acid molecule and this water clearly participates in hydrogen bonding to the carboxylic acid groups.

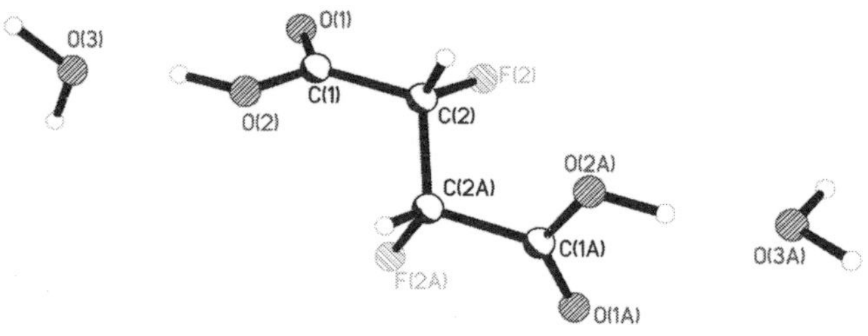

Figure 6. X-ray structure of threo-19.

The major by-product of the ozonolysis reaction of 13 was the vicinal difluorophenylpropionic acid 20 as a mixture of stereoisomers. The compound was purified by esterification with methanol to generate esters 21. These diastereoisomers could be separated by chromatography and then hydrolysis was achieved under acidic conditions, followed by recrystallisation as illustrated in Scheme 9 to generate racemic, but diastereomerically pure samples of erythro- and threo- 20.

Scheme 9. Strategy for the preparation of diastereoisomers of erythro- and threo- 20.

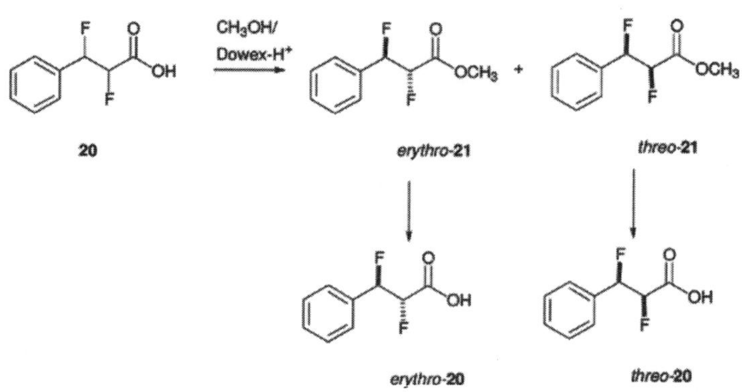

The 3JHH coupling constants of esters 21 remain small (2.8-3.6 Hz) and indicate a gauche relationship between these vicinal hydrogens as summarised in Figure 7. It follows that in each case the fluorines will be predominantly gauche to each other.

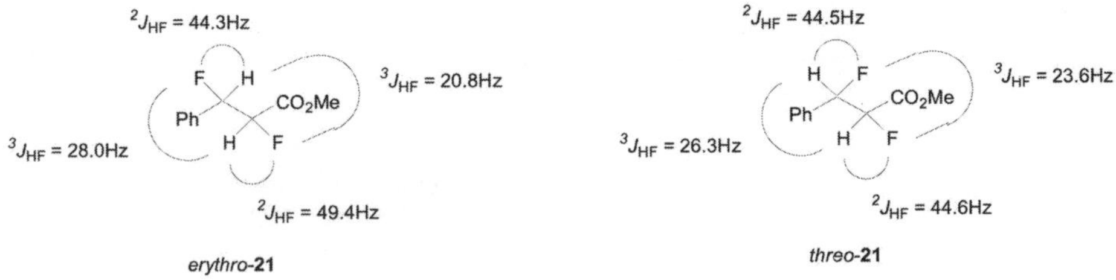

Figure 7. NMR (CDCl3, RT) coupling constants of erythro- and threo- 2,3-difluoro-3-phenylpropionates 21.

The observed values for erythro- 21 report a maximal 3JHF coupling constant for the β-fluorine (28 Hz), but an intermediate one for the α-fluorine (20.8 Hz). This suggests a conformational preference for rotamer c, which has a gauche vicinal fluorine relationship, over a (Figure 8). For the threo- 21 isomer, there are two vicinal 3JHF couplings of similar and large magnitude (26.3 and 23.6 Hz) suggesting that rotamer d, with two trans 3JHF relationships and again with the fluorines gauche, is the most significant contributor to the solution conformation.

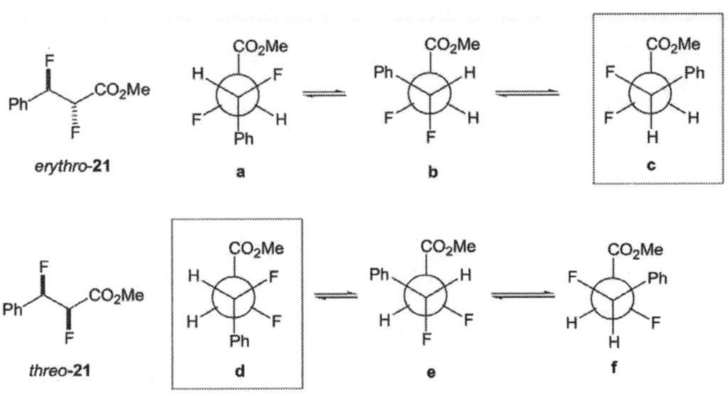

Figure 8. Newman projections showing the staggered conformations of erythro- and threo- 21.

In the solid state structure of threo-21 in Figure 9, the C-F bonds adopt a gauche relationship and the phenyl and ester groups are anti to each other. This is consistent with the preference for rotamer d found in solution. Attempts to crystallise erythro-21 as its free carboxylic acid resulted only in the formation of amorphous material and thus a comparison of solution and solid state structures was not possible for this isomer.

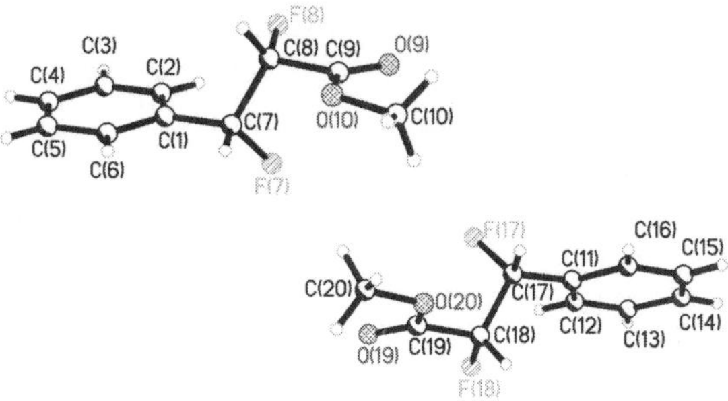

Figure 9. X-ray structure of methyl threo- 21.

Amides of 2,3-Difluorosuccinic Acid

It was an objective of this research to explore the conformational preferences of amides of 2,3-difluorosuccinamides, particularly as we have previously noticed a conformational presence in α-fluoroamides [30], where the C-F bond aligns anti and planar to the amide carbonyl as illustrated in Figure 10. This adds an additional conformational constraint to these amides with a barrier to rotation around the C(CO)-C(F) bond of around 7-8 kcal mol^{-1}. The preference of the C-F bond in α-fluoroamides to align anti periplanar to the carbonyl bond can be rationalized in terms of C-F bond and amide dipole relaxation as well as N-H...F hydrogen bonding [31].

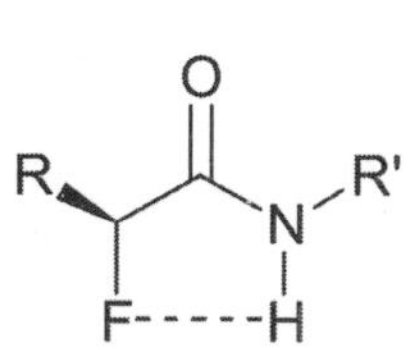

Figure 10. The preferred conformation of α-fluoroamides has the C-F and amide carbonyl anti-planar [30,31].

The solution and solid state structures of 2,3-difluorosuccinate benzylamides 22 have been evaluated. These compounds were prepared by a straightforward EDCI amide coupling between benzylamine and 2,3-difluorosuccinate 19 as shown in Scheme 10.

Scheme 10. The synthesis of stereoisomers of erythro- and threo- 22. These isomers could be separated by chromatography.

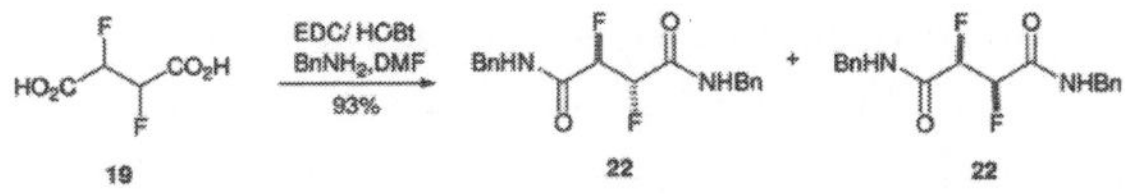

The diasteroisomers of 22 were separated by silica gel chromatography and recrystallisation of each allowed their X-ray structures to be compared. The structure of erythro-22 is illustrated in Figure 11.

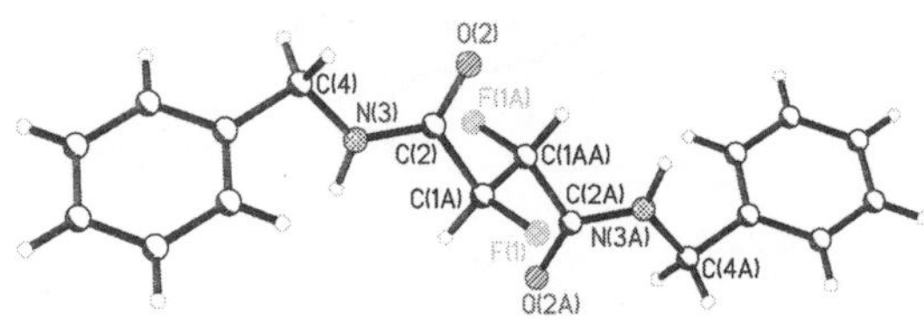

Figure 11. X-ray structure of erythro-22.

Erythro- 22 adopts an extended conformation of the main chain in which the C-F bonds are anti with respect to each other. In that conformation the large benzyl substituents point in opposite directions. The α-fluoroamide groups tend towards a syn-planar C-F...N-H conformation as it is typical for this functional group (Fig 10) with the C-F bonds only 23° off the plane. The carbonyls point in opposite directions and thus intramolecular hydrogen bonding is not possible. There is however strong intermolecular hydrogen bonding between the amide hydrogen and the carbonyl oxygen of adjacent molecules which is dominating the unit cell structure (Figure 12).

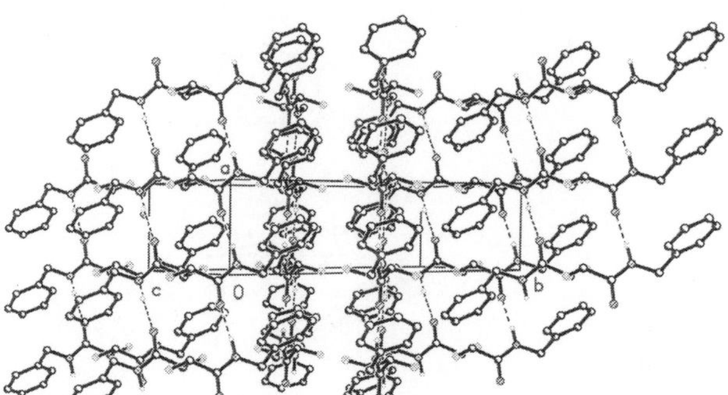

Figure 12. Crystal packing of erythro-22 clearly indicating intermolecular hydrogen bonding.

These intermolecular interactions apparently over-ride the stereoelectronic preference for the gauche arrangement of the C-F bonds, which is observed in solution (vide infra). So we conclude that the solid and solution state structures of erythro- 22 are quite different. By comparison with erythro-22, the crystal structure of threo- 22 in Figure 13 shows both C-F bonds perfectly syn planar with respect to the amide N-H bonds, consistent with the typical planar arrangement of the α-fluoroamide group (Fig 10). The vicinal fluorines are gauche to each other. In this case the solution and solid state structures appear to be much more similar.

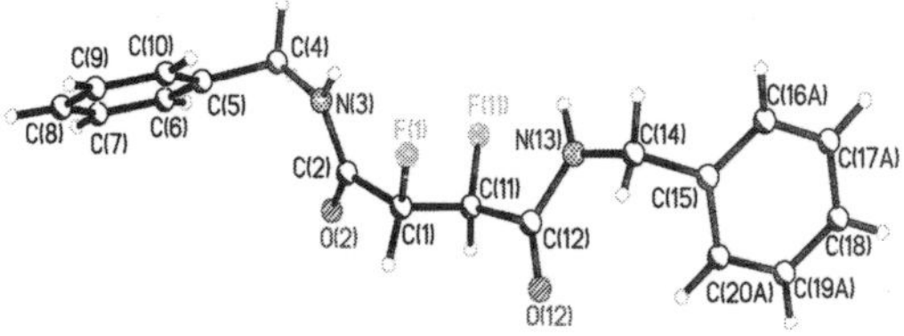

Figure 13. X-ray structure of threo-22.

In a recent Communication [18] we have reported the synthesis of peptides with 2,3-difluorosuccinic acid cores and revealed that such compounds adopt different conformations as a consequence of either the erythro or threo vicinal fluorine stereochemistry. That study highlighted both the solution and solid state conformations of the erythro and threo diastereoisomers of the bis-(S)-phenylalanine amides 23 as shown in Figure 14. The solution and solid state structures reinforced each other and the two diastereoisomers of 23 had preferred conformations where the fluorine atoms were again gauche to each other. This however gave very different shapes to the backbone connectivity in each diastereoisomer as illustrated in Figure 14.

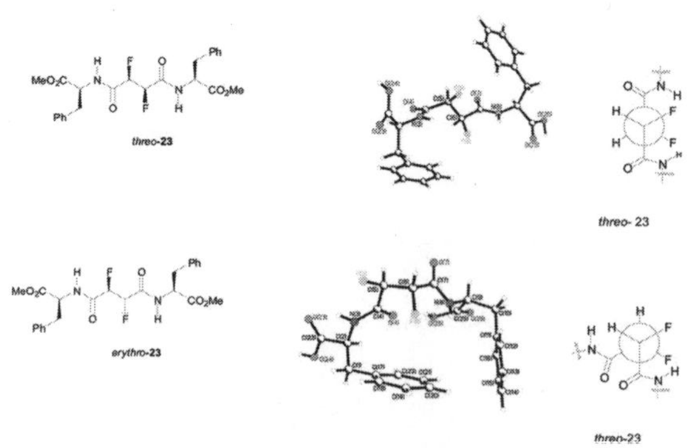

Figure 14. The conformations of erythro- and threo- 23 are very different as a consequence of each conformation preserving a vicinal fluorine gauche relationship [18].

NMR Studies of Vicinal Difluoro Diastereoisomers

The vicinal difluorosuccinates again give rise to second order NMR spectra due to the chemical equivalence but magnetic non equivalence of the fluorine and CHF methine hydrogen atoms similar to Figs 3. A comparison of the $^3J_{HH}$ and $^3J_{HF}$ coupling constants is outlined for the vicinal difluorosuccinate diastereoisomers 19 – 22, 24 and the 1,2-difluoro-1,2-diphenylethanes 13 in Figure 15.

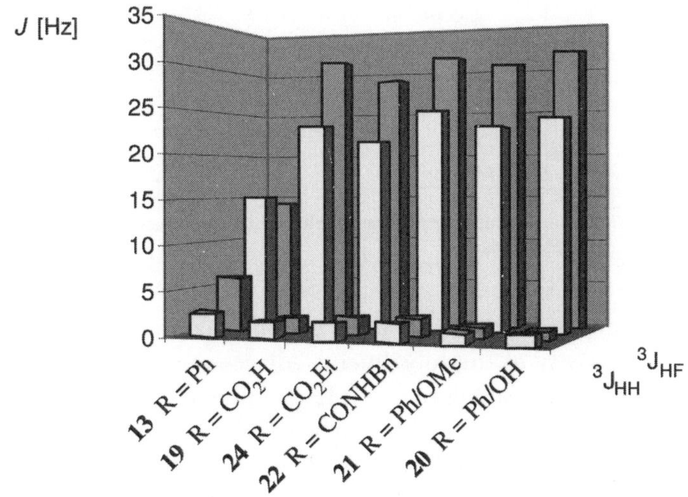

Figure 15. 3JHF and 3JHH coupling constants for the erythro (white) and threo (grey) diastereoisomers of the 2,3-difluorosuccinates 19–22, 24 as well as 1,2-difluoro-1,2-diphenylethanes 13. NMR spectra were recorded in CDCl3 with the exception of 2,3-difluorosuccinic acid, which was measured in CD3CN. The coupling constants were determined from second order spectra.

Interestingly, the 3JHF coupling constants are very similar to each other within each diastereoisomeric series and are essentially independent of the nature of the substituents attached to the carboxylate group. The only significant exception are the diastereoisomers of 1,2-difluoro-1,2-diphenylethanes 13 which have already been discussed in detail. For the 2,3-difluorosuccinate derivatives 19–22,24 the threo stereoisomers have larger 3JHF coupling constants than the erythro stereoisomers.

In order to interpret the data in Figure 15, it is again useful to consider the staggered conformations of each threo and erythro diastereoisomer as shown in Figure 16. Each rotational isomer has two 3JHF and two 3JHH coupling constants the overall magnitude of each being an average of the two.

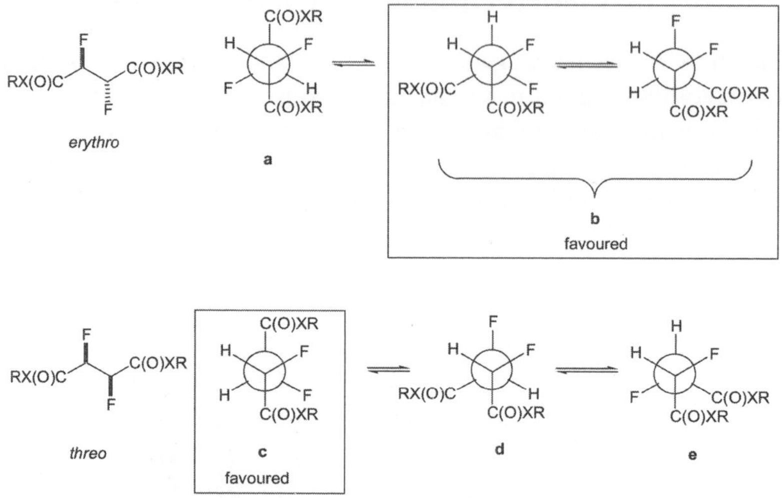

Figure 16. Newman projections of the three staggered conformations of the erythro and threo stereoisomers of the vicinaldifluoro compounds succinates.

The angular dependence of the 3JHF coupling constant is largely influenced by the electronegativity of the substituents adjacent to the coupling nuclei [32]. For related compounds, the full trans 3JHF coupling constant has been estimated to be approximately 32 Hz and the gauche 3JHF coupling constant is approximately 8 Hz [33]. With no conformational bias the average 3JHF coupling constants will be (16 Hz) for each of the diastereoisomers according to these values (Figure 17).

$$\bar{J}_{HF}(threo) = \frac{1}{3}\times\frac{J_t + J_t}{2} + \frac{1}{3}\times\frac{J_g + J_g}{2} + \frac{1}{3}\times\frac{J_g + J_g}{2} \sim 16Hz$$

$$\bar{J}_{HF}(erythro) = \frac{1}{3}\times\frac{J_g + J_g}{2} + \frac{1}{3}\times\frac{J_t + J_g}{2} + \frac{1}{3}\times\frac{J_t + J_g}{2} \sim 16Hz$$

Figure 17. The average coupling constant with no conformational bias. The limiting coupling constants Jg = 8 Hz and Jt = 32 Hz are estimated values.

The experimental 3JHF coupling constants are clearly different for the two diastereoiomeric series. The contributions of the different conformers can then be estimated from the observed 3JHF NMR coupling constants as illustrated in the equations in Figure 18.

$$J_{HF}(erythro) \quad 0\times\frac{J_g \quad J_g}{2} \quad 0.5\times\frac{J_t \quad J_g}{2} \quad 0.5\times\frac{J_t \quad J_g}{2} \sim 20Hz$$

$$J_{HF}(threo) \quad 1.0\times\frac{J_t \quad J_t}{2} \quad 0\times\frac{J_g \quad J_g}{2} \quad 0\times\frac{J_g \quad J_g}{2} \sim 32Hz$$

Figure 18. The observed $^3J_{HF}$ coupling constants are an average over the rotational isomers.

For the erythro diastereoisomer the enantiotopic conformers b (which will be equally populated), dominate the conformer profile. This is consistent with the observed average 3JHH values of (2–3 Hz) where in conformers b only H-H gauche relationships are found with no contributions from anti H-H couplings, which would raise this low value. The high value 3JHF of 32 Hz for the threo diastereoisomers is essentially a maximum value for a trans coupling constant indicating the dominant contribution from conformer c. This is also consistent with the observed average 3JHH values of (2–3 Hz) where in conformer c there are only H-H gauche relationships.

In overview the dominant conformers in each diastereoisomer series have structures which accommodate gauche relationships between the C-F bonds and these results suggest that the fluorine "gauche effect" is influencing the preferred conformations in solution. It is notable that the coupling constants for the 1,2-difluoro-1,2-diphenylethanes isomers 13 are different in the series and do not conform to the ratios described above.

Conclusion

In this article, we have described the synthesis and comparative structures of a series of diastereoisomers of vicinal difluoro compounds, which were generated by converting stilbenes to 1,2-difluoro,1-2-diphenylethanes 13 and then oxidation of the aryl rings to generate 2,3-difluorosuccinic acids and their derivatives. The preparative methods allowed the preparation of individual erythro or threo diastereoisomers. The tendency of the vicinal fluorines to adopt predominant gauche conformations in solution emerges from an analysis of vicinal $^3J_{HH}$ and $^3J_{HF}$ coupling constants of these molecules and reinforces earlier studies on the conformation of vicinal difluoro compounds. This is in line with the well described fluorine gauche effect. The only exception to this was found for the threo stereoisomer of 1,2-difluoro-1,2-diphenylethanes 13, where all of the data (ab initio, NMR and

X-ray) did not converge on a consensus structure. It emerges from this study that the stereoselective incorporation of vicinal fluorines can be used to influence the conformation of organic molecules. This is an attractive tool for the design of performance molecules in areas as diverse as pharmaceutical and medicinal chemistry research to materials science.

Acknowledgements

We thank the University of St Andrews for Studentship support (MS) and AMZS thanks the EPSRC for financial support. We also thank Professor Raymond Abrahams of the University of Liverpool for useful discussions on NMR interpretation.

References

1. Craig NC, Chen A, Suh KH, Klee S, Mellau GC, Winnewisser BP, Winnewisser M: J Am Chem Soc. 1997, 119:4789–4790.
2. Angelini G, Gavuzzo E, Segre AL, Speranza M: J Phys Chem. 1990, 94:8762–8766.
3. Tavasli M, O'Hagan D, Pearson C, Petty MC: Chem Comm. 2002, 1226–1227.
4. Merritt RF: J Am Chem Soc. 1967, 89:609–612.
5. Chia T, Yang NC, Chernick CL: J Am Chem Soc. 1964, 86:5021–5022.
6. Burmakov AI, Motnyak LA, Kunshenko BV, Alexeeva LA, Yagupolskii LM: J Fluorine Chem. 1981, 19:151–161.
7. Hudlicky M: J Fluorine Chem. 1983, 23:241–259.
8. Singh RP, Shreeve JM: J Fluorine Chem. 2002, 116:23–26.
9. Lal GS, Pez GP, Pesari RJ, Prozonic FM: J Chem Soc Chem Comm. 1999, 215–216. G. S. Lal GS, Pez GP, Pesaresi RJ, Prozonic FM, Cheng HS: J. Org. Chem., 1999, 64: 7048–7054.
10. Hulin B, Cabal S, Lopaze MG, Van Volkenburg MA, Andrews KM, Parker JC: Biorg Med Chem Lett. 2005, 15:4770–4773.
11. Caldwell CG, Chen P, He J, Parmee RE, Leiting B, Marsilio F, Patel RA, Wu JK, Eiermann GJ, Petrov A, He H, Lyons KA, Thornberry NA, Weber AE: Biorg Med Chem Lett. 2004, 14:1265–1268.

12. Hamatani T, Matsubara S, Matsuda H, Schlosser M: Tetrahedron. 1988, 44:2875–2881.

13. Lal GS, Labach E, Evans A: J Org Chem. 2000, 65:4830–4832.

14. Olah G: J Org Chem. 1979, 44:3872–3881.

15. Rozen S: J Org Chem. 1986, 51:3607–3611.

16. Hudlicky M: J Fluorine Chem. 1983, 23:241–259.

17. Norsikian S: Chem Eur J. 1999, 5:2055–2068.

18. Schüler M, O'Hagan D, Slawin AMZ: Chem Commun. 2005, 4324–4326.

19. Olah G, Nojima Keres MI: Synthesis. 1973, 780–783.

20. Ernet E, Haufe G: Synthesis. 1997, 953–956.

21. Barton DHR: J Chem Soc Perkin Trans 1. 1974, 739–742.

22. Harwood L: Polar Rearrangements. Oxford Chemistry Primers, Oxford University Press, Tokyo; 1992.

23. Abraham RJ, Loftus P: Tetrahedron. 1977, 33:1227–1234.

24. Frisch MJ, Trucks GW, Schlegel HB, Scuseria GE, Robb MA, Cheeseman JR, Montgomery JA, Vreven T Jr, Kudin KN, Burant JC, Millam JM, Iyengar SS, Tomasi J, Barone V, Mennucci B, Cossi M, Scalmani G, Rega N, Petersson GA, Nakatsuji H, Hada M, Ehara M, Toyota K, Fukuda R, Hasegawa J, Ishida M, Nakajima T, Honda Y, Kitao O, Nakai H, Klene M, Li X, Knox JE, Hratchian HP, Cross JB, Bakken V, Adamo C, Jaramillo J, Gomperts R, Stratmann RE, Yazyev O, Austin AJ, Cammi R, Pomelli C, Ochterski JW, Ayala PY, Morokuma K, Voth GA, Salvador P, Dannenberg JJ, Zakrzewski VG, Dapprich S, Daniels AD, Strain MC, Farkas O, Malick DK, Rabuck AD, Raghavachari K, Foresman JB, Ortiz JV, Cui Q, Baboul AG, Clifford S, Cioslowski J, Stefanov BB, Liu G, Liashenko A, Piskorz P, Komaromi I, Martin RL, Fox DJ, Keith T, Al-Laham MA, Peng CY, Nanayakkara A, Challacombe M, Gill PMW, Johnson B, Chen W, Wong MW, Gonzalez C, Pople JA Gaussian, Inc., Wallingford CT; 2004.

25. Kendall RA, Dunning TH Jr, Harrison RJ: J Chem Phys. 1992, 96:6796–6806.

26. Barone V, Cossi M: J Phys Chem A. 1998, 102:1995–2001.

27. Norsikian S: Chem Eur J. 1999, 5:2055–2068.

28. Aoyama T: J Chem Soc Perkin Trans 1. 1995, 15:1905–1912.

29. Nakajima M, Tomioka K, Koga K: Tetrahedron. 1993, 49:9735–9750.

30. Banks JW, Batsanov AS, Howard JAK, O'Hagan D, Rzepa HS, Martin-Santamaria S: J Chem Soc Perkin Trans 2. 1999, 2409–2411.
31. Reed AE, Curtiss LA, Weinhold F: Chem Rev. 1988, 88:899–926.
32. Abraham RJ, Cavalli L: Mol Phys. 1965, 9:67.
33. Ihrig AM, Smith SL: J Am Chem Soc. 1972, 94:34–41.

The Influence of Traffic and Wood Combustion on the Stable Isotopic Composition of Carbon Monoxide

M. Saurer, A. S. H. Prévôt, J. Dommen, J. Sandradewi, U. Baltensperger and R. T. W. Siegwolf

ABSTRACT

Carbon monoxide in the atmosphere is originating from various combustion and oxidation processes. Recently, the proportion of CO resulting from the combustion of wood for domestic heating may have increased due to political measures promoting this renewable energy source. Here, we used the stable isotope composition of CO ($\delta^{13}C$ and $\delta^{18}O$) for the characterization of different CO sources in Switzerland, along with other indicators for traffic and wood combustion (NO_x-concentration, aerosol light absorption at different wavelengths). We assessed diurnal variations of the isotopic composition *of CO at 3 sites during winter: a village site dominated by domestic heating, a site close to a motorway and a rural site. The isotope ratios of wood combustion emissions were studied at a test facility, indicating significantly lower $\delta^{18}O$ of CO from wood combustion compared to traffic emissions. At the village and the motorway site, we observed very pronounced diurnal $\delta^{18}O$-variations of CO with an amplitude of up to 8‰. Solving the isotope mass balance equation for three distinct sources (wood combustion, traffic, clean background air) resulted in diurnal patterns consistent with other indicators for wood burning and traffic. The average night-time contribution of wood-burning to total CO was 70% at the village site, 49% at the motorway site and 29% at the rural site based on the isotope mass balance. The results, however, depend strongly on the pure source isotope values, which are not very well known. We therefore additionally applied a combined CO/NO_x-isotope model for verification. Here, we separated the CO emissions into different sources based on distinct CO/NO_x emissions ratios for wood combustion and traffic, and inserted this information in the isotope mass balance equation. Accordingly, a highly significant agreement between measured and calculated $\delta^{18}O$ values of CO was found ($r = 0.67$, $p < 0.001$). While different proxies for wood combustion all have their uncertainties, our results indicate that the oxygen isotope ratio of CO (but not the carbon isotope ratio) is an independent sensitive tool for source attribution studies.*

Introduction

Carbon monoxide (CO) is an important trace gas of the atmosphere and widely studied due to its significant role in tropospheric chemistry (Crutzen and Zimmermann, 1991). In remote areas, CO is an important reactant of OH and thus influences the oxidation capacity of the troposphere. Also in polluted urban regions, CO is a significant OH sink (Dommen et al., 2002). However, CO reacts slower in the atmosphere compared to most other pollutants and can therefore be used as an overall indicator for the human activities related to emissions from combustion. When considering the isotope ratios of CO, this approach can be extended for the partitioning of emissions into different sources. It was shown in studies in a rural area (Schauinsland, Germany) (Gros et al., 2002) as well as in a remote area (Spitzbergen) (Röckmann et al., 2002) that the oxygen isotope ratio of CO is useful for the characterization of exceptional pollution events. The relatively low oxygen isotope content of CO from wood combustion ($\delta^{18}O \sim 16$‰) compared to CO from car emissions ($\delta^{18}O \sim 20$–24‰) enabled the detection of biomass burning events in these two studies. The method is not straightforward, however, because several source and sink effects have to be considered for explaining the $\delta^{18}O$ of CO and because the signals of the emissions are not very well known and may be variable (Tsunogai et al., 2003). Regarding CO emission from cars, diesel exhausts can have a much lower oxygen isotope composition of about 11‰ compared to the isotope ratio of atmospheric O_2 of 23.88‰ (Barkan and Luz, 2005), while even lower values down to 6‰ have been observed for cold gasoline engines due to fractionation effects (Kato Published by Copernicus Publications on behalf of the European Geosciences Union. M. Saurer et al.: Traffic and wood combustion influence on CO isotopes et al., 1999b). While such differences can hamper an unambiguous source apportionment, it should be considered that a traffic mix of many cars should have a more well defined isotopic composition, determined, e.g., to be 20.7‰±0.5‰ for Mainz, Germany (Kato et al., 1999b). The main source of atmospheric CO besides the above-mentioned combustion processes is the oxidation of methane and non-methane hydrocarbons, which produces CO with a very low 18O content (Gros et al., 2002). This effect is most important for aged air-masses and background air. On the other hand, the main sink effect, the reaction with OH to produce CO_2, involves a large inverse isotope effect, where the heavier 18O reacts more readily than 16O, resulting in a depletion of the remaining C18O in the atmosphere (Brenninkmeijer et al., 1999). Accordingly, a careful analysis of the isotope balance has to be made or additional tracers for constraining the isotope budget have to be measured, e.g. 14C of CO. Under conditions of heavy pollution, a simple two box-mixing model might be applicable by considering a source and background air (Kato et al., 1999a). Potentially, also $\delta^{13}C$ of CO may provide additional source information. The carbon isotopic composition of CO from automobile exhaust is known to be close to the isotopic compositions of fuels used (Stevens et al., 1972), while isotope fractionations during biomass burning may result in deviations of the isotope ratio from the original material (Kato et al., 1999b).

Emissions from wood combustion are not only important on the global scale, but are also relevant in regional pollution studies. Although the energy consumption from wood heating amounts to only a few percent compared to fossil fuel combustion in Switzerland, the emissions of carbonaceous particles from wood burning are significant and were apparently underestimated in the past. Furthermore, policies fostering the use of renewable sources of energy may result in increased use of wood for heating purposes in the future. A recent study based on 14C showed for an Alpine valley in winter that between 65% and 88% of carbonaceous matter of the ambient aerosol originate from non-fossil sources, most likely wood burning (Szidat et al., 2007). Also many VOC and OVOC concentrations are very strongly influenced by wood burning emissions (Gaeggeler et al., 2008). Besides traffic and industrial emissions, biomass burning contributes significantly to the organic aerosol mass even in a modern city like Zurich (Lanz et al., 2008; Szidat et al., 2006). During summer, forest and barbecue fires also result in a considerable fraction of the organic mass in urban areas (Lanz et al., 2007). A clear separation of the sources contributing to CO in the atmosphere could therefore further help in elucidating the role of wood combustion for the total pollution load.

In this study, we analysed 13C/12C and 18O/16O ratios of carbon monoxide at three sites, one influenced by domestic heating, one next to a motorway, and one at a rural location. The sampling was done at a high frequency (30 min intervals) over several days, to provide insights into short-term variability and diurnal cycles, using an automated sampling and analysis system. Data at such high temporal resolution are practically non-existent in the literature. Additional analyses were done in a controlled setting where emissions from a log wood stove were assessed at a test facility. The study was conducted in the framework of a project where aerosol emissions from wood burning were analysed in great detail, in both laboratory and field studies (AEROWOOD, Aerosols from Wood burning versus other sources, Sandradewi et al., 2008b; Weimer et al., 2008; Alfarra et al., 2007; Szidat et al., 2007). The isotope data were compared with other gas phase and aerosol parameters related to wood burning, mainly CO/NOx ratios and aerosol light absorption measured by an aethalometer. For evaluating different sources to CO, we used an isotope mass balance model, a model based on CO/NOx ratios and a combination of both. The main aim of the study presented here was to assess the use of high-frequency isotope data of CO for distinguishing between wood combustion and fossil fuel sources.

Materials and Methods

Sampling

Samples were collected during two campaigns in Southern Switzerland, investigating a village site (Roveredo, 46°14′18′ N, 9°07′45″ E, 298 m a.s.l.) and a site close to a motorway (Moleno, 46°16′46″ N, 8°59′49″ E, 254m a.s.l.), complemented by one measurement period in a rural area in Northern Switzerland (Paul Scherrer Institute, PSI, 47°32′15″ N, 8°13′37″ E, 335 m a.s.l.) and a wood stove experiment at the test facility of the EMPA (D¨ubendorf). Roveredo is located in an Alpine valley (Mesolcina) with often strong and stable temperature inversion prevailing during winter, influenced by local heating and traffic from a transit route (San Bernardino route). A large percentage of the people of Roveredo burn wood for domestic heating purposes (75% of the heating installations are fuelled by wood). An intensive measurement campaign took place here in January 2005, with sampling for isotopes of carbon monoxide from 11–13 January with a time-resolution of approximately 30 min. The Moleno site was located on a parking lot directly at the Gotthard motorway (average traffic density 19 700 vehicles per day), while the next village is in a distance of 800 m. Due to the direction and dimensions of the valley, the sun is more often reaching the valley bottom compared to the Roveredo site and therefore comparably more convection is taking place during daytime in winter. Sampling took place during February 2005, specifically on 8 to 9 February for the isotopes. The measurement period near PSI was conducted from 3 to 16 February. The PSI is located in a rural area with forests and agricultural fields nearby, influenced by minor emissions from traffic. During all campaigns, meteorological parameters as well as CO, NO_x and PM_{10} were continuously monitored.

The wood stove experiment at the test facility of EMPA was conducted during April 2005 (Weimer et al., 2008). For the isotope sampling, a log wood stove was fired with beech logs during 3 consecutive fires of approximately 30 min each. The wood stove was similar to those often used in households for domestic heating purposes, while beech is the main wood type (around 70 %) burnt in Southern Switzerland. Sampling lines were connected from the exhaust tube of the stove to three dilution stages passing the emissions to a suite of instruments assessing gas phase (CO, NOx, etc.) and particle properties. Sampling for the isotopes took place after an overall dilution by a factor of 1200 by adding compressed air, which had the main purpose to cool and dilute the effluent gas close to "real-world" conditions and allow condensation of semi-volatile particles in reasonable amounts, which is important for the aerosol measurements.

Analysis

Air sampling for the isotope analysis of CO during all campaigns was done with inhouse-built automated samplers (ASA's) that enable the time-programmed sampling of 33 samples of about 300 ml air in glass flasks (Theis et al., 2004). The air was drawn through magnesium perchlorate traps for drying at a rate of about 1 l/min, while the actual filling time was set to 3 min. The ASA's have been successfully applied for isotope analysis of CO_2, CH_4 and CO (Theis et al., 2004). After transporting the ASA's to the lab, they were attached to a modified Precon-and Gasbench-preparation unit (Thermo Finnigan, Bremen, Germany) for automated analysis. Isotopic carbon and oxygen isotope analysis was done in continuous-flow mode according to (Mak and Yang, 1998). In short, air samples passed through a liquid nitrogen trap and ascarite trap for complete removal of CO_2. Then CO was converted to CO_2 in a glass tube filled with Sch¨utze reagent. The evolving CO_2 was first trapped in a loop submersed in liquid nitrogen and then passed through a Poraplotcolumn before entering the mass-spectrometer (Delta-plus XL, Thermo Finnigan, Bremen, Germany). Isotope ratios are given as δ-values and expressed as relative deviations from the international standard, which is VPDB for carbon and VSMOW for oxygen (Saurer et al., 1998). As one oxygen-atom is added to CO in the Sch¨utze reactor, a careful calibration and correction is necessary regarding the $\delta^{18}O$ values (Mak and Yang, 1998). For this purpose, a standard CO gas with known isotopic composition was analysed in the same way as the samples, which then allowed calculating the isotopic composition of the oxygen of the Sch¨utze reagent by a mass balance calculation. The standard CO was calibrated vs. VSMOW by analysis of water with a high-temperature pyrolysis-unit and the analysis of intercomparison water samples provided by the International Atomic Energy Agency (Saurer et al., 1998). When filling a complete ASA (n=33) with a reference gas of 0.5 ppm CO in synthetic air (N_2/O_2), a standard deviation of 0.3‰ resulted for $\delta^{18}O$, and 0.2‰ for $\delta^{13}C$. Overall, CO isotopic values from 535 air samples are presented in this paper.

During the Roveredo campaign, aerosol light absorption was measured with a seven-wavelength aethalometer (Magee Scientific, λ=370, 470, 520, 590, 660, 880 and 950 nm) that was connected to a whole-air inlet without size-cut (Sandradewi et al., 2008b). This instrument collected the aerosols on quartz fibre filters at a flow rate of 2.5±0.1 lpm and recorded the data every 2 min. The absorption coefficient babs indicating the light attenuation through the aerosol layer on the filter was calculated according to Lambert-Beer's law (Sandradewi et al., 2008a). Based on the different wavelength dependence of babs for wood burning (wb) and traffic-related particles, the contribution of these two sources to the absorption coefficients was calculated. The obtained babs for pure wood burning at 470nm and pure traffic at 950 nm were used in a linear model to estimate the fractional contribution of particulate matter (PM) from wood burning to the total PM, expressed as PMwb/(PMwb+PMtraffic) (Sandradewi et al., 2008a).

Calculations

Isotope Mass Balance

Assuming a gas sample to be a mixture of compounds 1, 2, . . . N with concentrations or mass m_1, m_2, . . . m_N , with the total mass

$$m_{total} = \sum_{i=1}^{N} m_i, \tag{1}$$

assuming further that the isotopic composition of the element of interest of each compound is δ_1, δ_2,. . . δ_N, then the isotopic composition of the mixture is due to mass balance given as:

$$\delta_{mixture} \approx \frac{\sum_{i=1}^{N} m_i \delta_i}{\sum_{i=1}^{N} m_i} \tag{2}$$

This equation will be applied in the following to describe the isotopic composition of CO of an air sample as a mixture of different sources. Equation (2) would be mathematically correct when using just the mass of 12C16O isotopologues, but is also a good approximation when using the total mass of CO.

The Keeling-Plot Approach

A special case of the isotope mass balance is a gas sample taken as the mixture of 2 compounds, namely background air (δ_{bgd}, m_{bgd}), and a pollutant (δ_p, m_p). The isotope ratio of the gas mixture is then given by

$$\delta_{mixture} \approx \frac{m_{bgd}\delta_{bgd} + m_p\delta_p}{m_{bgd} + m_p} \tag{3}$$

This equation can be rearranged as:

$$\delta_{mixture} \approx \frac{m_{bgd}\delta_{bgd} + (m_{mixture} - m_{bgd})\delta_p}{m_{mixture}} = \frac{m_{bgd}(\delta_{bgd} - \delta_p)}{m_{mixture}} + \delta_p \tag{4}$$

Taking the background concentration, the background isotope ratio and the isotopic composition of the pollutant as constant, Eq. (4) is linear in 1/mmixture with y-intercept δp:

$$\delta_{mixture} \approx \frac{const.}{m_{mixture}} + \delta_p \tag{5}$$

This equation is very useful because it enables deriving the isotope value of the "pure" pollutant (δp) from a scatter plot of the measured isotope ratios of air samples as a function of the inverse of the measured concentrations, without any further knowledge required about mixing ratios or isotope ratio of the background (Pataki et al., 2003). With some limitations, Eq. (5) is also applicable for more complex mixtures, assuming e.g. the pollutant to be itself a mixture of two pollutants. The y-intercept can then be interpreted as the isotope ratio of the pollutant mixture, but the linearity of Eq. (5) does not strictly hold when the pollutant composition is variable.

Isotope Source Separation

Ideally, one would like to solve the isotope mass balance (Eqs. 1, 2) for the different contributions mi. For our case, we assume three sources – traffic (t), wood burning (wb) and background (bgd) –, contributing with mt, mwb and mbgd to total CO (mtotal), being characterized by the isotope ratios δ_t, δ_{wb}, δ_{bgd} and δ_{total}:

$$\begin{aligned} &\delta_{total} m_{total} - CO \approx \\ &\delta_{bgd} m_{mgd} - CO + \delta_t m_{t-CO} + \delta_{wb} m_{wb-CO} \end{aligned} \tag{6}$$

$$m_{total-CO} \approx m_{bgd-CO} + m_{t-CO} + m_{wb-CO} \tag{7}$$

While the mtotal and δtotal can be measured, the main difficulty here is that we have to know the pure source isotope values δbgd, δt and δwb, which also should be constant over time. Under certain conditions, we further assume a known and constant contribution of the CO background air, mbgd–CO=const. The validity of these four assumptions will be discussed later in detail. Accordingly, the isotope mass balance equations reduce to a system of two equations with two unknowns (mt–CO, mwb–CO) and can therefore be solved, for instance, for the contribution of wood burning to the total CO:

$$m_{wb-CO} \approx \frac{(\delta_{total} - \delta_t) m_{total-CO} + (\delta_t - \delta_{bdg}) m_{bdg-CO}}{(\delta_{wb} - \delta_t)}. \tag{8}$$

The fractional contribution of wood-burning is then given by mwb-CO/mtotal-CO.

CO-NO$_x$ Linear Model

In this paragraph, we consider the gas phase concentrations (not isotope ratios) of NO_x and CO, expressed for instance as partial pressures. Let's take a gas sample with concentrations c(NO_x) and c(CO), which is a mixture of background air (CO^{bgd}, NO_x^{bgd}), traffic emissions (CO^t, NO_x^t) and wood burning (CO^{wb}, NO_{x-}^{wb}). This may be expressed as

$$c(NO_x) = c\left(NO_x^{bgd}\right) + c(NO_x^t) + c(NO_x^{wb}) \tag{9}$$

$$c(CO) = c(CO^{bgd}) + c(CO^t) + c(CO^{wb}) \tag{10}$$

Without any additional assumptions, it is not possible to extract these 6 different contributions to the measured c(CO) and c(NOx). However, when assuming 1) that the background is known and constant, c(CObgd)=cconst(CObgd) and c(NOxbgd)=cconst(NOxbgd), and 2) that the ratios c(COt)/c(NOxt)=rt and c(COwb)/c(NOxwb)=rwb are known and constant, then there is a solution. The first assumption (constant background) may be valid depending on the actual weather conditions for a limited time period. The second assumption is based on distinct CO/NOx emission ratios for traffic emissions and wood burning (where the ratio is much larger for wood burning). By using these definitions and replacing c(NOxt) and c(NOxwb) in Eq. (10) with the help of the newly defined ratios, Eqs. 9 and 10 can be rewritten as:

$$c(NO_x) - c_{const}(NO_x^{bdg}) + \frac{c(CO^t)}{r_t} + \frac{c(CO^{wb})}{r_{wb}} \quad (11)$$

$$c(CO) - c_{const}(CO^{bdg}) + c(CO^t) + c(CO^{wb}), \quad (12)$$

which are two equations for the two unknowns $c(CO^t)$ and $c(CO^{wb})$. Solving for $c(CO^{wb})$, the concentration of CO originating from wood burning emissions, leads to:

$$c(CO^{wb}) = \frac{r_{wb}}{r_t r_{wb}} \left(c_{cont}(CO^{bgd}) - c(CO) + r_t[c(NO_x) - c_{const}(NO_x^{bgd})]\right) \quad (13)$$

and therefore based on (12) for traffic:

$$c(CO^t) = \frac{r_t}{r_t - r_{wb}\left(c(CO) - c_{const}(CO^{bgd}) + r_{wb}[c(NO_x) - c_{const}(NO_x^{bgd})]\right)} \quad (14)$$

Dividing these concentrations by c(CO) results in the fractional contributions of wood-burning and traffic to CO (or the contribution in % by multiplying with 100), which may be directly compared to the results from the isotope source separation mwb-CO/mtotal-CO and mt-CO/mtotal-CO. A graphical representation of the model will be shown later.

Combined CO/NOx Isotope Model

Based on the isotope mass-balance (Eq. 2), we describe the isotopic composition of CO of an air sample (δ_s) as a mixture of CO from background air (δ_{bgd}, m_{bgd-CO}), CO from traffic (δ_t, m_{t-CO}) and CO from wood burning (δ_{wb}, m_{wb-CO}):

$$\delta_s = \frac{\delta_{bgd} m_{bgd-CO} + \delta_t m_{t-CO} + \delta_{wb} m_{wb-CO}}{m_{bgd-CO} + m_{t-CO} + m_{wb-CO}} \tag{15}$$

Using the above CO/NOx model and the assumptions therein, the coefficients mt–CO and mwb–CO can be calculated for every air sample from the measurement of c(CO) and c(NOx) using the expressions derived in Eqs. (13 and 14) (the background concentrations are still taken as constant). When the isotope ratios of the three sources are known, Eq. (15) can thus be solved and this calculated value compared with the measured isotope ratio of CO of a gas sample. (A similar equation could also be written for the isotope mass balance of NOx, but this is not further discussed here due to the lack of isotope data of NOx). The purpose of the combined CO/NOx isotope model is the verification of the isotope mass balance (Eqs. 6–8).

Results

Wood Stove Experiment

Figure 1 shows $\delta^{18}O$ of CO plotted versus the inverse of the CO concentration 1/c(CO) from three fires of the combustion of beech logs in a controlled setting. While one fire lasted approximately 30 minutes, samples correspond to three minute filling time of the sample containers and were taken three times during each fire. CO concentrations were not constant during the fires, due to different combustion temperatures (Weimer et al., 2008). Additionally, the mixing of the combustion gas with the dilution gas (compressed air) has to be considered. The mixing line in Fig. 1 therefore should mainly be interpreted for the y-intercept, which corresponds to the pure wood-burning $\delta^{18}O$ signal according to the Keeling-plot equations (Eqs. 3–5), while the other endmember is determined by isotopic signal of the dilution air, which must have a relatively low $\delta^{18}O$ value. The y-intercept indicating the oxygen isotope signal from the combustion source was similar for all three batches, with an average of 16.3‰ ± 0.6‰. The corresponding $\delta^{13}C$-signal was –24.7‰ ± 1.0‰. The obtained value for $\delta^{18}O$ of CO from wood combustion is clearly lower than the values commonly observed for traffic emissions, which should be close to the value of atmospheric oxygen (O_2) of 23.88‰. Accordingly, a significant difference results between $\delta^{18}O$ of CO from wood heating and from traffic emissions (Fig. 1).

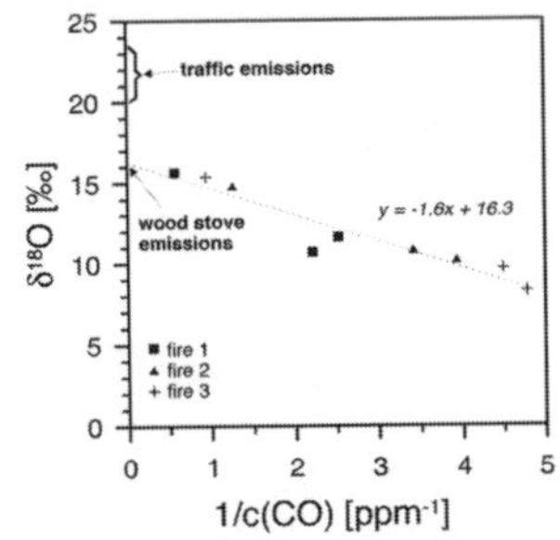

Figure 1. Oxygen isotope ratios of CO as a function of the inverse of the CO-concentration 1/c(CO) for the EMPA wood stove experiment, in comparison with the estimated range of values for traffic emissions.

Roveredo and Moleno: Data Overview

Isotope ratios and concentrations of CO are shown for a 48h period in winter in Roveredo (Fig. 2), the village site affected mainly by domestic heating and to a minor degree by traffic emissions. The $\delta^{18}O$ data show a very pronounced diurnal pattern with relatively low values in the night, a steep increase in the morning hours and high values during the day. The amplitude of this consistent signal over 2 days is almost 8‰. A diurnal pattern is also visible in c(CO), but with much higher noise (individual peaks) relative to the diurnal amplitude. In contrast, the $\delta^{13}C$ data do not show any diurnal variation. A relatively high pollution level is evident from the range of c(CO) from 300 to 1200 ppb and c(NO_x) from 10 to 140 ppb, which may be explained by the stable weather conditions with strong inversions and low wind speed during this period (Sandradewi et al., 2008b). The influence of NOx products on the NO_x signal of the conventional instrument used (Steinbacher et al., 2007) can be assumed to be negligible in winter at these concentrations. Only during night, the downslope winds were carrying relatively clean air and diluted the air at the bottom of the valley, where also the low emissions during the night contributed little as evident from low c(NO_x), in particular from 03:00 to 06:00a.m. (Prévôt et al., 2000b). CO/NO_x concentration ratios were strongly enriched during the night, when traffic was minimal. It is known that due to a lower combustion temperature, much more CO and less NO_x is emitted by wood combustion compared to traffic (Johansson et al., 2004; Kirchstetter et al., 1999). Therefore, the ratio CO/NO_x can be used as an indicator for the relative emission strength of wood burning and traffic. Qualitatively, it seems obvious that the low $\delta^{18}O$ values and high CO/NO_x in Roveredo at night reflect the influence of wood burning. This will be discussed in more detail in the following.

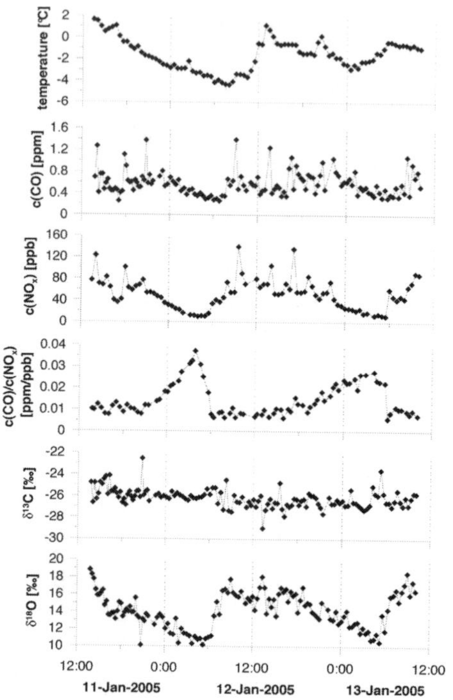

Figure 2. Temporal evolution of temperature, c(CO), c(NOx), c(CO)/c(NOx), carbon and oxygen isotope ratios of CO over 48h in winter in the village Roveredo, Switzerland.

For similar weather conditions as above (cold, low wind), the pollution situation and isotope values were assessed for a 24-h period in Moleno, a site close to a motorway, more than 600 m distant to the next small village (Fig. 3). Here, we observed the highest pollution during the night and early morning (not during the day as in Roveredo), which is probably caused by convection occurring during the day due to insolation and subsequent dilution of the polluted air, as reflected also in higher wind speeds around midday (data not shown). The concentrations of CO and NOx were higher than in Roveredo, with c(NOx) reaching 500 ppb in the morning rush hour. Two maxima per day were observed for c(NOx), clearly reflecting the traffic density and mixing during the day. The oxygen isotope data of CO also showed a diurnal signal (Fig. 3), following the pollution trends with higher values during the day than in the night, but with a smaller amplitude (about 4‰) than in Roveredo. The time course of the CO/NOx concentration ratio indicated very low values compared to Roveredo throughout all the measured period. The δ13C values of CO tended to be higher during the day than during the night, however, by less than 1‰.

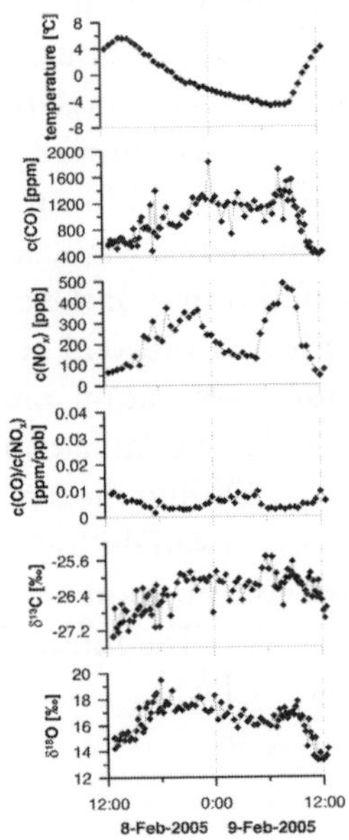

Figure 3. Temporal evolution of temperature, c(CO), c(NO_x), c(CO)/c(NO_x), carbon and oxygen isotope ratios of CO over 24 h in winter near a motorway (Moleno, Switzerland).

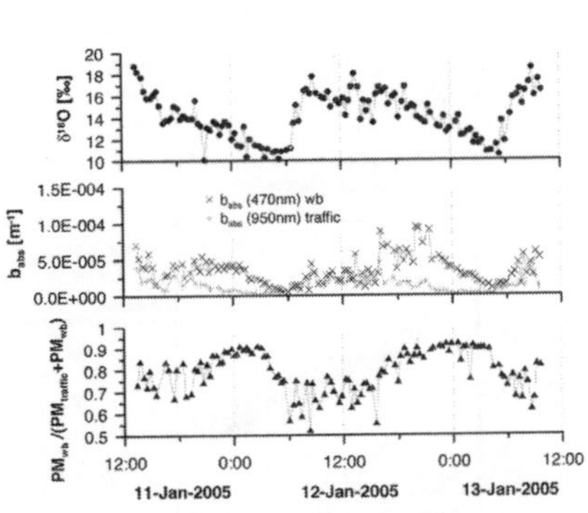

Figure 4. Comparison of $\delta^{18}O$ of CO with aethalometer results (aerosol light absorption coefficient babs and calculated proportion of PM from wood burning) for the measurement campaign in Roveredo.

Comparison with Indicators for Wood Combustion Versus Traffic

In Fig. 4, we show the data of $\delta^{18}O$ of CO for Roveredo together with parameters related to aerosol light absorption (b_{abs} at 470 nm and 950 nm), and thereof calculated contribution of wood burning to particulate matter $PM_{wb}/(PM_{wb}+PM_{traffic})$. $\delta^{18}O$ of CO shows an opposite diurnal pattern compared to the light absorption parameters, with either a maximum or a minimum in the early morning hours and a subsequent steep decrease or increase after the onset of traffic. Aerosol light absorption studies on wood smoke using an aethalometer showed that organic compounds in aerosols from wood combustion result in a strong UV absorption measured at low wavelengths, e.g at 470 nm, whereas absorption of traffic-originating particles dominates at higher wavelengths, e.g. at 950 nm (Jeong et al., 2004). Accordingly, the diurnal course of the absorption coefficient babs (470 nm)wb indicates a dominant influence of wood burning during night on the total particle load. Based on the source separation model and calculation of $PM_{wb}/(PM_{wb} + PM_{traffic})$ (Sandradewi et al., 2008a), a contribution of wood burning up to 90% during night was obtained. During the day, the relative PM_{wb} amount was 50–70%, with significant short-term variability. All parameters shown in Fig. 4 therefore indicate the maximum contribution of wood combustion to the pollution load at a similar time of the day. The linear correlation coefficient between the isotope ratio and $PM_{wb}/(PM_{wb} + PM_{traffic})$ is r = –0.58.

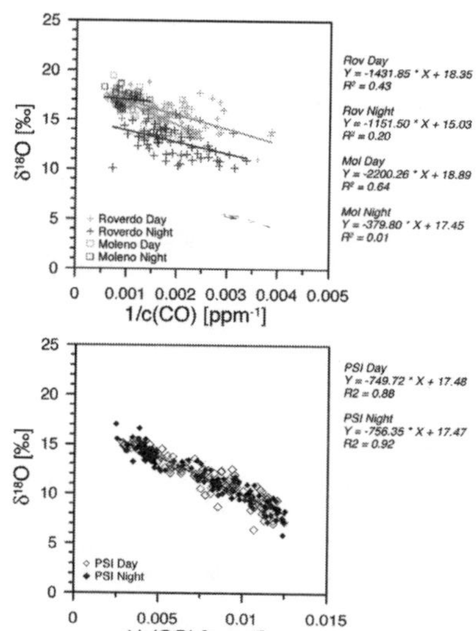

Figure 5. Keeling plots for Roveredo and Moleno (upper plot) and the rural site PSI (lower plot), separated into day and night values.

Keeling Plot Analysis

The relationship between CO-concentrations c(CO) and $\delta^{18}O$ was analysed by a Keeling-plot-approach for both sites (Fig. 5). However, Keeling-plots need some care in the interpretation in a situation where the air is originating from more than two sources (see also 2.3.2). Varying contributions of background, traffic and wood combustion emissions violate the assumption of a two end-member model. Indeed, the explained variance for the correlation of $\delta^{18}O_{CO}$ vs. 1/c(CO) is rather low for Roveredo (r^2=0.43 during the day and r^2=0.20 during the night), with y-intercepts of 18.35‰ and 15.03‰ during the day and night, respectively. In contrast, the same relationship is more linear for Moleno and accordingly yields a much higher correlation coefficient during the day (r^2=0.64), showing that the two-member mixing model may be more viable, with a y-intercept of 18.9‰, while the correlation breaks down in the night (r^2=0.01). The lower values for the isotopic signal of the pollution in Roveredo at night indicates the influence of wood combustion, which has a lower ^{18}O content than traffic, resulting in differences of up to 8‰ between the values for Roveredo and Moleno. The values do, however, not match the expected values for "pure" emissions (23.88‰ for traffic and 16.3‰ for wood combustion), but are lower. This discrepancy could have two reasons: either the presumed source values are not applicable or the assumptions for the application of the Keeling-plot approach are not well enough fulfilled.

Table 1. CO attributed to different sources as calculated with the isotope mass balance (Eqs. 6–8, parameters according to the base case scenario). Average values for the three investigated sites are shown, where "Day" corresponds to 09:00 a.m.–03:00 p.m. and "Night" to 09:00 p.m.–03:00 a.m.

Site	Day/night	CO from wood burning [%]	CO from traffic [%]	CO from background [%]
Roveredo	Day	35	37	28
Roveredo	Night	70	2	28
Moleno	Day	42	29	29
Moleno	Night	49	35	16
PSI	Day	27	36	37
PSI	Night	29	38	33

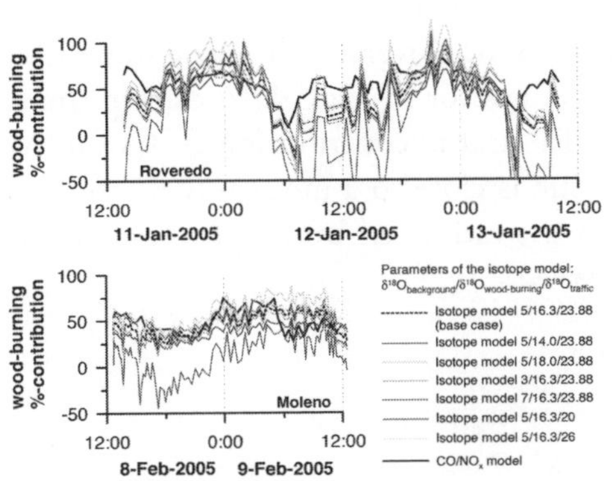

Figure 6. Calculated wood burning contribution to CO (m_{wb}) according to the isotope source separation model (Eq. 8) for different input parameters. The results according to the CO/NO_x split model are also shown.

Source Separation Based on Isotopes

Based on Eqs. (6–8), the relative contribution of wood burning to the total CO in an air sample can be estimated from the measurement of the concentration and oxygen isotope ratio of CO, assuming knowledge of the pure source isotope values. Since these source values have some uncertainty, a range of values is considered here. As base case, we take δ_t=23.88‰, δ_{wb}=16.3‰ (as discussed above) and δ_{bgd}=5‰. The isotope source value for the background is particularly uncertain, whereby it is known to be relatively depleted in the heavier isotope (Brenninkmeijer et al., 1999). As a sensitivity test, we used the low and high extreme values of δ_t =20‰ and 26‰, δ_{wb}=14‰ and 18‰ and δ_{bgd}=3‰ and 7‰. Further, a background value of $c(CO^{bgd})$=181 ppm was determined from a regression analysis under early morning conditions (see next section). Accordingly, we calculated the wood burning contribution m_{wb-CO} to the measured CO concentration for the Roveredo and Moleno campaigns with Eq. (8) (Fig. 6, Table 1). For the base case for Roveredo, we observed a clear diurnal pattern with an average value of mwb=70% during the night (09:00 p.m. to 03:00 a.m.), and much lower values from 0–50% during the day. A higher variability is observed during the day when traffic is more frequent. In Fig. 6, also calculated values below zero (which are not possible in reality) are shown for a better assessment of the influence of the source isotope values. Those combinations of source isotope values that result in negative mwb can obviously be considered as not plausible. It is apparent from Fig. 6 that the scenarios with relatively low $\delta^{18}O$ for the traffic emissions (δ_{bgd}=5‰; δ_{wb}=16.3‰; δ_t =20‰) and low $\delta^{18}O$ for the background CO (δ_{bgd}=3‰; δ_{wb}=16.3‰; δ_t =23.88‰) result in the most negative values of m_{wb} during the day. On the other hand, the scenario with a high value for the wood burning emissions (δ_{bgd}=3‰; δ_{wb}=18‰; δ_t =23.88‰) results in values above 100% during the night, which again is physically not possible. For the Moleno site, a diurnal pattern in m_{wb} is also visible, but less pronounced compared to the Roveredo site. Excluding the extreme scenarios as above, the estimated values for m_{wb} are 42% during the day and 49% during the night (Table 1). The traffic contribution to CO in Moleno is rather low, considering that the site is close to a motorway. Overall for both sites, a quite strong dependence of the estimated wood burning contribution on the source isotope values is observed.

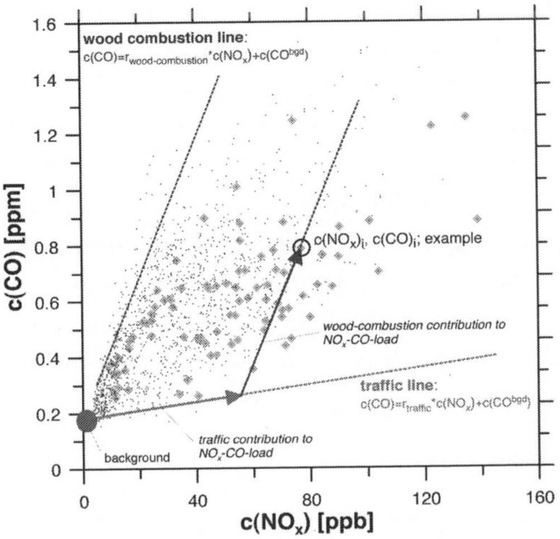

Figure 7. Scatter plot for CO and NO_x data from Roveredo (large symbols: data from 11–13 January 2005, small symbols: data from 1–31 January 2005), shown together with theoretical lines based on pure traffic and pure wood combustion CO/NO_x emission ratios. The arrows indicate how an actual sample $c(NO_x)_i$, $c(CO)_i$ can be expressed as linear combination of traffic and wood combustion emissions.

CO/NO_x-Model

In Fig. 7, a scatter plot of c(CO) versus $c(NO_x)$ is shown for Roveredo for winter conditions. As mentioned above, the CO/NO_x concentration ratio of wood burning emissions is much higher than for traffic emissions. The range of values of $c(CO)/c(NO_x)$ observed in actual air samples depends on the contribution of these two sources and the admixture of background (relatively) clean air. The lowest ratios are observed when traffic emissions dominate, which is the case in the morning rush hour (06:00–07:00 a.m.), while the highest ratios are observed after midnight (02:00–03:00 a.m.), when traffic is minimal and heating dominates. Accordingly, we extracted emissions ratios for pure traffic (r_t) and pure wood burning (r_{wb}) by calculating the CO/NO_x ratios, using the average of either the lowest 50% of values (for r_t) or the highest 50% values (for r_{wb}) of the respective 1 h-time slots. This calculation was done after subtracting a background of 181 ppb for CO and 5 ppb for NO_x determined from the y-intercept of the regression line for the morning rush hour conditions. We obtained r_t=0.0015ppm/ppb and r_{wb}=0.025 ppm/ppb. These pure emission ratios correspond to the slopes indicated in Fig. 7 and represent lower and upper boundaries for the possible CO and NO_x compositions. It is now possible extracting the individual contributions of wood burning and traffic with a linear model indicated by the arrows in Fig. 7, which is equivalent to the derivation in Sect. 2.3.4. Uncertainties in this approach are involved with the assumption of a constant background of $c(CO^{bgd})$=181 ppb and c(NO_x^{bgd})=5 ppb) and the way of derivation of values for r_t and r_{wb}. While the first point seems plausible for the short time period where the model actually will be applied (2 days), the second point clearly leaves some uncertainty because pure emissions never really occur. Several additional tests were performed to check the plausibility of the r_t and r_{wb}-values. CO/NO_x concentration measurements during summer, for instance, were calculated to get an independent estimate of traffic emissions (because heating should be negligible in summer). The value obtained for r_t was higher (0.003 ppm/ppb), however, the original calculation was kept, because the traffic mix (diesel versus gasoline vehicles) may also be different in summer. Calculations in the following are shown for a range of r_b and r_{wb}.

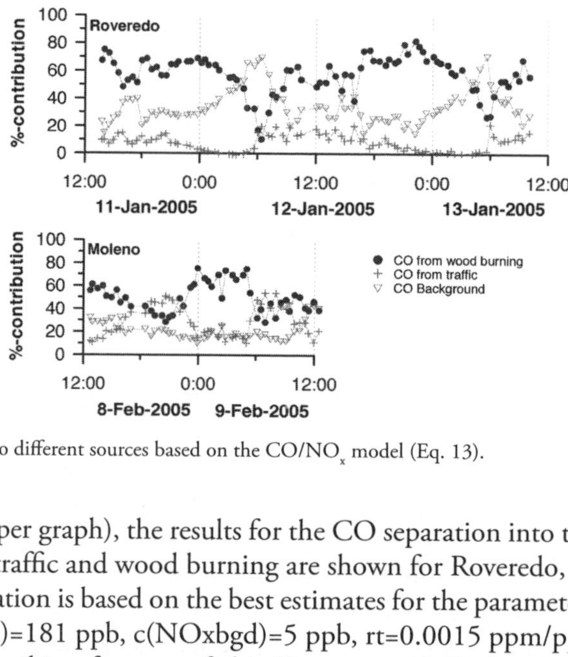

Figure 8. CO split into different sources based on the CO/NO_x model (Eq. 13).

In Fig. 8 (upper graph), the results for the CO separation into the three sources background, traffic and wood burning are shown for Roveredo, for January 11 to 13. The separation is based on the best estimates for the parameters as discussed above: c(CObgd)=181 ppb, c(NOxbgd)=5 ppb, rt=0.0015 ppm/ppb, rwb=0.025 ppm/ppb. The striking features of this Figure are: a) High values for the contribution of CO from wood burning during most of the day (67% from 09:00 pm–03:00 a.m., maximum value 82%), b) a low value of this contribution during a short period in the morning due to admixture of clean background air (during low CO conditions), c) traffic contributions of not more than 20% during the day, and even lower during the night. The wood burning contribution as shown in Fig. 8 is also indicated in Fig. 6 for comparison with the calculations based on the isotope mass balance. There is a good agreement between mwb calculated with the CO/NOx-model and the base case scenario in the isotope-only model, particularly during the night, whereas during the day the mwb-calculation with the CO/NOx model is higher. Further, the CO/NOx model seems to result in less short-term variability.

The influence of varying emission ratios rt and rwb on the %-contribution of CO from wood burning is shown in Table 2 (based on the CO/NOx-separation). Both ratios were changed by a factor of 3. It is obvious from this Table that both increasing rt at constant rwb as well as increasing rwb at constant rt result in a relatively minor reduction in the calculated %-contribution from wood burning of 2–7%. Higher deviations were observed for daytime hours, in particular for the morning rush hour, with differences up to 25% for different emission ratios.

Table 2. CO originating from wood burning for Roveredo (09:00 p.m.–03:00 a.m., 11–13 January 2005) calculated with the CO/NOx model for varying emissions ratios for traffic (r_t) and wood burning (r_{wb}).

r_t [ppm/ppb]	r_{wb} [ppm/ppb]	CO from wood burning [%]
0.001	0.02	68.8
0.002	0.02	67.6
0.003	0.02	66.3
0.001	0.04	67.2
0.002	0.04	64.1
0.003	0.04	61.0
0.001	0.06	66.4
0.002	0.06	63.0
0.003	0.06	59.4

The calculation of the wood burning contribution to CO for Moleno calculated with the CO/NOx-model also yielded high values at night (Fig. 8, lower graph), but these high values were prevalent for a shorter time period compared to Roveredo. The night (09:00 p.m.–03:00 a.m.) average was 57% calculated with this model and thus somewhat higher than with the isotope mass balance (49%, see Table 1). The contribution of traffic was higher in Moleno compared to Roveredo for both day and night according to the CO/NOx-model, while this was only the case during the night according to the isotope mass balance.

Combined CO/NO$_x$-Isotope Model

The calculated CO split into different sources with the CO/NO$_x$-model (Fig. 8) was then used as input for the combined CO/NO$_x$-isotope model (according to Eq. 15). Taking the base case for the pure source δ^{18}O values (23.88‰ for traffic, 5‰ for background CO and 16.3‰ for wood burning), the isotopic composition of the air samples was calculated and compared to the measured values

(Fig. 9). Generally, there is a good agreement between measured and modelled data for Roveredo ($r = 0.67$, $p < 0.001$), although modelled values are somewhat low during the day (06:00 a.m. to 06:00 p.m.). Nevertheless, the diurnal pattern is well reproduced by the model. The combined CO/NO_x isotope model was also applied for Moleno, using the same parameters as in Roveredo. Here, an even better agreement between measured and modelled values was observed ($r=0.70$, Fig. 9, lower panel).

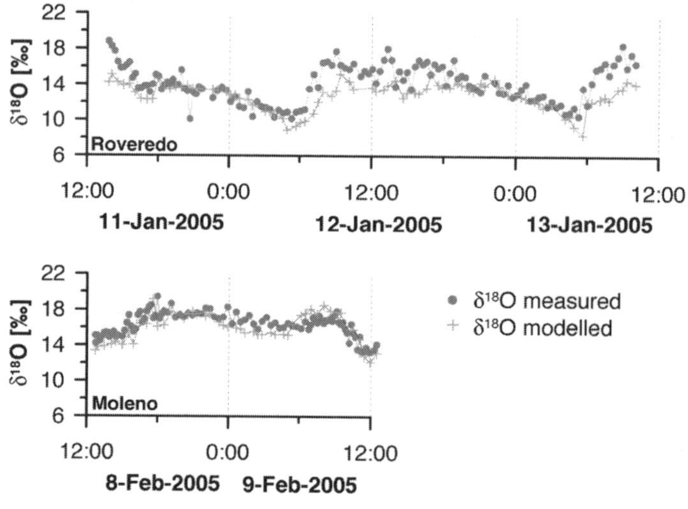

Figure 9. Comparison of calculated and measured $\delta^{18}O$ (model based on the combination of CO/NO_x-separation and isotope mass balance, Eq. 15).

Besides the uncertainty in the CO/NOx separation due to uncertain emission ratios, the agreement between modelled and measured values depends strongly on the assumed isotope values for the different sources. We assessed the performance of the combined CO/NOx-isotope model and its dependence on the used values of the pure isotope sources with a sensitivity analysis for Roveredo (Fig. 10). We varied the isotope value of CO from wood combustion from 8–20‰ and the background isotope composition from 1–9‰, setting the traffic isotope value to either 20‰ or 23.88‰ (the lower value derived from Kato et al. (1999b) for an average traffic mix). As indicators of the performance of the model, we used the correlation coefficient between modelled and measured values (r), the slope of this correlation and the difference (offset) between model and data (expressed as square root of the sum of the squared differences). Obviously, the first two parameters should be as high as possible, while the latter one should be as low as possible for best model performance. It can be seen in Fig. 10a–c (i.e. for δ18Otraffic=23.88‰) that r is highest for relatively low δ18Owb (10–14‰), while the offset is lowest for high δ18Owb (16–20‰), which makes it difficult to find the optimum δ18Owb. This holds for all δ18Obgd, but the mismatch between the optima regarding r and the offset is lowest for relatively high δ18Obgd. The slope criterion, however, indicates that too high values for δ18Obgd are not appropriate, because then the slope would be too low. The slope would be highest for high δ18Owb. Overall, it therefore seems that the values used so far (from literature and the wood-stove experiment) are close to optimum, because it is not possible to improve both r and the offset by changing δ18Owb, while changing δ18Obgd would either increase the offset (when decreasing δ18Obgd) or decrease the slope (when increasing δ18Obgd). The same analysis for δ18Otraffic=20‰ (Fig. 10d–f) shows slightly lower values for r and the slope, independent from the values of the other sources, indicating lower model performance.

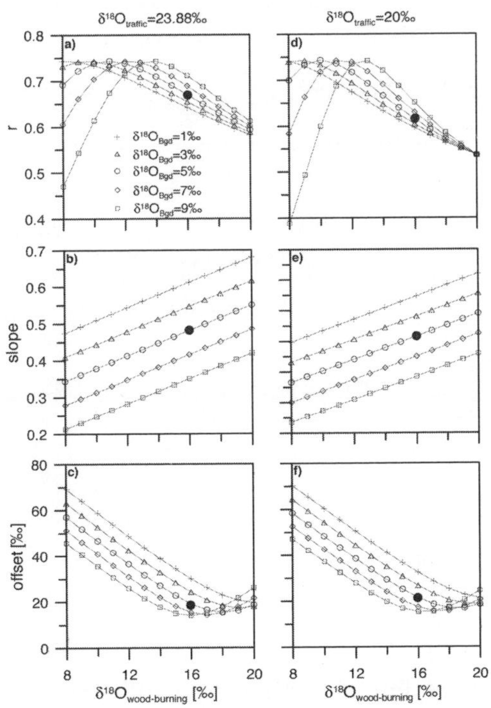

Figure 10. Sensitivity analysis of the combined CO/NO_x-isotope model (Eq. 15). The model performance is given as a function of the assumed values of $\delta^{18}O$ of wood combustion, showing the correlation coefficient for the regression between measured and modelled values (r), the slope and the offset between measurements and model (square root of sum of squared differences). Different lines in each panel refer to different values for the $\delta^{18}O$ of the background. The black dot indicates the values used as the base case scenario. Figure a–c refer to $\delta^{18}O_{traffic}$=23.88‰ and Fig. d–f to $\delta^{18}O_{traffic}$=20‰.

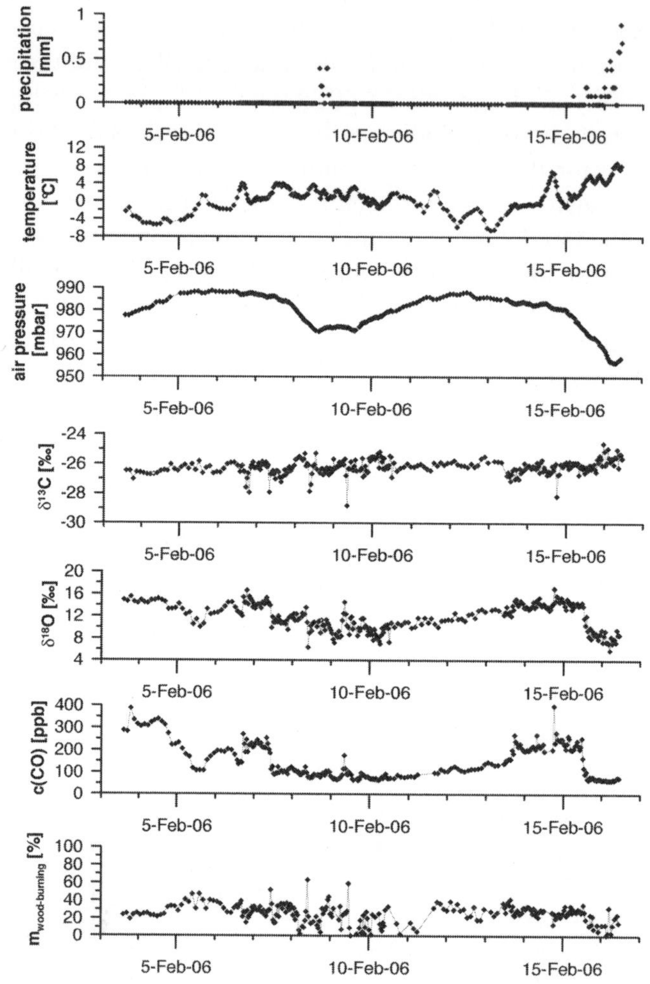

Figure 11. Meteorological parameters, concentration and isotope ratios of CO, and calculated contribution of wood-combustion to CO according to the isotope separation model (Eq. 8) over a period of 18 days in winter at a rural site (PSI).

Rural Site (PSI)

For a better assessment of diurnal cycles and the influence of meteorological conditions on the isotope variations of CO, an 18-day period was analysed at a rural site (PSI) during winter 2006. The investigated period was characterized by dry and cold conditions, with temperatures mostly around or below the freezing point (Fig. 11). Precipitation only occurred two times (on 8 and 16 February), both times associated with a drop in air pressure and increase in temperature, indicating the passage of a warm front. Further, a drop in the CO concentration c(CO) as well as the $\delta^{18}O$ of CO was observed, more strongly in the second event (Fig. 11). While the change in pressure was a gradual one occurring over several hours, the CO drop was almost instantaneous at the passage of the front, which obviously removed the local pollution and replaced it with relatively clean air. Over the whole 18-day period, $\delta^{18}O$-values of CO closely followed the course of c(CO). Between the first and the second rain event, c(CO) as well as $\delta^{18}O$ of CO increased slowly due to accumulation of pollutants in the inversion layer. The Keeling-plot revealed a tight linear relationship between 1/c(CO) and $\delta^{18}O$ for this site with r 2=0.88 for day values and r 2=0.92 for night values, while the y-intercept was relatively low (17.48‰ and 17.47‰ for day and night, respectively), see Fig. 5. The carbon isotope ratio of CO did not change during the rain events and did not show systematic longer-term variations either, but rather seemed to be influenced by small local disturbances. There is a significant negative correlation between c(CO) and $\delta^{13}C$ (r = -0.37), which is much weaker, however, than the respective correlation for $\delta^{18}O$ (r = 0.88).

Table 3. Correlation coefficients (r) between meteorological conditions and the isotope ratios of CO for the rural site PSI. Correlations for the hour of the day are also shown. Bold numbers are significant according to t-test (p<0.001).

	$\delta^{13}C$[‰]	$\delta^{18}O$ [‰]
Hour GMT	−0.16	0.07
Wind direction (DEG)	0.04	−0.13
Wind speed (m/s)	**0.27**	**−0.68**
Wind speed N-S component (m/s)	0.12	**−0.58**
Windspeed E-W component (m/s)	**0.24**	**−0.55**
Air pressure (mbar)	**−0.30**	**0.61**
Temperature (°C)	0.09	**−0.42**
Rainfall (MM)	0.17	**−0.31**

There are no clear diurnal cycles observed for δ13C and δ18O of CO at the rural site PSI and accordingly there is no correlation between the hour of day and the isotope signals (Fig. 11, Table 3). A correlation analysis with meteorological conditions showed that wind speed and air pressure can best explain the δ18O variations (Table 3). The same analysis for δ13C also showed a significant influence of these meteorological parameters. Further, the passage of weather fronts (represented by temperature and rainfall) had a strong impact on δ18O.

We applied the isotope separation model (Eq. 8) to the PSI data, assigning the exact same δ18O values for the pure sources (referring to the base case above, δbgd=5‰; δwb = 16.3‰; δt = 23.88‰) and further using 55 ppb as background

CO concentration. On average, we found a contribution of wood burning mwb to total CO of 23.7% (1σ =12.0%), with no clear trends over the investigated period, but relatively high variability during low CO concentration periods (Fig. 11, Table 1). Interestingly, some anomalously high mwb values appear to be related to low δ13C values the reason for which is unknown. Diurnal cycles in calculated mwb are absent, except maybe for 7 to 10 February. The CO/NOx separation model could not be applied at the PSI site, unfortunately, due to the lack of NOx measurements.

Discussion and Conclusions

For a successful source apportionment with stable isotopes of CO, it is necessary to carefully characterize the isotopic signatures of the pure emissions. Regarding wood burning, the value of 16.3‰ obtained in our wood-stove experiment agrees very well with other estimates in the literature, namely 18‰ ± 1‰ (Stevens and Wagner, 1989) and 16.3‰ (Gros et al., 2002). In a controlled wood burning experiment, a larger range of values was reported, with low values down to 10‰ for the smoldering phase, but the set-up used may be more representative for open fires (Kato et al., 1999b). This study indicated that under poor combustion conditions (e.g. low temperature or low oxygen), when the highest CO emissions occur, the highest isotope fractionation is involved resulting in the lowest $\delta^{18}O$ values. Regarding traffic CO emissions, values could be close to the value of atmospheric O_2 under ideal combustion conditions (hot engine) and in the absence of fractionations. A range of values was, however, reported for car emissions, where fuel type (Kato et al., 1999) and the use of catalysts (Huff and Thiemens, 1998) showed to have an impact on the $\delta^{18}O$ of the emitted CO. In particular, diesel emissions were reported to produce significantly lower values than the atmospheric value of about 23.88‰. Depending on the car mix and the relative number of trucks, variable source values are to be expected. Kato et al. (1999a) observed an average source value of 20.7‰ ± 0.5‰ for Mainz, Germany, sampled over an annual course and estimated the contribution of diesel exhaust to total CO as 14%. Based on the above numbers, the difference between the wood-burning and the traffic signal could be as low as 20.7–18.3=2.4‰ or as high as 23.88‰–10‰=13.88‰, but nevertheless appears to be significant under all scenarios. Regarding uncertainties with sources, it should also be noted that other sources than wood burning and traffic could have an influence, like industrial emissions, although these are probably of minor importance at the sites studies here. For δ13C, the difference between the biomass burning and fossil fuel combustion signal is clearly smaller. Car emissions are expected to be isotopically lighter than the value of –24.7‰ found in our assessment of the wood stove emissions, possibly by about 1–4‰ (Widory et al., 2004). This small difference is a drawback for the successful use of carbon isotopes in source apportionment, which is also confirmed by the small variability of $\delta^{13}C$ observed in our study.

Oxygen isotope values of CO at the village site Roveredo observed during the day in winter were much higher, when traffic was more frequent, compared to the evening and the night, when heating emissions dominated, reflecting the difference between the two sources. Other indicators of wood burning showed similar diurnal courses, namely the CO/NOx concentration ratios and light absorption coefficients of aerosols collected on filters. The latter data in the low UV-range are known to be correlated to wood burning signature as organic compounds in wood smoke aerosols result in a strong UV absorption. Recent investigations also show a strong correlation between the absorption coefficients and 14C data of carbonaceous aerosols collected during the same period and therefore enabled a quantitative calibration of the aethalometer data (Sandradewi et al., 2008a). 14C is related to the ratio of fossil to non fossil sources and therefore depends to a large degree on the ratio of wood burning to traffic (Szidat et al., 2007). The radiocarbon data showed that the carbonaceous particles during the evening in winter in Roveredo originate to a major part from wood combustion, up to 90% during the night. Similar estimates for particulate matter originating from wood burning were obtained by the aethalometer model. The significant correlation between δ18O of CO and PMwb/PMtotal indicates an influence of wood burning for all these parameters.

By high-resolution sampling, a very pronounced diurnal cycle in δ18O of CO (not in δ13C) could be detected. The diurnal variations are, however, not only caused by variations in the sources, but also by turbulence and convection and mixing with clean, background air from higher levels of the atmosphere. The diurnal pattern is strongly enhanced by the alternating occurrence of static and turbulent conditions typical for valleys (Prót et al., 2000a). This is similar for the two polluted sites studied, even when the dilution occurs by downwind-slopes at night in Roveredo, or by convection during the day in Moleno. These results are in strong contrast with a prolonged series of measurements done in flat terrain at a rural location. First, the source strength is smaller for this site. The emissions are farther away and have more time to mix. Weaker inversions over the flatter terrain also allow longer transport distances and thus mixing of source regions compared to the valley. The sites in the valley are thus more affected by close pollution sources and subject to a higher range of c(CO) (300–1500 ppb) compared to the rural site (70 to 400 ppb), and second there is not this distinct alternating pattern of static and turbulent conditions when a typical wintertime stratus cloud deck is covering the flat terrain in Switzerland. Accordingly, no significant diurnal signal in δ18O of CO was observed on flat terrain at the site PSI, but a strong influence of the synoptic weather patterns, reflected in the high correlations to wind speed and atmospheric pressure (Table 3). This again emphasizes the important influence of background air compared to local conditions. This holds in particular, as CO in background air with increasing age of the air masses is affected by fractionation processes that result in a very low 18O-content (Brenninkmeijer et al., 1999).

For a quantitative analysis including the effect of the admixture of background air, a Keeling-plot approach is often useful. This calculation, however, is based on the assumption of a two end-member mixing model (Eqs. 3–5). In fact, there is no strong relationship observed between δ18O and 1/c(CO) for both Roveredo and Moleno, in particular during the night. The correlation coefficients are lower than in other published studies (Brenninkmeijer et al., 1999), reflecting the fact that there are more than two components to be considered at the studied sites. As we have shown, there are large diurnal variations in the contribution of wood-burning and traffic to the CO load and accordingly there does not exist a unique y-intercept, i.e. source isotope signal, which results in the scatter observed in Fig. 5. The y-intercepts therefore have to be interpreted with caution. They are lower during the night, indicating the increased influence of wood-burning. At the traffic-dominated site Moleno, the y-intercept during the day is highest (18.9‰), but still far from the "ideal" traffic value of 23.88‰. This is due to the influence of still some wood-burning emissions at this location, and probably also a value for pure traffic emissions closer to about 20‰ as determined by Kato et al. (1999b). The conditions at the study site are apparently not ideal for applying the Keeling-plot calculation. Another uncertainty that cannot completely be ruled out is the influence of variations in the background isotopic composition. For the rural site PSI, we observed a much stronger relationship between 1/c(CO) and δ18O (r2=0.88–0.92), but no significant difference between day and night Keeling plots. The closer relationship of the two parameters compared to Roveredo and Moleno indicates less variability in the pollution mix reaching this site, which is reasonable considering that it is farther away from direct sources. Still the interpretation of the y-intercepts in the order of 17.5‰ is not straightforward, as this value reflects the mixture of pollutants reaching PSI. The value is in between the wood-burning-dominated night-time value of Roveredo (15.0‰) and the traffic-dominated day-time value of Moleno (18.9‰), probably showing also a mixture of the two at this less polluted site.

Solving the isotope mass balance equation with 3 source terms for the wood burning contribution to CO for Roveredo and Moleno (Fig. 6) resulted in a plausible diurnal pattern of traffic dominated emissions during the day and wood burning dominated emissions during the night. However, a strong sensitivity of the model result to the input parameters was also apparent. For the base case scenario, we inferred a mwb of 60–80% for Roveredo and 40–60% for Moleno during night. Nevertheless, a quantitative analysis is still hampered by the incomplete knowledge of the source isotope values. We found that changes of only a few permil in the assumed source isotope values may easily result in changes of 20 or more percent in the calculated mwb. We investigated a range of values for the three different sources and could to some degree constrain possible source isotope values with the aid of the simple criterion that mwb should not be smaller than 0% and not larger than 100% during any time of the investigated period. In this way, we could show that δbgd should be close to 5‰, while δt could not clearly be discriminated between 20‰ and 23.88‰.

For an independent evaluation of this result and to more clearly constrain the isotope budget, we additionally considered the CO/NOx concentration ratio. The gas phase concentrations c(CO) and c(NOx) are rather simple to measure and are routinely monitored at many stations for air quality control. However, there is a similar problem as above in that the "pure" CO/NOx emission ratios for wood burning and for traffic are not very well confined or may be variable over time. We approached this problem by a detailed analysis of CO/NOx concentration ratios at different times of the day, assigning the highest values to wood-emissions and the lowest values to traffic emissions. The calculated mwb with this model agrees remarkably well with the independent estimation based on the isotope source separation model (Fig. 6). The two models yield similar diurnal patterns for Roveredo and Moleno and similar maximum night-time contributions of about 70–80% in Roverdo and 50–60% in Moleno. Results from 14C measurements indicated even higher contributions of non-fossil derived carbonaceaous aerosols in the evening, which could be due to higher emission ratios of wood burning to traffic for carbonaceous aerosols compared to CO (Szidat et al., 2007). The surprisingly high contribution of non-fossil carbonaceous aerosol in Moleno was explained by the influence of heating sources in the surrounding villages. At a rural site (PSI), we observed much lower wood burning contributions (20–30%) to CO – a result that was expected because wood-heating in this area is not known as very significant. Still, these findings confirm that wood burning emissions during winter may be a much more widespread phenomenon in industrialized regions than previously assumed. In Zurich, a city approximately 30 km distant from the PSI site, biomass burning emissions were estimated to contribute between 25% and 40% to the total carbonaceous aerosol load in winter (Szidat et al., 2006; Lanz et al., 2008), similar to the value of 20–30% inferred here for CO at the rural site.

Finally, with the CO/NOx isotope model (Eq. 15), we applied a novel combination of isotope and gas phase measurements to tackle the problem that the source isotope values are not known very precisely. This combined model performed well on the "best estimates" (base case) for the pure isotope source values, for two different sites with different pollution mixtures and different advection of clean background air. The strengths of the approach are twofold: (1) It confirms indirectly the usefulness of the CO/NOx split model (Eqs. 9–14), because it would not work on two such different sites as Roveredo and Moleno if the emission ratios would not be appropriate. By including the isotope part, it is therefore possible to confirm the split-model (assuming here that the isotope source values are appropriate). The CO/NOx split model can then be applied to much larger datasets and other stations because these data are widely available (2) Testing the isotope source values is possible by a sensitivity analysis (Fig. 10). This test indicated that our best estimate values are actually close to optimum, because changing the source values would either reduce the correlation coefficient between measurements and calculations or would result in an offset between the two. These two applications of the model are not independent from each other as the first assumes knowledge of the source isotope values to test the CO/NOx-split model, while the second assumes the CO/NOx-split model to be correct for testing variations of the isotope source values. The good agreement between model and measurements in Fig. 9 is a strong indication for the usefulness of the combined model, but we would consider it a proof of concept rather than a strict validation. More studies with better definition of the source isotope values would clearly be helpful to verify the approach. This could be done with the help of Keeling-plot analyses in situations where one pollutant dominates. Once the source values are better characterized the isotope mass balance equation may be more widely applied. We think that the CO/NOx model and the isotope mass balance are complementary and confirm each other in a situation where there are uncertainties in both.

Acknowledgements

We thank Jan Eitel for support during field measurements and M. Mohr and D. Schreiber (EMPA, D¨ubendorf, Switzerland) for providing access to the wood stove experiment at the test facility of EMPA. We thank the Office of Nature and Environment of Canton Graub¨unden for providing data from the monitoring stations and the Swiss Federal Office for the Environment (FOEN) for funding the AEROWOOD project.

Edited by: J. Kaiser

References

1. Alfarra, M. R., Prót, A. S. H., Szidat, S., Sandradewi, J., Weimer, S., Lanz, V. A., Schreiber, D., Mohr, M., and Baltensperger, U.: Identification of the mass spectral signature of organic aerosols from wood burning emissions, Environ. Sci. Technol., 41, 5770– 5777, 2007.

2. Barkan, E. and B. Luz: High precision measurements of 17O/16O and 18O/16O ratios in H2O, Rapid Comm. Mass Spectrom., 19(24), 3737–3742, 2005.

3. Brenninkmeijer, C. A. M., Röckmann, T., Bräunlich, M., Jockel, P., and Bergamaschi, P.: Review of progress in isotope studies of atmospheric carbon monoxide, Chemosphere, 1, 33–52, 1999.

4. Crutzen, P. J. and Zimmermann, P. H.: The changing photochemistry of the troposphere, Tellus, 43, 136–151, 1991.

5. Dommen, J., Prévôt, A.S.H., Neininger, B., and Baumle, M.: Characterization of the photooxidant formation in the metropolitan area of Milan from aircraft measurements, J. Geophys. Res., 107, 8197, doi:8110.1029/2000JD000283, 2002.

6. Gaeggeler, K., Prévôt, A. S. H., Dommen, J., Legreid, G., Reimann, S., and Baltensperger, U.: Residential wood burning in an Alpine valley as a source for oxygenated volatile organic compounds, hydrocarbons and organic acids, Atmos. Environ., 42, 8278– 8287, doi:10.1016/j.atmosenv.2008.07.038, 2008.

7. Gros, V., Jöckel, P., Brenninkmeijer, C.A.M., Röckmann, T., Meinhardt, F., and Graul, R.: Characterization of pollution events observed at Schauinsland, Germany, using CO and its stable isotopes, Atmos. Environ., 36, 2831–2840, 2002.

8. Huff, A. K. and Thiemens, M. H.: 17O/16O and 18O/16O isotope measurements of atmospheric carbon monoxide and its sources, Geophys. Res. Lett., 25, 3509–3512, 1998.

9. Jeong, C.-H., Hopke, P. K., Kim, E., and Lee, D.-W.: The comparison between thermal-optical transmittance elemental carbon and Aethalometer black carbon measured at multiple monitoring sites, Atmos. Environ., 38, 5193–5204, 2004.

10. Johansson, L. S., Leckner, B., Gustavsson, L., Cooper, D., Tullin, C., and Potter, A.: Emission characteristics of modern and old-type residential boilers fired with wood logs and wood pellets, Atmos. Environ., 38, 4183–4195, 2004.

11. Kato, S., Akimoto, H., Bröckmann, T., and Brenaunlich, M., Rininkmeijer, C. A. M.: Measurements of stable carbon and oxygen isotopic compositions of

CO in automobile exhausts and ambient air from semi-urban Mainz, Germany, Geochem. J., 33, 73– 77, 1999a.

12. Kato, S., Akimoto, H., Röckmann, T., Bräunlich, M., and Brenninkmeijer, C. A. M.: Stable isotopic compositions of carbon monoxide from biomass burning experiments, Atmos. Environ., 33, 4357–4362, 1999b. Atmos. Chem. Phys., 9, 3147–3161, 2009
13. Kirchstetter, T. W., Harley, R. A., Kreisberg, N. M., Stolzenburg, M. R., and Hering, S. V.: On-road measurement of fine particle and nitrogen oxide emissions from light-and heavy-duty motor vehicles Atmos. Environ., 33, 2955–2968, 1999.
14. Lanz, V. A., Alfarra, M. R., Baltensperger, U., Buchmann, B., Hueglin, C., and Prévôt, A. S. H.: Source apportionment of submicron organic aerosols at an urban site by factor analytical modelling of aerosol mass spectra, Atmos. Chem. Phys., 7, 1503– 1522, 2007, http://www.atmos-chem-phys.net/7/1503/2007/.
15. Lanz, V. A., Alfarra, M. R., Baltensperger, U., Buchmann, B., Hueglin, C., Szidat, S., Wehrli, M. N., Wacker, L., Weimer, S., Caseiro, A., Puxbaum, H., and Prévôt, A. S. H.: Source attribution of submicron organic aerosols during wintertime inversions by advanced factor analysis of aerosol mass spectra, Environ. Sci. Technol., 42, 214–220, 2008.
16. Mak, J. E. and Yang, W. B.: Technique for analysis of air samples for 13C and 18O in carbon monoxide via continuous-flow isotope ratio mass spectrometry, Anal. Chem., 70, 5159–5161, 1998.
17. Pataki, D. E., Ehleringer, J. R., Flanagan, L. B., Yakir, D., Bowling, D. R., Still, C. J., Buchmann, N., Kaplan, J. O., and Berry, J. A.: The application and interpretation of Keeling plots in terrestrial carbon cycle research, Global Biogeochem. Cy., 1022, doi:10.1029/2001GB001850, 2003.
18. Prévôt, A. S. H., Dommen, J., and Bäumle, M.: Influence of road traffic on volatile organic compound concentrations in and above a deep Alpine valley, Atmos. Environ., 34, 4719–4726, 2000a.
19. Prévôt, A. S. H., Dommen, J., Bäumle, M., and Furger, M.: Diurnal Variations of volatile organic compounds and local circulation systems in an Alpine valley, Atmos. Environ., 34, 1413–1423, 2000b.
20. Röckel, P., Gros, V., Bröckmann, T., Jäunlich, M., Possnert, G., and Brenninkmeijer, C. A. M.: Using 14C, 13C, 18O and 17O isotopic variations to provide insights into the high northern latitude surface CO inventory, Atmos. Chem. Phys., 2, 147–159, 2002, http://www.atmos-chem-phys.net/2/147/2002/.
21. Sandradewi, J., Prévôt, A. S. H., Szidat, S., Perron, N., Alfarra, M. R., Lanz, V. A., Weingartner, E., and Baltensperger, U.: Using aerosol light absorption

measurements for the quantitative determination of wood burning and traffic emission contributions to particulate matter, Environ. Sci. Technol., 42, 3316–3323, 2008a.

22. Sandradewi, J., Prévôt, A. S. H., Weingartner, E., Schmidhauser, R., Gysel, M., and Baltensperger, U.: A study of wood burning and traffic aerosols in an Alpine valley using a multi-wavelength Aethalometer, Atmos. Environ., 42, 101–112, doi:110.1016/j.atmosenv.2007.1009.1034, 2008b.
23. Saurer, M., Robertson, I., Siegwolf, R., and Leuenberger, M.: Oxygen isotope analysis of cellulose: An interlaboratory comparison, Anal. Chem., 70, 2074–2080, 1998.
24. Steinbacher, M., Zellweger, C., Schwarzenbach, B., Bugmann, S., Buchmann, B., Ordonez, C., Prevót, A. S. H., and Hueglin, C.: Nitrogen oxide measurements at rural sites in Switzerland: Bias of conventional measurement techniques, J. Geophys. Res., 112, D11307, doi:10.1029/2006JD007971, 2007.
25. Stevens, C. M., Walling, D., Venters, A., Ross, L. E., Engelkem, A, and Krout, L.: Isotopic composition of atmospheric carbon monoxide, Earth. Planet. Sc. Lett., 16, 147–165, 1972.
26. Stevens, C. M. and Wagner, A. F.: The role of isotope fractionation effects in atmospheric chemistry, Z. Naturforsch., 44A, 376–384, 1989.
27. Szidat, S., Jenk, T. M., Synal, H.-A., Kalberer, M., Wacker, L., Hajdas, I., Kasper-Giebl, A., and Baltensperger, U.: Contribution of fossil fuel, biomass-burning, and biogenic emissions to carbonaceous aerosols in Zurich as traced by 14C, J. Geophys. Res., 111, D07206, doi:10.1029/2005JD006590, 2006.
28. Szidat, S., Prót, A. S. H., Sandradewi, J., Alfarra, M. R., ev^Synal, H. A., Wacker, L., and Baltensperger, U.: Dominant impact of residential wood burning on particulate matter in Alpine valleys during winter, Geophys. Res. Lett., 34, L05820, doi:10.1029/2006GL028325, 2007.
29. Theis, D. E., Saurer, M., Blum, H., Frossard, E., and Siegwolf, R. T. W.: A portable automated system for trace gas sampling in the field and stable isotope analysis in the laboratory, Rapid Commun. Mass Spectrom., 18, 2106–2112, 2004.
30. Tsunogai, U., Hachisu, Y., Komatsu, D. D., Nakagawa, F., Gamo, T., and Akiyama, K.: An updated estimation of the stable carbon and oxygen isotopic compositions of automobile CO emissions, Atmos. Environ., 37, 4901-4910, 2003.
31. Weimer, S., Alfarra, M. R., Schreiber, D., Mohr, M., Prévôt, A. S. H., and Baltensperger, U.: Organic aerosol mass spectral signatures from wood

burning emissions: Influence of burning conditions and wood type, J. Geophys. Res., 113, D10304, doi:10.1029/2007JD009309, 2008.

32. Widory, D., Roy, S., Le Moullec, Y., Goupil, G., and Cocherie, A.: The origin of atmospheric particles in Paris: a view through carbon and lead isotopes, Atmos. Environ., 38, 953–961, 2004.

Reduction of Arenediazonium Salts by Tetrakis(Dimethylamino) Ethylene (TDAE): Efficient Formation of Products Derived from Aryl Radicals

Mohan Mahesh, John A. Murphy, Franck LeStrat and Hans Peter Wessel

ABSTRACT

Tetrakis(dimethylamino)ethylene (TDAE 1), has been exploited for the first time as a mild reagent for the reduction of arenediazonium salts to aryl radical intermediates through a single electron transfer (SET) pathway. Cyclization of the aryl radicals produced in this way led, in appropriate substrates, to syntheses of indolines and indoles. Cascade radical cyclizations of aryl radicals derived from arenediazonium salts are also reported. The relative ease of removal of the oxidized by-products of TDAE from the reaction mixture makes the methodology synthetically attractive.

Keywords: cyclization, electron transfer, indole, indoline, radical

Introduction

Arenediazonium salts have long proved useful as sources of aryl radicals in many reactions featuring carbon-carbon (e.g., Meerwein [1], Pschorr [2–3], Gomberg [3] reactions) and carbon-heteroatom bond (e.g., Sandmeyer [4]) formation. The radical-polar crossover reaction [5–15] of arenediazonium salts, developed in our group since 1993, also features aryl radical intermediates and is a more recent addition to these reactions. It involves a novel splicing of radical and polar reactions in one pot, employing tetrathiafulvalene (TTF, 4a, Scheme 1) as electron donor. A number of functionalised heterocycles [5–17] such as dihydrobenzofurans, indolines and indoles have been synthesized using this methodology and the radical-polar methodology has been employed successfully in the total synthesis [10] of aspidospermidine (15), the alkaloid of the Aspidosperma genus (Scheme 2). In line with our interests in generating aryl radicals by reduction of arenediazonium salts with tetrathiafulvalenes [5–16] 4 and dithiadiazafulvalenes [16–17] (DTDAF) 6 (Scheme 1) and later by electrochemical means [18], we were keen to compare the outcomes of these reactions of diazonium salts with those arising from the use of alternative neutral organic electron donors [19–20]. An interesting member of these alternative reagents is the commercially available and economically attractive tetrakis(dimethylamino)ethylene (TDAE, 1). This paper describes the results of our investigations on reactions of TDAE as a neutral organic electron donor with arenediazonium salts.

TDAE (1), has been widely exploited as a strong electron donor [21–59] to electron-poor aliphatic and benzylic halides, notably those derived from organofluorine sources. Burkholder, Dolbier and Médebielle reported [28] that the electrochemical oxidation of TDAE in acetonitrile occurs in two reversible one-electron oxidation steps, to TDAE+• 2 and TDAE++ 3 at −0.78 V and −0.61 V vs saturated calomel electrode (SCE). Recently, TDAE-promoted reduction of electron-deficient o- and p-nitrobenzyl chlorides [44–47], 1,2-bis(bromomethyl)arenes [48], mono and trichloromethyl azaheterocycles [49–50], 2-(dibromomethyl) quinoxaline [51], α-bromoketones [52] have been reported. Vanelle and coworkers recently reported a photoinduced reduction of p-nitrobenzaldehyde in the presence of TDAE [53].

Scheme 1. Aza- and thia-substituted electron donors.

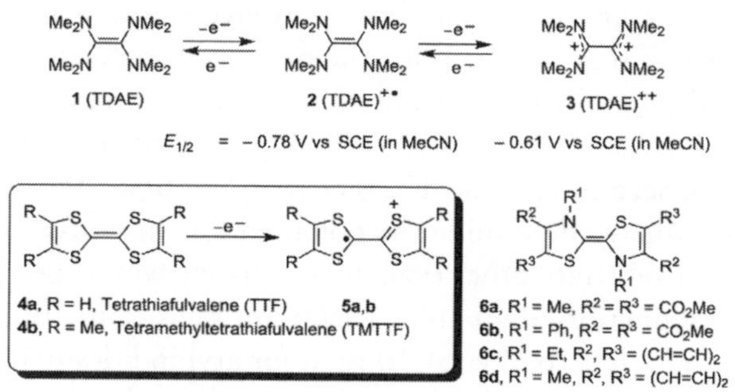

Scheme 2. Radical-polar crossover reaction of arenediazonium salts by TTF.

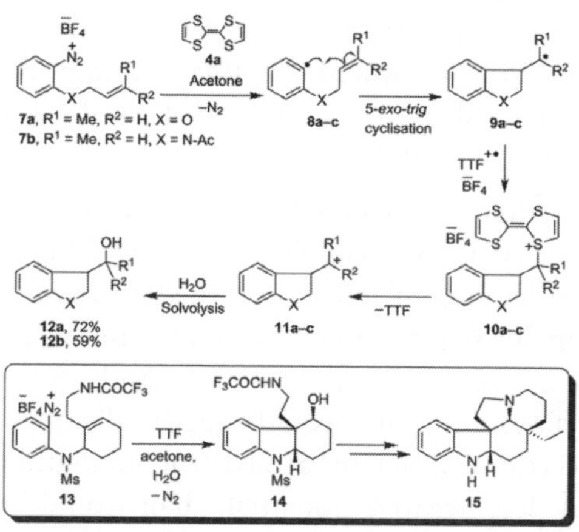

The utility of TDAE as a strong electron donor in specific organometallic reactions, such as the chromium-mediated allylation of aldehydes and ketones [54–55] and the palladium-catalyzed reductive homo-coupling of aryl halides to afford the corresponding biaryls [56–59] illustrate further versatility of the reagent.

The fact that formation of aryl radicals had never been reported using TDAE meant that we were keen to compare its reactions with those of the structurally related TTF (4a). Thus, as shown in Scheme 2, the radical-cation of TTF intercepts intermediates with the formation of C-S bonds in the radical-polar crossover reaction; would the analogous chemistry be seen with TDAE where no sulfur atoms are present?

The experiments were of heightened interest because of the recent report by Andrieux and Pinson [60] on the standard reduction potential of the phenyl radical (formed by electrochemical reduction of the arenediazonium cation) to the phenyl anion (+0.05 V vs SCE). Thus, it had long been noted that cyclic voltammetry of aryl halides, particularly iodides, can give rise to a single two-electron wave in the reductive part of the cycle. The first electron converts the aryl iodide to the corresponding aryl radical, while the second electron transforms the aryl radical to an aryl anion. The two-electron single reductive wave arises because the transfer of the second electron is easier than the first. Andrieux and Pinson reasoned that in order to determine the potential for the conversion of aryl radical to aryl anion, a substrate other than an aryl halide would need to be used. As the one-electron reduction of an arenediazonium salt occurs [60] at much more positive potentials (E p 0.16 V vs SCE) than for aryl iodides [61] (E p –2.2 V vs SCE), this gives a much better chance to observe a second reductive peak in cyclic voltammetry and to determine the potential for conversion of aryl radicals to aryl anions. In the event, their study [60] showed two reductive peaks for benzenediazonium tetrafluoroborate (E p 0.16 V and –0.64 V vs SCE). Through detailed analysis, Andrieux and Pinson showed that this second peak was consistent with reduction of the aryl radical to aryl anion and derived a value for the standard potential of this step as E 0 = +0.05 V.

The reduction potentials determined by Andrieux and Pinson would be consistent with the chemistry that we had observed using TTF, in that TTF had been able to achieve the easier step of reducing arenediazonium salts to aryl radicals, but not the more difficult step (aryl radicals to aryl anions). The redox potentials associated with TTF are +0.32 V and +0.71 V vs SCE [62] so, even transferring one electron to the diazonium salts would superficially appear difficult, but it is well known that electron-transfer by a mediator in solution [63] is frequently more easily achieved (less negative potential) than would be expected from the bare electrochemical data. In the light of these facts, and given that TDAE is a much more powerful donor than TTF (by about 1.1 V for the transfer of the first electron), there is a danger that aryl radicals formed from arenediazonium salts using this reagent would be further converted into aryl anions, if the second electron transfer were sufficiently rapid. Therefore, we proposed to examine cyclization reactions of aryl radicals produced in this way to investigate this point.

Results and Discussion

Preparation of indolines

Our initial studies reacted TDAE with simple arenediazonium salt 16 (Scheme 3Scheme 3). On addition of TDAE to a solution of the arenediazonium tetrafluoroborate salt 16 in acetonitrile [Table 1, entries (i) and (ii)], or, alternately, on addition of the arenediazonium salt to excess TDAE (2.5 equiv) [Table

1, entry (iii)] the reaction mixture underwent effervescence as it turned from deep red to pale orange. In each case, the reaction yielded an inseparable mixture of indolines 20a and 20b in approximately equal yield. In entry (iii) of that table, these yields were estimated from a calibrated NMR determination; following this, the mixture was subjected to epoxidation with mCPBA, leading to isolation of 20b and the epoxide 20c (not shown in Scheme 3) derived from 20a. From this series of experiments, the oxidized product of TDAE, namely octamethyloxamidinium bis(tetrafluoroborate) (21) was isolated in up to 64% yield as an off-white powder. The structure of the salt 21 was characterized by NMR studies and also by mass spectrometry.

Scheme 3. Studies on the reductive radical cyclization of arenediazonium salt 16 by TDAE.

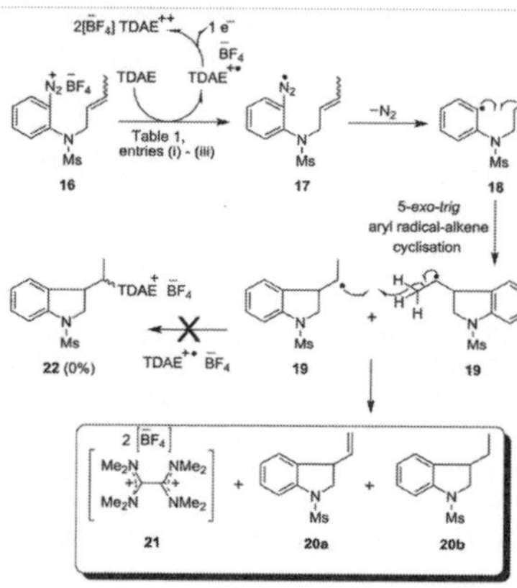

Table 1. Reductive radical cyclization of arenediazonium salt 16 by TDAE.

Entry	Equivalents of TDAE	Solvent	Temperature (°C)	Time	Yield (%) of		
					20a	**20b**	**21**
(i)	1.0	Acetonitrile	25	24 h	11	11	50
(ii)	1.0	Methanol	−40 to 25	24 h	4	5	64
(iii)	2.5	Acetonitrile	25	10 min	12	17	–[a]

[a][TDAE]$^{++}$ 2[BF_4^-] salt **21** was not isolated from the reaction

The formation of the indolines 20a and 20b could then be envisaged through an intermolecular radical disproportionation reaction of two cyclised radical intermediates 19 as explained in Scheme 3Scheme 3. No evidence was seen for the formation of salt 22 although the yields of the products 20 were not high. Non-observation of 22 illustrates that TDAE+•, unlike TTF+•, does not provide an efficient termination of radical processes, and the low yields of isolated compounds could be consistent with radical chemistry where efficient termination was lacking.

With this as guidance to our thinking, the remaining substrates below were designed to provide internal termination routes for the radical chemistry.

One way to achieve clean termination of the radical process would be by providing a radical leaving group adjacent to the cyclised radical 19. Appropriate groups might be sulfide, sulfoxide and sulfonyl groups [64–65]. Accordingly, arenediazonium salts 31a–d were prepared bearing appropriate terminal radical leaving groups (Scheme 4) and treated with 1 equivalent of TDAE under different solvent conditions and temperature. As expected, the aryl radical generated from the reduction of the arenediazonium salts, underwent facile self-terminating 5-exo-trig aryl radical-alkene cyclization to afford the indolines 20a, 32 as the sole products in very high yields (Table 2). Owing to the sensitive nature of the arenediazonium salts 31a–d, they were usually generated in situ from the corresponding amines 30a–d by treatment with nitrosonium tetrafluoroborate. One of the notable features of these cyclizations is the ease of purification of the product from the reaction mixture. The oxidized product of TDAE, namely octamethyloxamidinium bis(tetrafluoroborate) (21), was easily removed either by filtration or by simple work-up with water.

Scheme 4. Preparation of the arenediazonium salts 31a–d. Reagents and conditions: (a) 23, NaH, THF, 0 °C, 0.5 h, then TBDPS-Cl, 0 °C to 25 °C, 4 h, 85%; (b) 25, DIAD, PPh3, THF, 0 °C to 25 °C, 12 h, 80% (26a), 64% (26d); (c) TBAF, THF, 25 °C, 1.5 h, 85% (27a), 0.5 h, 97% (27d); (d) PBr3, DCM, 0 °C to 25 °C, 1 h, 99% (28a), 93% (28d); (e) PhSH, NaH, THF, 0 °C to 25 °C, 1 h, then 28, 25 °C, 5 h, 98% (29a), 94% (29d); (f) Cu(acac)2, NaBH4, EtOH, 25 °C, 15 h, 70% (30a), SnCl2·2H2O, MeOH, 65 °C, 4 h, 91% (30d); (g) NOBF4, CH2Cl2, -20 °C to -10 °C, 1.5 h, 98% (31a), 100% (31d); (h) NaIO4, MeOH/H2O, r.t., 1 h 15 min, 92%; (i) SnCl2, MeOH, 65 °C, 4 h, 71%; (j) NOBF4, CH2Cl2, -20 °C to -10 °C, 1.5 h, 100%; (k) NaIO4, MeOH/H2O (1:1), r.t., 72 h, 82%; (l) SnCl2·2H2O, MeOH, 3.5 h, 83%; (m) NOBF4, CH2Cl2, -20 °C to -10 °C, 1.5 h, 100%.

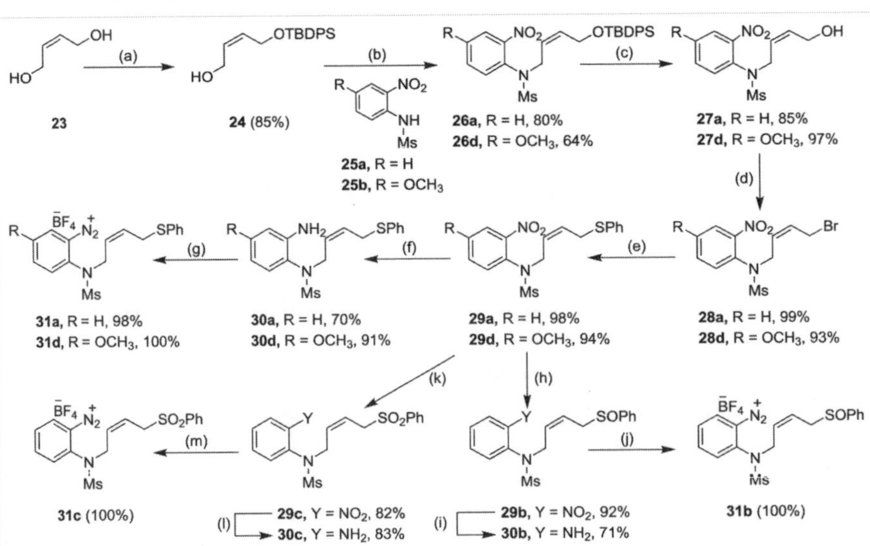

Table 2. Reductive radical cyclization of arenediazonium salts 31a–d by TDAE.

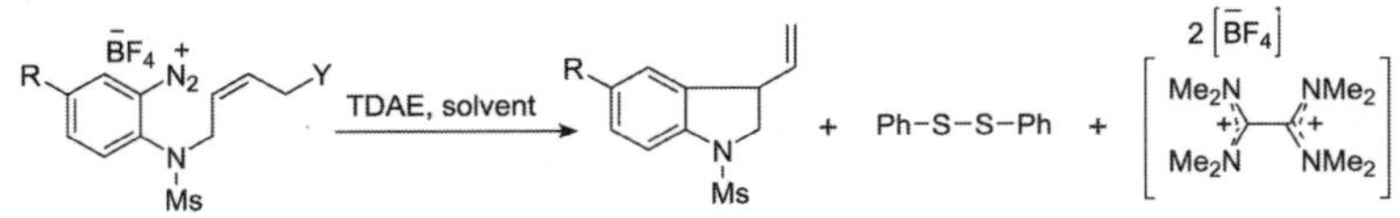

31a, R = H, Y = SPh
31b, R = H, Y = SOPh
31c, R = H, Y = SO_2Ph
31d, R = OCH_3, Y = SPh

20a, R = H
32, R = OCH_3

33

21

Entry	Diazonium Salt	Equivalents of TDAE	Solvent	Temperature (°C)	Time	Isolated Yield (%) of		
						20a/32	**33**	**21**
(i)	**31a**	1.0	Acetonitrile	25	5 min	88	82	12
(ii)[a]	**31a**	0	Acetonitrile	25	5 min	0	0	0
(iii)[b,d]	**31a**	3.0	Acetonitrile	25	5 min	85	39	–[c]
(iv)	**31a**	1.0	Acetone	25	5 min	84	78	17
(v)	**31a**	1.0	Methanol	0 to 25	2 h	81	34	71
(vi)[d]	**31b**	1.0	Acetone	25	5 min	81	0	24
(vii)[d]	**31c**	1.0	Acetone	25	5 min	63	0	41
(viii)[d]	**31d**	1.0	Acetone	25	5 min	88	55	35

[a]Control reaction performed in the absence of TDAE reagent. [b]This experiment was conducted by adding a solution of the diazonium salt in dry MeCN to a solution of TDAE in dry MeCN, while the other experiments in this table all featured addition of the TDAE to the diazonium salt. [c]$[TDAE]^{++}\ 2[BF_4^-]$ salt **21** was not isolated from the reaction. [d]In these experiments, the arenediazonium salts **31a–d** were made *in situ* from their corresponding amines **30a–d**, while all the other experiments in this table featured direct use of arenediazonium salt.

Cascade Cyclizations

To determine the scope of the TDAE-mediated reduction of arenediazonium salts, we sought to extend this methodology to more complex substrates, namely 42 and 44. Pleasingly, the arenediazonium salts 42 and 44, prepared in situ from the amines 41 and 43 respectively upon treatment with 1 equivalent of TDAE, underwent facile cascade radical cyclizations to afford the bicyclized product 47 in 85% and 77% yield respectively (Scheme 5). The ability of TDAE to mediate such efficient cascade cyclizations via two C-C bond formations reactions in one pot from the aryl radical 45 was significant considering the fact that our previous studies on similar substrates by TTF [5,8,16–17] and TMTTF [16] had shown competitive trapping of the intermediate alkyl radical 46 by TTF+• or TMTTF+•.

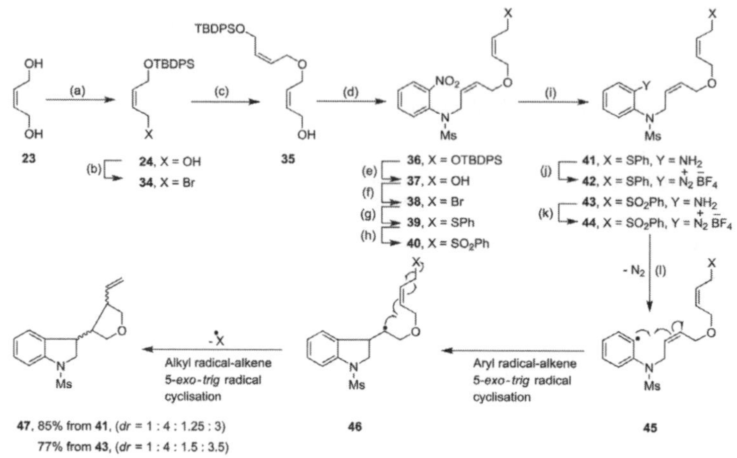

Scheme 5. Cascade radical cyclizations of arenediazonium salts 42 and 44 by TDAE. Reagents and conditions: (a) 23, NaH, THF, 0 °C, 0.5 h, then TBDPS-Cl, 0 °C to 25 °C, 4 h, 85%; (b) NBS, Me2S, CH2Cl2, -30 °C, 0.5 h, then 24, -30 °C to r.t., 3 h, 95%; (c) 23, NaH, THF, 0 °C to r.t., 1 h, then 34, 72 h, 61%; (d) N-(2-nitrophenyl)methanesulfonamide (25a), DIAD, PPh3, THF, 0 °C to r.t., 1 h, 98%; (e) TBAF, THF, r.t., 20 min, 90%; (f) NBS, PPh3, CH2Cl2, -25 °C, 20 min, then 37, -25 °C to r.t., 40 min, 96%; (g) PhSH, NaH, THF, 0 °C to r.t., 1 h, then 38, r.t., 12 h, 86%; (h) NaIO4, H2O, MeOH, r.t., 76 h, 73%; (i) SnCl2, CH3OH, H2O, reflux, 3.5 h to 4 h, 95% (41), 78% (43); (j) NOBF4, CH2Cl2, -15 °C to 0 °C, 1.5 h; (k) NOBF4, CH2Cl2, -15 °C to 0 °C, 1.5 h; (l) TDAE (1.0 equiv), acetone, 10 min, r.t., 85% 47, 46% 33, 52% 21 in two steps, from 41; 77% 47, 46% 21 in two steps, from 43.

Preparation of Indoles

Following the successful implementation of the methodology on the synthesis of indolines, we next sought to harness the aryl radicals in the synthesis of indoles by a radical-based addition-elimination strategy [66–67]. However, our initial attempts in this area upon cyclization of arenediazonium salt 49a were not fruitful as the reactions afforded a mixture of the exocyclic alkene 50a and the alcohol 52a [Table 3, entry (i)]. We expected that the aryl radical 53a generated by the reduction of arenediazonium salt 49a by TDAE would undergo 5-exo-trig radical cyclization onto the vinyl bromide to afford the alkyl radical intermediate 54a, from which Br• would be eliminated to afford the exocyclic alkene 50a (Scheme 6). Such alkenes tautomerise easily to the corresponding indoles (in this case 51a) in the presence of a trace of acid. Alcohol 52a can arise by 1,2-bromine shift [66,68–69] from radical 54a followed by either (a) loss a hydrogen atom from the resulting benzylic radical 56a in collision with another radical or, less likely, (b) oxidation of 56a by electron transfer to an arenediazonium cation. Loss of a proton from the cation so formed would yield bromoalkylindole 57a and subsequent hydrolysis would result in the alcohol 52a. However, when the same reaction was re-examined in anhydrous DMF as the solvent with 1.5 equivalents of

TDAE, it afforded the unstable exocyclic alkene 50a as the sole product, which after treatment with p-toluenesulfonic acid tautomerised to the indole 51a in an overall 68% yield in three steps from 48a. Adopting the optimized procedure, the diazonium salts 49b–d on treatment with 1.5 equivalents of TDAE in anhydrous DMF yielded the indoles 51b–d in 63%, 43% and 64% yields (in three steps from the corresponding aryl amines 48b–d) respectively (Scheme 7). The indole 51d bearing a fused 9-membered ring was of particular interest to us because the important anticancer drugs vinblastine (58a) and vincristine (58b) contain such a system.

Table 3. Initial optimization studies of cyclization of arenediazonium salt 49a.

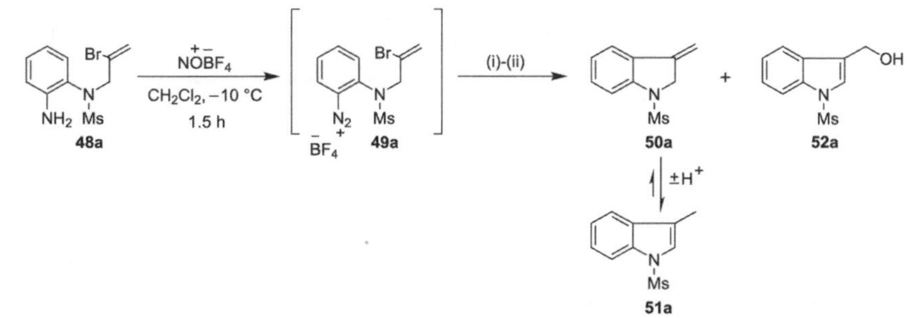

Entry	Diazonium Salt	Equivalents of TDAE	Solvent	Temperature (°C)	Time	Isolated Yield[a] (%) of 51a	52a
(i)	**49a**	1.0	Acetone	25	30 min	39[b]	40
(ii)	**49a**	1.5	DMF (anhydrous)	25	10 min	68[c]	0

[a]All isolated yields were calculated on the basis of the quantity of the starting aryl amine **48a**. [b]The product **51a** was isolated directly from the reaction mixture after auto-tautomerisation of **50a** to indole **51a** during storage of reaction mixture prior to flash chromatography. [c]The product **51a** was obtained by treatment of the intermediate exocyclic alkene **50a** with *p*-toluenesulfonic acid monohydrate in dichloromethane at r.t. for 12 h.

Scheme 6. TDAE-mediated radical based addition-elimination route to indoles.

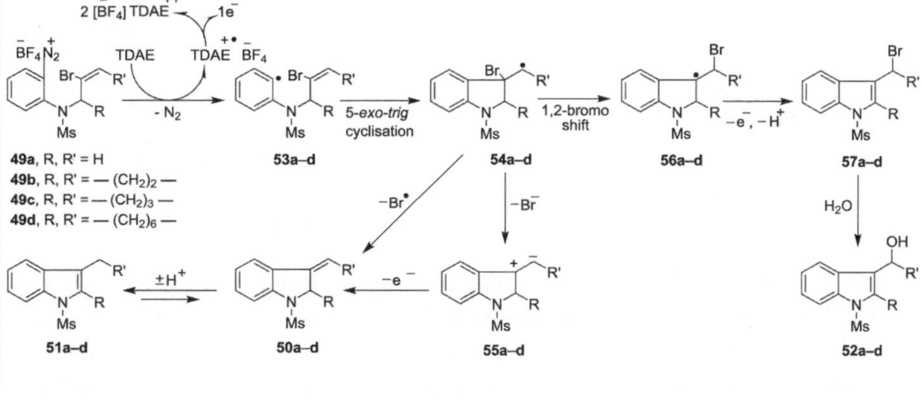

Scheme 7. Cyclization of the arenediazonium salts 49b–d by TDAE. Reagents and conditions: (a) NOBF4, CH2Cl2, -10 °C to 0 °C, 1.5 h; (b) TDAE (1.5 equiv), anhydrous DMF, r.t., 10 min; (c) p-toluenesulfonic acid monohydrate, CH2Cl , r.t., 12 h, 63% (51b, in three steps from 48b), 43% (51c, in three steps from 48c); (d) NOBF4, CH2Cl2, -10 °C to 0 °C, 1.5 h; (e) TDAE (1.5 equiv), anhydrous DMF, r.t., 10 min, 74% 50d; (f) p-toluenesulfonic acid monohydrate, CH2Cl2, r.t., 12 h, 64% (overall yield in three steps, from 48d).

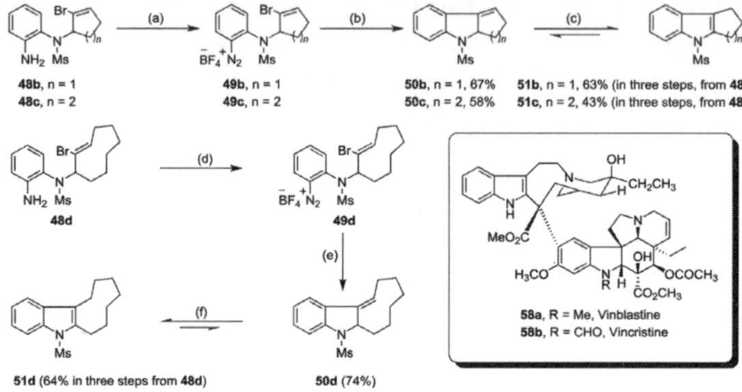

Aryl C-C Bond Formation

As a final extension of this methodology, we probed the feasibility of this methodology in aryl-aryl C-C bond formation reactions. Accordingly, the diazonium salt 62 was prepared from indoline (59), and treated with one equivalent of TDAE in acetone as solvent. The reaction mixture instantaneously turned deep red, with accompanying effervescence of nitrogen, and afforded the tetracyclic sulfonamide 65 in 60% yield along with indole (63) and indole sulfonamide 64 in 33% and 5% yield respectively (Scheme 8).

Scheme 8. Cyclization of the arenediazonium salt 62 by TDAE. Reagents and conditions: (a) 2-Nitrobenzenesulfonyl chloride, DMAP, pyridine, 0 °C then 110 °C, 27 h, 57%; (b) H2, Pd/C, EtOAc, 3 h, 98%; (c) 61, NOBF4, CH2Cl2, -10 °C, (more ...)

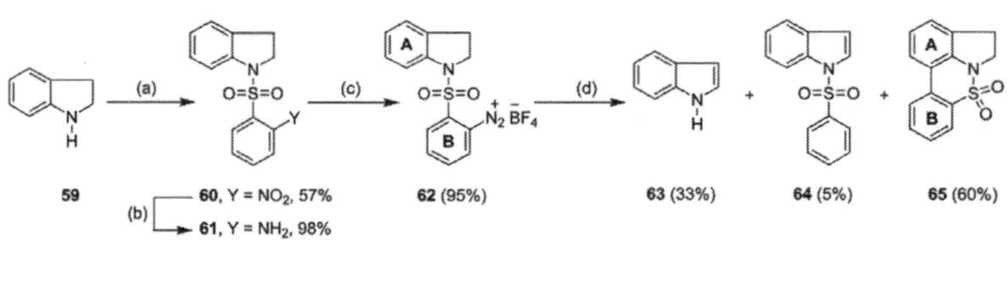

Initial SET from TDAE to the arenediazonium salt 62 afforded the aryl radical 67, with release of molecular nitrogen. The aryl radical 67 would be expected [3,11,70] to cyclise onto the aryl ring A either through 5-exo or 6-endo cyclization. The radicals 68 and 69 could interconvert. Alternatively, aryl radical 67 could also undergo direct formation of the radical 69. Rearomatization from 69 might then occur through a number of pathways; in one of these, the radical intermediate 69 would lose a proton to yield the radical anion 70 which, upon oxidation by loss of single electron to the starting diazonium salt 62, would result in the formation of the tetracyclic sulfonamide 65 (Scheme 9).

Scheme 9. Mechanism for the formation of the tetracyclic sulfonamide 65.

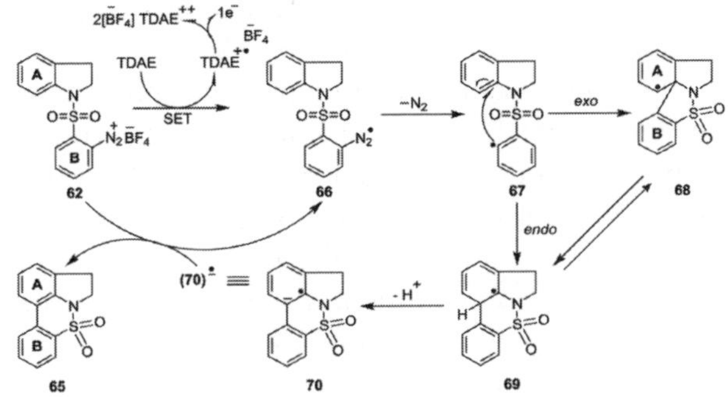

Indole (63) and indole sulfonamide 64 can be formed via the indolinyl radical intermediate 71 (Scheme 10). The formation of the indolinyl radical 71 could be envisaged by abstraction of the hydrogen atom (1,5-hydrogen translocation) by the aryl radical 67 from the carbon atom in the α-position to the nitrogen atom of the indoline nucleus within the same molecule. The indolinyl radical 71 might follow pathway A and undergo radical fragmentation to the intermediate 75 which would eventually tautomerise to indole (63). The precedent for this radical fragmentation of the sulfonyl group comes from the previous work of our group [71], where a similar indolinyl radical underwent a radical cleavage of N-S bond to eliminate the sulfonyl group.

Scheme 10. Possible mechanism for the formation of indole (63) and indole sulfonamide 64.

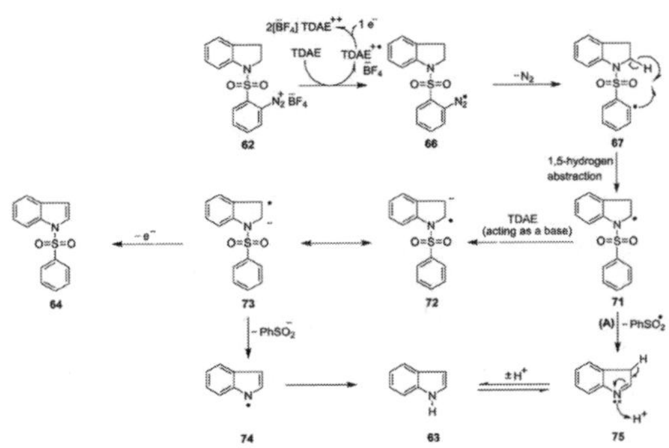

Indole sulfonamide 64 could be explained by deprotonation of radical 71 by TDAE to form radical-anion 72, followed by electron loss. Alternatively, removal of hydrogen atom through reaction with another radical could afford 64.

Conclusion

We have reported a mild and direct method for generation of aryl radicals by reduction of arenediazonium salts using TDAE as a neutral ground-state organic electron donor. Additionally, we have described the utility of the aryl radicals in the construction of indolines and indoles by intramolecular radical cyclization of aryl radicals onto appropriately placed alkenes bearing terminal radical leaving groups. The presence of a suitable radical leaving group like a sulfide, sulfoxide or sulfone is necessary for the self-termination of the 5-exo-trig radical cyclization reactions to avoid competing intermolecular radical side-reactions. The TDAE-mediated radical-based addition-elimination route for the construction of indole ring systems warranted anhydrous reaction conditions for greater efficiency. A preliminary study on TDAE-mediated aryl-aryl C-C bond formation reaction has also been discussed. TDAE possesses a distinct advantage over other organic reducing agents as the oxidized products of TDAE are water soluble – thus the purification process is highly convenient. Further extensions of this methodology in the construction of several heterocyclic ring systems and complex synthetic targets for natural product synthesis are currently in progress in our laboratory.

Acknowledgements

We thank Universities UK (CVCP) for the award of Overseas Research Scholarship (ORS), University of Strathclyde for an International Research Scholarship, and F. Hoffmann-La Roche, Ltd., Basel for a CASE award to M.M.; EPSRC for funding; EPSRC National Mass Spectrometry Service, Swansea for the mass spectral analysis and Dr. John A. Parkinson, University of Strathclyde for NOE NMR experiments.

References

1. Meerwein H, Buchner E, van Emster K. J Prakt Chem. 1939;152:237.
2. Abramovitch R A. Adv Free-Radical Chem. 1966;2:87.
3. Sainsbury M. Tetrahedron. 1980;36:3327.
4. Hanson P, Hammond R C, Goodacre P R, Purcell J, Timms A W. J Chem Soc, Perkin Trans 2. 1994:691.

5. Bashir N, Patro B, Murphy J A. In: Advances in Free Radical Chemistry, Vol. 2. Zard S Z, editor. Stamford, CT, U.S.A.: JAI Press; 1999. pp. 123–150.
6. Murphy J A. The Radical-Polar Crossover Reaction. In: Renaud P, Sibi M P, editors. Radicals in Organic Synthesis. Vol. 1. Weinheim, Germany: Wiley-VCH; 2001. pp. 298–315.
7. Lampard C, Murphy J A, Lewis N. J Chem Soc, Chem Commun. 1993:295.
8. Fletcher R J, Lampard C, Murphy J A, Lewis N. J Chem Soc, Perkin Trans 1. 1995:623.
9. Murphy J A, Rasheed F, Gastaldi S, Ravishanker T, Lewis N. J Chem Soc, Perkin Trans 1. 1997:1549.
10. Callaghan O, Lampard C, Kennedy A R, Murphy J A. J Chem Soc, Perkin Trans 1. 1999:995.
11. Lampard C, Murphy J A, Rasheed F, Lewis N, Hursthouse M B, Hibbs D E. Tetrahedron Lett. 1994;35:8675. doi: 10.1016/S0040-4039(00)78469-3.
12. Murphy J A, Rasheed F, Roome S J, Lewis N. Chem Commun. 1996:737.
13. Fletcher R, Kizil M, Lampard C, Murphy J A, Roome S J. J Chem Soc, Perkin Trans 1. 1998:2341.
14. Kizil M, Lampard C, Murphy J A. Tetrahedron Lett. 1996;37:2511.
15. Fletcher R J, Hibbs D E, Hursthouse M, Lampard C, Murphy J A, Roome S J. Chem Commun. 1996:739.
16. Koizumi T, Bashir N, Murphy J A. Tetrahedron Lett. 1997;38:7635.
17. Koizumi T, Bashir N, Kennedy A R, Murphy J A. J Chem Soc, Perkin Trans 1. 1999:3637.
18. LeStrat F, Murphy J A, Hughes M. Org Lett. 2002;4:2735.
19. Murphy J A, Khan T A, Zhou S-z, Thomson D W, Mahesh M. Angew Chem, Int Ed. 2005;44:1356.
20. Murphy J A, Zhou S-z, Thomson D W, Schoenebeck F, Mahesh M, Park S R, Tuttle T, Berlouis L E A. Angew Chem, Int Ed. 2007;46:5178.
21. Wiberg N. Angew Chem, Int Ed Engl. 1968;7:766.
22. Fritsch J M, Weingarten H, Wilson J D. J Am Chem Soc. 1970;92:4038.
23. Kolomeitsev A, Médebielle M, Kirsch P, Lork E, Röschenthaler G-V. J Chem Soc, Perkin Trans 1. 2000:2183.
24. Carpenter W, Bens E M. Tetrahedron. 1970;26:59.
25. Winberg H E, Downing J R, Coffman D D. J Am Chem Soc. 1965;87:2054.

26. Wiberg N, Buchler J W. Angew Chem, Int Ed Engl. 1962;1:406.
27. Wiberg N, Buchler J W. Chem Ber. 1963;96:3223.
28. Burkholder C, Dolbier W R, Jr, Médebielle M. J Org Chem. 1998;63:5385.
29. Burkholder C, Dolbier W R, Jr, Médebielle M, Ndedi A. Tetrahedron Lett. 1998;39:8853.
30. Burkholder C, Dolbier W R, Jr, Médebielle M, Aït-Mohand S. Tetrahedron Lett. 2001;42:3077.
31. Médebielle M, Keirouz R, Okada E, Ashida T. Synlett. 2001:821.
32. Aït-Mohand S, Takechi N, Médebielle M, Dolbier W R., Jr Org Lett. 2001;3:4271.
33. Takechi N, Aït-Mohand S, Médebielle M, Dolbier W R., Jr Tetrahedron Lett. 2002;43:4317.
34. Takechi N, Aït-Mohand S, Médebielle M, Dolbier W R., Jr Org Lett. 2002;4:4671.
35. Médebielle M, Kato K, Dolbier W R., Jr Tetrahedron Lett. 2003;44:7871.
36. Médebielle M, Hohn S, Okada E, Myoken H, Shibata D. Tetrahedron Lett. 2005;46:7817.
37. Prakash G K S, Wang Y, Hu J, Olah G A. J Fluorine Chem. 2005;126:1361.
38. Xu W, Dolbier W R., Jr J Org Chem. 2005;70:4741.
39. Peng W, Zhao J, He P, Zhu S. Synlett. 2006:296.
40. Peng W, Zhao J, Zhu S. J Fluorine Chem. 2006;127:360.
41. Peng W, Zhao J, Zhu S. Synthesis. 2006:1470.
42. Pooput C, Dolbier W R, Jr, Médebielle M. J Org Chem. 2006;71:3564.
43. Médebielle M, Keirouz R, Okada E, Shibata D, Dolbier W R., Jr Tetrahedron Lett. 2008;49:589.
44. Giuglio-Tonolo G, Terme T, Médebielle M, Vanelle P. Tetrahedron Lett. 2003;44:6433.
45. Giuglio-Tonolo G, Terme T, Médebielle M, Vanelle P. Tetrahedron Lett. 2004;45:5121.
46. Giuglio-Tonolo G, Terme T, Vanelle P. Synlett. 2005:251.
47. Amiri-Attou O, Terme T, Vanelle P. Synlett. 2005:3047.
48. Nishiyama Y, Kawabata H, Kobayashi A, Nishino T, Sonoda N. Tetrahedron Lett. 2005;46:867.

49. Montana M, Crozet M D, Castera-Ducros C, Terme T, Vanelle P. Heterocycles. 2008;75:925.
50. Montana M, Terme T, Vanelle P. Tetrahedron Lett. 2006;47:6573.
51. Montana M, Terme T, Vanelle P. Tetrahedron Lett. 2005;46:8373.
52. Nishiyama Y, Kobayashi A. Tetrahedron Lett. 2006;47:5565.
53. Amiri-Attou O, Terme T, Médebielle M, Vanelle P. Tetrahedron Lett. 2008;49:1016.
54. Kuroboshi M, Goto K, Mochizuki M, Tanaka H. Synlett. 1999:1930.
55. Kuroboshi M, Tanaka M, Kishimoto S, Goto K, Mochizuki M, Tanaka H. Tetrahedron Lett. 2000;41:81.
56. Kuroboshi M, Waki Y, Tanaka H. Synlett. 2002:637.
57. Kuroboshi M, Waki Y, Tanaka H. J Org Chem. 2003;68:3938.
58. Park S B, Alper H. Tetrahedron Lett. 2004;45:5515.
59. Kuroboshi M, Takeda T, Motoki R, Tanaka H. Chem Lett. 2005;34:530.
60. Andrieux C P, Pinson J. J Am Chem Soc. 2003;125:14801.
61. Pause L, Robert M, Savéant J-M. J Am Chem Soc. 1999;121:7158.
62. Ashton P R, Balzani V, Becher J, Credi A, Fyfe M C T, Mattersteig G, Menzer S, Nielsen M B, Raymo F M, Stoddart J F, et al. J Am Chem Soc. 1999;121:3951.
63. Simonet J, Pilard J-F. Electrogenerated Reagents. In: Lund H, Hammerich O, editors. Organic Electrochemistry. 4. New York: Marcel Dekker Inc.; 1991. pp. 1163–1225. See in particular pp 1171 ff.
64. Lacôte E, Delouvrié B, Fensterbank L, Malacria M. Angew Chem, Int Ed. 1998;37:2116. doi: 10.1002/(SICI)1521-3773(19980817)37:15<2116::AID-ANIE2116>3.0.CO;2-L.
65. Wagner P J, Sedon J H, Lindstrom M J. J Am Chem Soc. 1978;100:2579.
66. Murphy J A, Scott K A, Sinclair R S, Martin C G, Kennedy A R, Lewis N. J Chem Soc, Perkin Trans 1. 2000:2395.
67. Murphy J A, Scott K A, Sinclair R S, Lewis N. Tetrahedron Lett. 1997;38:7295.
68. Freidlina R Kh, Terent'ev A B. Adv Free-Radical Chem. 1980;6:1.
69. Freidlina R Kh. Adv Free-Radical Chem. 1965;1:211.
70. Bowman W R, Heaney H, Jordan B M. Tetrahedron. 1991;47:10119. 71. Bommezijn S, Martin C G, Kennedy A R, Lizos D, Murphy J A. Org Lett. 2001;3:3405.

Copyrights

Index

A

B

C

D

E

F

G

H

I

Q

R

S

T

U

V